Nitrogen Management in Crop Production

Nitrogen Management in Crop Production

Editor: Mark Taylor

STATES
ACADEMIC PRESS
www.statesacademicpress.com

States Academic Press,
109 South 5th Street,
Brooklyn, NY 11249, USA

Visit us on the World Wide Web at:
www.statesacademicpress.com

ISBN: 978-1-63989-384-3 (Hardback)

Cataloging-in-Publication Data

Nitrogen management in crop production / edited by Mark Taylor.
p. cm.
Includes bibliographical references and index.
ISBN 978-1-63989-384-3
1. Nitrogen in agriculture. 2. Crops and nitrogen. 3. Nitrogen fertilizers.
I. Taylor, Mark.
S587.5.N5 N58 2022
631.84--dc23

Table of Contents

Preface

Nitrogen is an essential plant nutrient which is considered as a mineral element. Nitrogen deficiency reduces the fertility of the soil and affects the crop production severely. Plants absorb nitrogen through their roots in the form of nitrate and ammonium ions. It plays a crucial role in photosynthesis and is responsible for the activation of various enzymes. Nitrogen based fertilizers are the most important means of nitrogen management. They play an important role in the areas with prolonged wet field conditions which makes the soil nitrogen deficient. However, excess nitrogen in the soil can cause nitrogen pollution. Alternative nitrogen sources such as organic residues and biological nitrogen fixation could be used to manage nitrogen effectively and increase the crop yield while reducing the harmful effects on the environment. The various studies that are constantly contributing towards advancing technologies and evolution of this field are examined in detail. Also included herein is a detailed explanation of the various concepts and applications of nitrogen management in crop production. For all those who are interested in crop production, this book can prove to be an essential guide.

This book is a result of research of several months to collate the most relevant data in the field.

When I was approached with the idea of this book and the proposal to edit it, I was overwhelmed. It gave me an opportunity to reach out to all those who share a common interest with me in this field. I had 3 main parameters for editing this text:

1. Accuracy - The data and information provided in this book should be up-to-date and valuable to the readers.

2. Structure - The data must be presented in a structured format for easy understanding and better grasping of the readers.

3. Universal Approach - This book not only targets students but also experts and innovators in the field, thus my aim was to present topics which are of use to all.

Thus, it took me a couple of months to finish the editing of this book.

I would like to make a special mention of my publisher who considered me worthy of this opportunity and also supported me throughout the editing process. I would also like to thank the editing team at the back-end who extended their help whenever required.

Editor

1

Nitrogen Transformations Associated with N_2O Emissions in Agricultural Soils

Ling Zhang and Xiaojun Liu

Abstract

Nitrogen (N) is one of the most important plant nutrient, and its availability and transformations are vital for net primary production. Soil N transformations include mineralization, nitrification and denitrification processes. Nitrogen mineralization transforms organic N into inorganic N, providing available N for crops. Both nitrification and denitrification are microbe-driven processes associated with nitrous oxide (N_2O) emissions. N_2O emissions from agricultural soils decrease N fertilization efficiency and potentially induce global warming. The mitigation of soil N_2O emissions in agricultural practice is essential for sustainable development of agriculture considering the environmental effect of N_2O. Various strategies have been proposed for the mitigation of N_2O emissions. Nitrification inhibitors have been demonstrated to be useful in decreasing soil N_2O emissions, including the application of nitrification inhibitors, such as dicyandiamide (DCD) and 3,4-dimethylpyrazole phosphate (DMPP). Recently, biological nitrification inhibitors have also attracted researchers' attention, which may be more environment-friendly. In addition, biochar commonly used as soil ameliorant to improve soil quality and C sequestration could also mitigate soil N_2O emissions. Once all effective strategies would be widely implemented, more environment-friendly agriculture could be expected.

Keywords: fertilizer efficiency, global change, mitigation, nitrogen emissions, nutrition

1. Introduction

Nitrogen (N) is one of the most important elements for plant growth and hence it is vital for ecosystem's primary production [1]. In the background of global climate change, understanding soil N balance in agricultural soils is important for sustaining the development of modern agriculture and mitigation of climate change considering the substantial contribution made by agriculture to climate change [2–4].

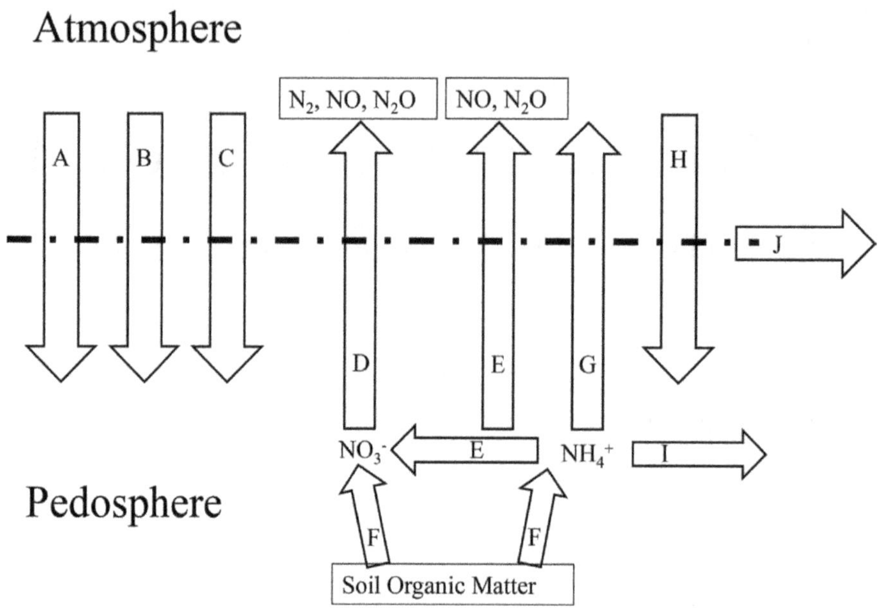

Figure 1. Simple diagram showing N cycle between atmosphere and agricultural soil. Note: (A) N deposition; (B) N fixation; (C) fertilization; (D) denitrification; (E) nitrification; (F) organic N mineralization; (G) plant uptake; (H) litter decomposition; (I) soil immobilization; (J) N volatilization, leaching and crops harvest.

In general, soil N mainly originates from atmospheric N deposition (both dry and wet deposition) (**Figure 1**) and biological fixation in natural ecosystems [5]. In agricultural soils, however, N input via anthropogenic activities (e.g., fertilization practice) has been considered as the most important way [6]. Fertilization in agricultural practice could introduce many kinds of N into soil ecosystems. For example, both organic and inorganic N, in liquid or particle status, could often be observed in agricultural fertilization practice. In addition to N input by anthropogenic activities, N output via crops harvest or human-induced runoffs with irrigation could occur simultaneously in agricultural soils (**Figure 1**).

Moreover, the complicated N transformations going on in soil ecosystems also account for large part of N output from soils [7]. For example, N mineralization associated with transforming organic N into inorganic N might induce more N leaching with runoffs or underground water; nitrification process could increase N loss from soils via gas emissions [2, 7, 8] (**Figure 1**).

Thereby, understanding the process and mechanisms underlining N transformations in agricultural soils would be vital for more efficient fertilization practice and agriculture management.

2. Nitrogen transformations in agricultural soils

2.1. Overview of soil N transformations

Except for N transformation process ongoing during litter decomposition process at the litter-soil interface, N was also transformed by processes including mineralization, nitrification, denitrification, and so on, in agricultural soils [9].

While organic N accounts for larger part of total soil N, only soil inorganic N (mainly including ammonium and nitrate N) derived from mineralization of organic N is readily available for plant. Thereby, organic N mineralization is an important factor controlling levels of soil inorganic N availability, which is vital for crop production. In agricultural soils, most crop residuals were harvested by agricultural activities, while some of them were returned into soils as roots, rhizospheric deposition or aboveground litter. During decomposition of these organic matters, most inorganic N would be released into surrounding soils via mineralization of organic N. In these processes, N uptake by plant from soils was returned into soils again, during N mineralization process.

Nitrogen produced by mineralization process, including both ammonium and nitrate N, was readily available for crop growth. If there was no leaching or loss via runoffs, all available N will be adsorbed by plant roots and used for growth. However, nitrification and denitrification processes (both produce greenhouse gas N_2O) also occurred between ammonium and nitrate N, causing new balance among different N components [2, 4, 7].

In addition, soil ammonium N could also be immobilized by soil particles, and both important inorganic N may loss with surface runoff, further decreasing its availability level. Before mineralized by microbes, soluble organic N was also reliable to be removed by runoffs.

2.2. General methods in studying soil N transformations

2.2.1. Net and gross N transformation rates

Due to complicated interactions among all soil N transformation process, net N transformation rates have been widely studied [10, 11]. Once plant roots were excluded, changes in soil N levels could be considered as results of transformations. Specifically, when soil mineralization rates were examined, both ammonium and nitrate concentrations were determined both at the beginning and at the end of the study. Net N mineralization rates were considered as changes in concentration of both inorganic N per given time unit [11, 12]. The equations used for the calculation were as follows:

Net N mineralization rate:

$$N_{\text{mineralized}} = \frac{\left[(\text{Nitrate}_f + \text{Ammonium}_f) - (\text{Nitrate}_0 + \text{Ammonium}_0)\right]}{T_{days}}. \tag{1}$$

where $N_{mineralized}$ = net N mineralization rate, expressed as mg N kg^{-1} soil day^{-1}, Ammonium$_f$ = final ammonium concentration, expressed as mg NH_4^+-N kg^{-1} soil, Ammonium$_0$ = initial ammonium concentration, as mg NH_4^+-N kg^{-1} soil, T_{days} = incubation time, days.

Similarly, net nitrification rates and ammonification rates were calculated as follows:

Net nitrification rate:

$$N_{nitrified} = \frac{(\text{Nitrate}_f - \text{Nitrate}_0)}{T_{days}}, \tag{2}$$

Net ammonification rate:

$$N_{ammonification} = \frac{(\text{Ammonium}_f - \text{Ammonium}_0)}{T_{days}}, \tag{3}$$

where $N_{nitrified}$ = net nitrification rate, expressed as mg NO_3^--N kg^{-1} soil day^{-1}, $Nitrate_f$ = final nitrate concentration, expressed as mg NO_3^--N kg^{-1} soil, $Nitrate_0$ = initial nitrate concentration, as mg NO_3^--N kg^{-1} soil, $Ammonium_f$ = final ammonium concentration, expressed as mg NH_4^+-N kg^{-1} soil, $Ammonium_0$ = initial ammonium concentration, as mg NH_4^+-N kg^{-1} soil, T_{days} = incubation time, days.

Meanwhile, with the development of isotope labeling strategy, more studies have been conducted to determine the gross N transformation rates in agriculture and forestry soils [13–16]. By labeling N in ammonium or nitrate N, N element could be traced during the complicated transformation process. Thereby, the gross N transformation rates could be obtained using the isotope labeling method.

2.2.2. *In situ and ex situ studies on N transformations*

Studies on N transformations could also be sorted by study place or site, into *in situ* or *ex situ* studies. Laboratory soil incubation studies were widely used to examine N transformations in environmental science [10, 17]. During laboratory incubation, the environmental factors could be easily altered to check their role played on N transformation rates. Thereby, studies conducted in incubators are more operable than those *in situ*. For example, incubation studies could be manipulated with different soil water content, incubated temperature, or even aeration status. However, not all N transformation studies could be conducted in the laboratory. In agricultural ecosystems, the dynamics of soil N might be important for fertilization practice in field crops research and management. In this condition, laboratory incubation studies are no longer applied. Instead, *in situ* soil core incubation is more suitable. Similar to laboratory incubations, both concentrations of soil inorganic N at the beginning and at the monitored date should be determined of the soil column. Moreover, the soil column incubated *in situ* for study should be isolated from the surrounding soils to prevent potential N uptake by roots of crops.

In recent years, ion-exchange resins (Unibest PST-1, Unibest, Bozeman, MT, USA) have been widely used for *in situ* studies on soil N availability [18, 19]. In these studies, incubated soils cores were isolated from surrounding soil. Exchangeable anion/cation resin was used to capture any inorganic N moved into or out of the incubated soil core [19]. Changes in soil inorganic N in soil core and the resin relative to that of the initial soil core with time were considered as net N mineralization rate. Similarly, changes in soil nitrate N with time were considered as net nitrification rate.

3. Nitrogen transformations associated with N_2O emissions

3.1. Illustration of N_2O emissions

3.1.1. *Soil N_2O production*

N_2O is one important component of greenhouse gas emitted from soil. The global warming potential of N_2O is much larger than that of methane and carbon dioxide (CH_4 and CO_2). According to the latest report, even though not so much as CH_4 and CO_2 in atmospheric environment, the global warming potential of N_2O is 265 times that of CO_2, while CH_4 is only 28 times [3]. Thereby, N_2O might have contributed substantially to global warming considering its larger global warming potential [3, 8]. Importantly, most of the atmospheric N_2O was emitted from soils, especially those fertilized by N fertilization in agricultural ecosystems. Hence, studying the way N_2O is produced, emitted and adsorbed in agricultural soils would be important for mitigation of soil N_2O emissions and hence for mitigation of atmospheric N_2O in this climate changing world.

Soil N_2O has been demonstrated to be the by-product or production of N transformation process by previous studies [20]. Specifically, the main process producing N_2O in soil ecosystems, namely nitrification and denitrification process, transformed ammonium N to nitrate N, and nitrate N to N_2, respectively [21, 22]. The processes associated with N_2O production could be described by hole in the pipe model (HIP, **Figure 2**) [20]. The model showed that N_2O was produced by complete nitrification process and was produced during denitrification process. In both processes, nitric oxide (NO) was also produced by both nitrification and denitrification processes, which was also considered as an important GHG [23–25]. Recently, in addition to N_2O, NO has also attracted increasing attention due to their role played in altering atmospheric chemistry and global climate [26–29].

3.1.2. *Measurement of soil N_2O emissions*

Methods being developed for the measurement of soil N_2O emissions rate including two different sets. This first one was used to measure soil N_2O emission rate in laboratory incubation studies. Different to that conducted *in situ*, incubation studies monitored soil N_2O emission based on soil mass rather than soil surface area. Thereby, those studies measured N_2O emission rate based on soil mass per time. Equation developed for the calculation of N_2O emission rate of incubated soil could be described as follows [17, 30, 31]:

$$E = P \times V \times \frac{dc}{dt} \times \frac{1}{RT} \times M \times \frac{1}{m} \times \frac{1}{t} \tag{4}$$

where E refers to emission rates of soil N_2O (ng g^{-1} h^{-1}), P is standard atmospheric pressure (Pa), V is headspace volume of the incubation flask (cm^3), c is the concentration of N_2O (ppb),

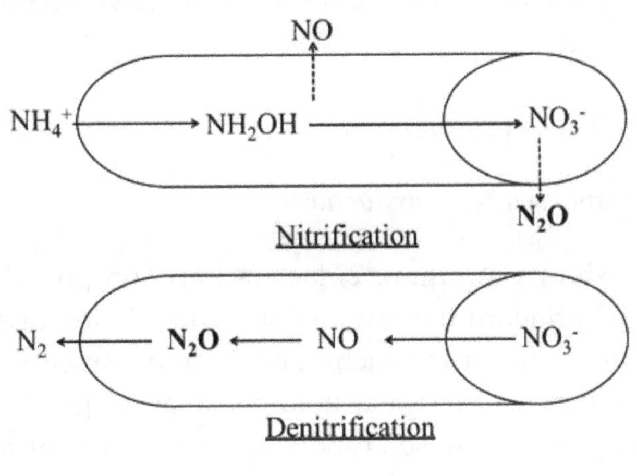

Figure 2. Simple diagram showing the production process of soil N_2O.

t is the time between two sample collections (h), R is the universal gas constant, T is the absolute air temperature (K), M is the molecular mass of N_2O (g mol^{-1}), and m is incubated soil mass by dry weight basis (g).

Relative to laboratory incubation studies, *in situ* studies could capture effects of various factors on soil N_2O emissions in natural environment. These studies generally calculated soil N_2O emission rate based on soil surface area. In general, *in situ* measurements could be conducted using static opaque chamber/gas chromatography method. To be specific, circular or square grooved collars should be buried into soil, with groove filled with water to seal the gas collection chamber [32, 33]. When soil N_2O emission rate would be measured, open-bottom cylindrical or cubic PVC gas sampling chamber would be fit into the groove. To exclude the potential effects of temperature variation during gas collection process, gas chambers were usually wrapped by foam and aluminum foil. Inside each chamber, battery-powered fans were used during gas accumulation process to mix air samples. Gas samples were usually collected manually or automatically using single-use syringes or sir bags, respectively. After gas collection, gas chromatograph with electron capture detector was used to measure N_2O concentrations. Soil N_2O emission rates were usually determined by the equation as follows [31–33]:

$$F = P \times V \times \frac{d\,N_2O}{dt} \times \frac{1}{RT} \times M \times \frac{1}{A} \times \frac{M_n}{M} \tag{5}$$

where F refers to soil N_2O emission rates (mg N_2O m^{-2} h^{-1}), P is the standard atmospheric pressure (Pa), V and A are the volume (m^3) and interior bottom area (m^2) of the gas collection chamber, R stands for universal gas constant, T is the absolute air temperature (K) when the gas sample was aspirated and M_n and M are the molecular masses of N and N_2O (g mol^{-1}), respectively.

When cumulative emissions were needed for study purpose, total soil N_2O emissions within a given time could be obtained by multiplying average soil N_2O emission rate and the corresponding time span [17, 33].

It should be noted that both methods were used to obtain the net soil N_2O emission rates. During measurement, soil might be source or sink of N_2O depends on soils used for studies. However, when results were positive based on two equations, it could be determined that soils were emitting N_2O. Similarly, when values were negative, soil could be adsorbing N_2O in the corresponding studies.

3.2. Factors impacting soil N_2O productions

3.2.1. General factors impacting soil N_2O productions

According to the model shown in **Figure 2**, factors impacting nitrification and denitrification could also be able to influence the production and emission of soil N_2O. As have been reported by previous studies, factors impacting nitrification process including quantity and quality of soil N input, soil moisture (water holding capacity) [34, 35], soil temperature [10, 17, 30], irrigation and tillage practices, soil type, soil oxygen concentration, dissolved organic C availability (controlling substrate availability of soil microbes) [17, 36], additives for soil amelioration, vegetation or crop types [37], land use change and soil pH [36, 38]. While soil N provided substrate for transformation process producing soil N_2O, other factors regulated N_2O production process mainly via indirect effects on soil microbial activities.

For example, soil temperature is the key factor controlling microbial activities. Since both nitrification and denitrification processes have been demonstrated as being driven by nitrification and denitrification bacteria, soil temperature could impact both transformation processes via its effects on bacterial activities. According to previous studies, nitrification process preferred temperature between 25 and 35°C, while it will be inhibited when soil temperature decreased below 5°C or increased above 50°C [39]. Moreover, the favorable temperature for denitrification falls within 30 and 67°C [40].

In addition, soil acidification levels as shown by soil pH are also important for microbial activities. To be specific, the ratio of N_2O in denitrification process would increase with relatively lower soil pH, indicating enhancement of denitrification bacteria activities. Similarly, activities of some nitrification bacteria would also increase in response to lower pH.

Soil aeration also controls the quantity of N_2O by nitrification or denitrification process. In well-ventilated soil environment, nitrification process could be complete while denitrification process posed at stage producing N_2O. Since complete nitrification process was also accompanied by N_2O production, under the same environmental conditions, more N_2O would be produced in upland soils relative to flooded soils.

3.2.2. *Factors impacting N_2O production in agricultural soils*

In agricultural soils with intensive anthropogenic disturbance, soil N_2O productions became more complicated compared with those in natural soils [4]. Agricultural practice generally including fertilization, tillage, water regime [34], and so on, all of which could alter soil physical and chemical properties, impacting soil N_2O productions [4].

First, agricultural soils received much more N input via fertilization, increasing N availabilities for nitrification and denitrification process [2, 4, 8]. Except for increasing N availability directly, fertilization types, quantity of fertilizations, fertilization method and the time when soil was fertilized together regulated N transformation process. In general, N fertilization including inorganic N or organic N input in agricultural management. In organic agriculture, activities of denitrification bacteria were higher, potentially facilitating the denitrification process, decreasing soil N_2O productions [41]. Indeed, soil N_2O emissions were found lower in organic agriculture than conventional agriculture in another study by Phillips [42].

Second, farmland with decreased-tillage or non-tillage management potentially enhance the accumulation of soil organic C [43]. Due to the balance between C and N regulated by C to N ratio, increased soil organic C might be accompanied by increased fixation of soil N and hence less N_2O emissions from agricultural field [36]. However, increased soil N_2O emissions were also observed in studies on decreased tillage farmland [34, 44, 45]. Moreover, soil tillage could also impact soil aeration conditions and indirectly regulate soil N_2O emissions via effects on microbial activities. Thereby, soil tillage and other disturbance management may be important in impacting soil N_2O emissions [34, 45, 46].

Water regime, especially that in paddy field, plays an important role in controlling soil N_2O production and emissions [47–49]. When paddy field was flooded, soil aeration was inhibited, anaerobic microenvironment was hence cultivated. In this kind of condition, N_2 became the main production of denitrification process, soil could be considered as the sink of N_2O, as almost no emissions were observed. Meanwhile, flooded soil was generally favorable to methanogens, which were associated with soil CH_4 production. Indeed, flooded soil, generally in paddy field, has been demonstrated to be the main source of CH_4 [47]. However, during drainage time, paddy field was not flooded any more, especially during time when soils were humid (i.e., wetting and drying cycles), aerobiotic soil environment was formed, and soil N_2O emission rate could reach a peak [31]. In this kind of environment, soil is not completely dry but experiencing wetting and drying cycles, allowing more oxygen in soil pores, increasing the production of N_2O [50]. Yan et al. [51] studied the correlations between soil water content and soil N_2O emissions, reporting the largest soil N_2O emission rate when soil water content was equivalent to water holding capacity. Thereby, soil water content and soil aeration condition interact in impacting soil N_2O emissions in agricultural soil.

3.3. Mitigation of N_2O emissions from agricultural soils

Various strategies have been developed for the mitigation of soil N_2O emissions, especially in agricultural ecosystems. Agricultural management including fertilization, tillage, crop rotations, and so on has been employed in mitigation of soil GHG, especially N_2O emissions. Not only because N_2O induce global warming, but also N losses that accompanied the production and emission process of N_2O. To increase fertilization efficiency, the economy of agriculture management, and the benefit for environment, more efficient mitigation strategies are still needed. Presently, nitrification inhibitor has been widely used in agricultural management and has been demonstrated to be much more effective.

Nitrification inhibitor interrupted the transformation process from ammonium N to nitrate N, which could decrease N losses from soil (emissions or leaching, for example, Marsden et al. [52]) and increase N availability level for crops and hence the adsorption of ammonium N. Using nitrification inhibitors could potentially decrease soil emissions via interruptions on both nitrification and denitrification processes simultaneously. Chemical nitrification inhibitor and biological nitrification inhibitor are two important choices in recent studies.

3.3.1. Chemical nitrification inhibitors

Chemical nitrification inhibitors are human-synthesized materials. By decreasing soil N loss induced by nitrification and denitrification process, nitrification inhibitors could enhance efficiency of N fertilization. Thereby, nitrification inhibitors were also used as additives for N fertilizers. Nitrogen fertilizers with these additives (or similar additives like urease inhibitors, etc.) were usually used and called as enhanced efficiency nitrogen fertilizers [53].

Dicyandiamide (DCD) and 3,4-dimethylpyrazole phosphate (DMPP) are two most widely used nitrification inhibitors [54]. Both inhibitors interrupted the oxidation of ammonium N, limiting the important step occurred in nitrification process. Even though both inhibitors could be used with N fertilization and were effective in mitigation of soil N_2O emissions, DMPP has been demonstrated to be less phytotoxic and used at lower rate relative to DCD in general. The efficacy of both inhibitors in mitigation of soil N_2O emissions could depend on temperature, soil chemical and physical characteristics, and so on. However, the difference in mitigation efficacy could also have been induced by the mobility of inhibitors in soil environment.

However, it should be noted that attentions should be paid to the negative effect on soil N availability [55] or food security [54] induced by applications of nitrification inhibitors. The first negative effect is the potentially increased ammonia (NH_3) volatilization induced by nitrification inhibitors [55]. Indeed, nitrification inhibitors decreased the rate of nitrification and denitrification process, potentially prolonged the retention time of ammonium N in soil environment, increasing the possibilities of more ammonia volatilization. Increased ammonia emissions would on one hand decrease the efficacy of N fertilization practice, and importantly, on the other hand, have economic and environmental consequences [53] considering their potential driving effect on soil N_2O emissions.

To reduce potentially increased ammonia volatilization after application of nitrification inhibitors, manufactures developed urease inhibitor N-(n-butyl) thiophosphoric triamide (NBPT) [56, 57]. Application of NBPT has been proved effective in reducing pasture soil urease activity and mitigating ammonia volatilization [58, 59]. The combination of nitrification inhibitor and NBPT could also decrease the yield-scaled N_2O emissions relative to treatments only with fertilizer in banana plantations [56].

The second caution is the grain yield and quality following alteration in soil N components [54]. While it is rational to expect that grain N concentration may increase in response to soil with longer N retention time, there are studies demonstrating no such effect in grain N following DMPP application [54]. A recent study on banana plantations in tropical China reported decreased yield-scaled N_2O emission, but banana yield showed no significant difference between N fertilization treatment and N with inhibitors [56]. Thereby, the efficiency of N fertilization practice with nitrification inhibitors might be crops or vegetation-type dependent, which should be considered in future applications.

There are also cautions on N fertilizer types that could be used combining with the application of nitrification inhibitors. To maintain the efficiency of nitrification inhibitors, only several N fertilizers could be widely used with nitrification inhibitors. Due to higher cost of theses fertilizers, wide adoption in agricultural practice became more difficult [59]. Recently, new compounds with higher nitrification inhibitor efficiencies have been developed, such as 3,4-dimethylpyrazole succinic (DMPSA), which is more stable when applied combined with other basic fertilizers (e.g., calcium ammonium nitrate) at basic conditions [54, 59].

3.3.2. Biological nitrification inhibitors

In tropical grassland and forest ecosystems, nitrification rate was found much slower relative to that in other similar soils. In further studies, it was found that some plant species synthesized important organic compounds and released these compounds into surrounding soils via roots [60]. These compounds were found being able to inhibit nitrification process, imposing similar effects on nitrification process as chemical nitrification inhibitors. They were called biological nitrification inhibitors as they were not human synthesized. Thereby, biological nitrification inhibitors are organic materials that have similar negative effects on nitrification process exuded by plant.

Biological nitrification inhibitors generally including phenolic compounds, alkaloid, isothiocyanate and terpenoid [60, 61]. For example, chemicals produced by *Arbutus unedo*, including phenolic compounds gallocatechin and catechin were able to decrease soil N_2O emissions [62]. In tropical grassland of Africa, both *Brachiaria humidicola* and *Brachiaria decumbens* were found be able to release biological nitrification inhibitors by producing linoleic, and so on, which enable them to survive in the low-N south Africa Savannas [63].

Compared with chemical nitrification inhibitors, biological inhibitors were environment-friendly in their producing process and application area. However, due to varying ability in synthesizing these compounds, further studies in how to cultivate species with stronger ability in producing these compounds are still needed.

3.3.3. *Other gradients incorporated into agricultural soils*

Other efforts have also been tried in decreasing soil N_2O emissions. In agricultural and forest soils, biochar has been used to improve soil quality and C sequestration [64–67]. Biochar is produced by slow pyrolysis of crop residues, household garbage, poultry litter, wood chips, or some other similar materials at high temperature (pyrolysis temperature generally between 400 and 600°C) without oxygen [67]. The physical characteristics of biochar enable it to be an ideal soil ameliorant. By applying biochar, soil aeration could be improved significantly, providing more oxygen and hence enhancing soil microbial activities. Due to its special physical characteristics, biochar could also prevent soil N leaching by adsorbing nitrate N (temporary immobilization, which would not impose negative effect on plant nutrient availability forever).

Acknowledgements

We acknowledge funding support by National Natural Science Foundation of China (Award number: 41501317), China Postdoctoral Science Foundation (Award number: 2017M612153), and Key Project of Jiangxi Education Department (Award number: GJJ160348).

Author details

Ling Zhang* and Xiaojun Liu

*Address all correspondence to: lingzhang09@126.com

Jiangxi Key Laboratory of Silviculture, Co-Innovation Center of Jiangxi Typical Trees Cultivation and Utilization, College of Forestry, Jiangxi Agricultural University, Nanchang, China

References

[1] Galloway JN, Townsend AR, Erisman JW, Bekunda M, Cai Z, Freney JR, et al. Transformation of the nitrogen cycle: Recent trends, questions, and potential solutions. Science. 2008;**320**(5878):889-892

[2] Bouwman AF. Direct emission of nitrous oxide from agricultural soils. Nutrient Cycling in Agroecosystems. 1996;**46**(1):53-70

[3] IPCC. Agriculture, forestry and other land use (AFOLU). In: Climate Change 2014: Mitigation of Climate Change; 2014.

[4] Freney JR. Emission of nitrous oxide from soils used for agriculture. Nutrient Cycling in Agroecosystems. 1997;**49**(1):1-6

[5] Vitousek PM, Aber JD, Howarth RW, Likens GE, Matson PA, Schindler DW, et al. Human alteration of the global nitrogen cycle: Sources and consequences. Ecological Applications. 1997;**7**(3):737-750

[6] Zheng X, Fu C, Xu X, Yan X, Huang Y, Han S, et al. The Asian nitrogen cycle case study. AMBIO: A Journal of the Human Environment. 2002;**31**(2):79-87

[7] Zhu T, Zhang J, Cai Z. The contribution of nitrogen transformation processes to total N_2O emissions from soils used for intensive vegetable cultivation. Plant and Soil. 2011; **343**(1-2):313-327

[8] Zou J, Lu Y, Huang Y. Estimates of synthetic fertilizer N-induced direct nitrous oxide emission from Chinese croplands during 1980-2000. Environmental Pollution. 2010;**158**(2):631-635

[9] Mosier AR, Zhaoliang Z. Changes in patterns of fertilizer nitrogen use in Asia and its consequences for N_2O emissions from agricultural systems. Nutrient Cycling in Agroecosystems. 2000;**57**(1):107-117

[10] Jiang L, Zhang L, Deng B, Liu X, Yi H, Xiang H, et al. Alpine meadow restorations by non-dominant species increased soil nitrogen transformation rates but decreased their sensitivity to warming. Journal of Soils and Sediments. 2017;**17**(9):2329-2337

[11] Robertson GP, Coleman DC, Bledsoe CS, Sollins P. Standard Soil Methods for Long-Term Ecological Research. New York, Oxford: Oxford University Press; 1999. 462 pp

[12] Zhang L, Zhang Y, Zou J, Siemann E. Decomposition of *Phragmites australis* litter retarded by invasive *Solidago canadensis* in mixtures: An antagonistic non-additive effect. Scientific Reports. 2014;**4**:5488

[13] Masse J, Prescott CE, Müller C, Grayston SJ. Gross nitrogen transformation rates differ in reconstructed oil-sand soils from natural boreal-forest soils as revealed using a ^{15}N tracing method. Geoderma. 2016;**282**:37-48

[14] Lang M, Li P, Han X, Qiao Y, Miao S. Gross nitrogen transformations in black soil under different land uses and management systems. Biology and Fertility of Soils. 2015;**52**(2): 233-241

[15] Chen Z, Zhang J, Xiong Z, Pan G, Müller C. Enhanced gross nitrogen transformation rates and nitrogen supply in paddy field under elevated atmospheric carbon dioxide and temperature. Soil Biology and Biochemistry. 2016;**94**:80-87

[16] Lan T, Han Y, Roelcke M, Nieder R, Cai Z. Effects of the nitrification inhibitor dicyandiamide (DCD) on gross N transformation rates and mitigating N_2O emission in paddy soils. Soil Biology and Biochemistry. 2013;**67**:174-182

[17] Hu Y, Zhang L, Deng B, Liu Y, Liu Q, Zheng X, et al. The non-additive effects of temperature and nitrogen deposition on CO_2 emissions, nitrification, and nitrogen mineralization in soils mixed with termite nests. Catena. 2017;**154**:12-20

[18] Jones RO, Chapman SK. The roles of biotic resistance and nitrogen deposition in regulating non-native understory plant diversity. Plant and Soil. 2011;**345**(1-2):257-269

[19] Parker SS, Schimel JP. Soil nitrogen availability and transformations differ between the summer and the growing season in a California grassland. Applied Soil Ecology. 2011;**48**(2):185-192

[20] Wrage N, Velthof GL, van Beusichem ML, Oenema O. Role of nitrifier denitrification in the production of nitrous oxide. Soil Biology and Biochemistry. 2001;**33**(12-13):1723-1732

[21] Li X, Sørensen P, Olesen JE, Petersen SO. Evidence for denitrification as main source of N_2O emission from residue-amended soil. Soil Biology and Biochemistry. 2016;**92**:153-160

[22] Zhang J, Cai Z, Cheng Y, Zhu T. Denitrification and total nitrogen gas production from forest soils of eastern China. Soil Biology and Biochemistry. 2009;**41**(12):2551-2557

[23] Erickson HE, Perakis SS. Soil fluxes of methane, nitrous oxide, and nitric oxide from aggrading forests in coastal Oregon. Soil Biology and Biochemistry. 2014;**76**:268-277

[24] Huang Y, Li D. Soil nitric oxide emissions from terrestrial ecosystems in China: A synthesis of modeling and measurements. Scientific Reports. 2014;**4**:7406

[25] Loick N, Dixon ER, Abalos D, Vallejo A, Matthews GP, McGeough KL, et al. Denitrification as a source of nitric oxide emissions from incubated soil cores from a UK grassland soil. Soil Biology and Biochemistry. 2016;**95**:1-7

[26] Hickman JE, Huang Y, Wu S, Diru W, Groffman PM, Tully KL, et al. Non-linear response of nitric oxide fluxes to fertilizer inputs and the impacts of agricultural intensification on tropospheric ozone pollution in Kenya. Global Change Biology. 2017;**23**(8):3193-3204

[27] Zhang Y, Lin F, Jin Y, Wang X, Liu S, Zou J. Response of nitric and nitrous oxide fluxes to N fertilizer application in greenhouse vegetable cropping systems in southeast China. Scientific Reports. 2016;**6**:20700

[28] Liu S, Lin F, Wu S, Cheng J, Sun Y, Jin Y, et al. A meta-analysis of fertilizer-induced soil NO and combined with N_2O emissions. Global Change Biology. 2016;**23**(6):2520-2532

[29] Yao Z, Liu C, Dong H, Wang R, Zheng X. Annual nitric and nitrous oxide fluxes from Chinese subtropical plastic greenhouse and conventional vegetable cultivations. Environmental Pollution. 2015;**196**:89-97

[30] Deng BL, Li ZZ, Zhang L, Ma YC, Li Z, Zhang WY, et al. Increases in soil CO_2 and N_2O emissions with warming depend on plant species in restored alpine meadows of Wugong Mountain, China. Journal of Soils and Sediments. 2016;**16**(3):777-784

[31] Liu S, Zhang L, Liu Q, Zou J. Fe(III) fertilization mitigating net global warming potential and greenhouse gas intensity in paddy rice-wheat rotation systems in China. Environmental Pollution. 2012;**164**:73-80

[32] Zhang L, Ma X, Wang H, Liu S, Siemann E, Zou J. Soil respiration and litter decomposition increased following perennial forb invasion into an annual grassland. Pedosphere. 2016;**26**(4):567-576

[33] Zhang L, Wang H, Zou J, Rogers WE, Siemann E. Non-native plant litter enhances soil carbon dioxide emissions in an invaded annual grassland. PLoS One. 2014;**9**(3):e92301

[34] Liu XJ, Mosier AR, Halvorson AD, Reule CA, Zhang FS. Dinitrogen and N_2O emissions in arable soils: Effect of tillage, N source and soil moisture. Soil Biology and Biochemistry. 2007;**39**(9):2362-2370

[35] Zheng X, Wang M, Wang Y, Shen R, Gou J, Li J, et al. Impacts of soil moisture on nitrous oxide emission from croplands: A case study on the rice-based agro-ecosystem in Southeast China. Chemosphere - Global Change Science. 2000;**2**(2):207-224

[36] Huang Y, Zou J, Zheng X, Wang Y, Xu X. Nitrous oxide emissions as influenced by amendment of plant residues with different C:N ratios. Soil Biology and Biochemistry. 2004;**36**(6):973-981

[37] Zhu T, Zhang J, Meng T, Zhang Y, Yang J, Müller C, et al. Tea plantation destroys soil retention of NO_3^- and increases N_2O emissions in subtropical China. Soil Biology and Biochemistry. 2014;**73**:106-114

[38] Homyak PM, Kamiyama M, Sickman JO, Schimel JP. Acidity and organic matter promote abiotic nitric oxide production in drying soils. Global Change Biology. 2016;**23**(4):1735

[39] Weil RR, Brady NC. Nitrogen and sulfur economy of soils. In: Brady RWaN, editor. The Nature and Properties of Soils. Cambridge: Pearson; 2017. pp. 584-642

[40] Huang YZ, Feng Z, Fuzhu Z. Study on loss of nitrogen fertilizer from agricultural fields and countermeasure. Journal of Graduate School, Academia Sinica. 2000;**17**(2):49-58

[41] Kramer SB, Reganold JP, Glover JD, Bohannan BJ, Mooney HA. Reduced nitrate leaching and enhanced denitrifier activity and efficiency in organically fertilized soils. Proceedings of the National Academy of Sciences of the United States of America. 2006;**103**(12): 4522-4527

[42] Phillips RL. Organic agriculture and nitrous oxide emissions at sub-zero soil temperatures. Journal of Environmental Quality. 2007;**36**(1):23-30

[43] Lu F, Wang X, Han B, Ouyang Z, Duan X, Zheng H, et al. Soil carbon sequestrations by nitrogen fertilizer application, straw return and no-tillage in China's cropland. Global Change Biology. 2009;**15**(2):281-305

[44] Li C, Frolking S, Butterbach-Bahl K. Carbon sequestration in arable soils is likely to increase nitrous oxide emissions, offsetting reductions in climate radiative forcing. Climatic Change. 2005;**72**(3):321-338

[45] Six J, Ogle SM, Conant RT, Mosier AR, Paustian K. The potential to mitigate global warming with no-tillage management is only realized when practised in the long term. Global Change Biology. 2004;**10**(2):155-160

[46] Yagioka A, Komatsuzaki M, Kaneko N, Ueno H. Effect of no-tillage with weed cover mulching versus conventional tillage on global warming potential and nitrate leaching. Agriculture, Ecosystems and Environment. 2015;**200**:42-53

[47] Zou J, Huang Y, Jiang J, Zheng X, Sass RL. A 3-year field measurement of methane and nitrous oxide emissions from rice paddies in China. Global Biogeochemical Cycles. 2005;**19**(2):153-174

[48] Wang JY, Jia JX, Xiong ZQ, Khalil MAK, Xing GX. Water regime–nitrogen fertilizer–straw incorporation interaction: Field study on nitrous oxide emissions from a rice agroecosystem in Nanjing, China. Agriculture, Ecosystems and Environment. 2011;**141**(3-4):437-446

[49] Cai Y, Chang SX, Ma B, Bork EW. Watering increased DOC concentration but decreased N_2O emission from a mixed grassland soil under different defoliation regimes. Biology and Fertility of Soils. 2016;**52**(7):987-996

[50] Davidson EA. Sources of nitric oxide and nitrous oxide following wetting of dry soil. Soil Science Society of America Journal. 1992;**56**(1):95-102

[51] Yan X, Shi S, Du L, Guangxi X. N_2O emission from paddy soil as affected by water regime. Acta Pedologica Sinica. 2000;**37**(4):482-489

[52] Marsden KA, Marín-Martínez AJ, Vallejo A, Hill PW, Jones DL, Chadwick DR. The mobility of nitrification inhibitors under simulated ruminant urine deposition and rainfall: A comparison between DCD and DMPP. Biology and Fertility of Soils. 2016;**52**(4): 491-503

[53] Linquist BA, Liu L, van Kessel C, van Groenigen KJ. Enhanced efficiency nitrogen fertilizers for rice systems: Meta-analysis of yield and nitrogen uptake. Field Crops Research. 2013; **154**:246-254

[54] Huérfano X, Fuertes-Mendizábal T, Fernández-Diez K, Estavillo JM, González-Murua C, Menéndez S. The new nitrification inhibitor 3,4-dimethylpyrazole succinic (DMPSA) as an alternative to DMPP for reducing N_2O emissions from wheat crops under humid Mediterranean conditions. European Journal of Agronomy. 2016;**80**:78-87

[55] Lam SK, Suter H, Mosier AR, Chen D. Using nitrification inhibitors to mitigate agricultural N_2O emission: A double-edged sword? Global Change Biology. 2017;**23**(2):485-489

[56] Zhu T, Zhang J, Huang P, Suo L, Wang C, Ding W, et al. N_2O emissions from banana plantations in tropical China as affected by the application rates of urea and a urease/nitrification inhibitor. Biology and Fertility of Soils. 2015;**51**(6):673-683

[57] Abalos D, De Deyn GB, Kuyper TW, van Groenigen JW. Plant species identity surpasses species richness as a key driver of N_2O emissions from grassland. Global Change Biology. 2014;**20**(1):265-275

[58] Shi X, Hu H-W, Kelly K, Chen D, He J-Z, Suter H. Response of ammonia oxidizers and denitrifiers to repeated applications of a nitrification inhibitor and a urease inhibitor in two pasture soils. Journal of Soils and Sediments. 2017;**17**(4):974-984

[59] Guardia G, Cangani MT, Andreu G, Sanz-Cobena A, García-Marco S, Álvarez JM, et al. Effect of inhibitors and fertigation strategies on GHG emissions, NO fluxes and yield in irrigated maize. Field Crops Research. 2017;**204**:135-145

[60] Sun L, Lu Y, Yu F, Kronzucker HJ, Shi W. Biological nitrification inhibition by rice root exudates and its relationship with nitrogen-use efficiency. The New Phytologist. 2016; **212**(3):646-656

[61] Bending GD, Lincoln SD. Inhibition of soil nitrifying bacteria communities and their activities by glucosinolate hydrolysis products. Soil Biology and Biochemistry. 2000; **32**(8):1261-1269

[62] Castaldi S, Carfora A, Fiorentino A, Natale A, Messere A, Miglietta F, et al. Inhibition of net nitrification activity in a Mediterranean woodland: Possible role of chemicals produced by *Arbutus unedo*. Plant and Soil. 2009;**315**(1):273-283

[63] Subbarao GV, Rondon M, Ito O, Ishikawa T, Rao IM, Nakahara K, et al. Biological nitrification inhibition (BNI)—Is it a widespread phenomenon? Plant and Soil. 2007;**294**(1):5-18

[64] Liimatainen M, Martikainen PJ, Maljanen M. Why granulated wood ash decreases N_2O production in boreal acidic peat soil? Soil Biology and Biochemistry. 2014;**79**:140-148

[65] Ameloot N, Sleutel S, Das KC, Kanagaratnam J, de Neve S. Biochar amendment to soils with contrasting organic matter level: Effects on N mineralization and biological soil properties. GCB Bioenergy. 2015;**7**(1):135-144

[66] Case SDC, McNamara NP, Reay DS, Stott AW, Grant HK, Whitaker J. Biochar suppresses N_2O emissions while maintaining N availability in a sandy loam soil. Soil Biology and Biochemistry. 2015;**81**:178-185

[67] Lehmann J, Gaunt J, Rondon M. Bio-char sequestration in terrestrial ecosystems—A review. Mitigation and Adaptation Strategies for Global Change. 2006;**11**(2):395-419

2

Controlled-Release Fertilizers as a Means to Reduce Nitrogen Leaching and Runoff in Container-Grown Plant Production

Jianjun Chen and Xiangying Wei

Abstract

Container-grown plants refer to plants produced in confined volume filled with substrates. The substrates endogenously have limited nutrients and low water-holding capacity. Plants grown in the containers must be fertilized and watered frequently varying from daily to weekly. Frequent fertilization and irrigation can result in nutrient leaching and/or runoff. Since nitrogen (N) is a key component of the majority of fertilizers, container plant production has been viewed as a source of N leaching and/or runoff. The leaching and runoff, if in large quantities on a year-round basis, could affect surface and ground water quality. Application of controlled-release fertilizers (CRFs) has been reported to have less N leaching than plants fertilized with water-soluble fertilizers (WSFs). However, there are different types of CRFs with different compositions and longevities on the market. Container plants also differ greatly in their growth and development and in N requirement. Thus, production of high-quality container plants with minimum N leaching using CRFs still remains challenging. This article is intended to discuss characteristics of container plant production and N leaching and runoff during production, and to document that CRF application can reduce N leaching and/or runoff. Certain requirements for future development of CRFs are also discussed.

Keywords: container-grown plants, controlled-release fertilizers, nitrogen leaching and runoff, nitrate

1. Introduction

Container-grown plants refer to those grown from seedlings, liners, rooted cuttings or grafted plants in containers or pots filled with substrates to marketable sizes or harvestable stages. Substrates or growing media are comprised of peat, perlite, soil, vermiculate or other organic components in different proportions. Many plants can be produced in containers including floriculture, nursery, fruit and vegetable crops. According to the United States Department of Agriculture (USDA) National Agriculture Statistics Service [1], floriculture crops are ornamental plants without woody stems, such as annual and perennial bedding and garden plants, cut flowers, cut cultivated greenery, potted flowering plants, tropical foliage plants and unfinished propagative material. Nursery crops are finished ornamental plants and trees with woody stems that are used for outdoor landscaping. Nursery crops also include ornamental vines, turfgrass sod and other groundcovers. Fruit and vegetable crops can also be produced in containers. Container fruit crops commonly include apple, blueberry, cherry, citrus, fig, orange, peach, pear and plum trees. Container vegetables include basil, beet, carrot, cucumber, ginger, lettuce, radish, onion, strawberry and tomato.

Container crop production has become increasingly popular over the past 50 years [2, 3]. This is because container plant production has several advantages over traditional field production: (1) container plants are grown in substrates, not in soil, their production does not rely on arable land; (2) container sizes, substrate types and pH, pest, disease, water and nutrient management are easier to control or modify in container plant production than field production [4]; (3) plants grown in containers have a greater fine root mass compared to field-grown plants [5, 6]. Root surface area of holly plants (*Ilex x attenuata* Ashe 'East Palatka') grown in containers increased more than twofold than those grown in ground, and plant leaf dry weight and total top dry weight were 22.5 and 15% greater, respectively, when grown in containers [5]; (4) container plants are more convenient for moving and shipping, allowing more operational flexibility and improving shipping efficiency; (5) containerization allows growers to sell plants throughout the year regardless of soil conditions or plant growth stage, which increases productivity per unit area; (6) container-grown plants exhibit much less transplant shock and higher survival rates after transplanting compared to field-grown plants [7]; (7) plant spacing for containers ranges from 17,300 to 247,000 plants per hectare in nurseries and 99,000–865,000 plants per hectare in greenhouse production compared to 1480–12,360 plants per hectare in field production [2], thus, much more plants are produced per hectare by container production and more profit is made per unit area and (8) container-grown plants can be consolidated to provide space for growing additional plants after inventories are sold. However, such consolidation will not be possible for field-grown plants. More plants per unit area of container-grown crops means higher revenue compared with field production [8].

Currently, approximately 90% of greenhouse, nursery and floriculture crops in the USA are produced in containers [9]. The floriculture and nursery industries are strong and fast-growing sectors of US agriculture. Together, it accounts for a total of $11.7 billion in sales in 2009, a 10.7% increase since 1998. Floriculture and nursery crops comprise almost 30% of the specialty crops grown in the USA [10]. Since floriculture and nursery crops are used largely for decoration of the surrounding environment, they are produced in every state in the USA. The leading floriculture and nursery states are California, Florida, Michigan, Texas and New York [11]. The floriculture and nursery industries generate 170,000 jobs worth $3.78 billion to California's economy [12]. Floriculture and nursery crops are among the largest agricultural commodity groups in Florida. According to the Census of Horticulture Specialties for 2014

[13], there were over 2069 commercial nursery and greenhouse farms in Florida, with total sales of $1.796 billion, and $3.291 billion in capital assets in land, buildings and equipment.

2. Nitrogen loss during container plant production

The rapid increase in container plant production, however, has been under increasing scrutiny because of potential contamination of surface and/or ground water by nutrient elements, particularly nitrogen (N). In Europe, extremely high NO_3—N concentrations, up to 2000 kg N/ha, were found in soil depth of 100 cm underlying commercial greenhouses [14]. In Connecticut, US, NO_3—N accumulation over 2300 kg/ha was recorded in soil under decades-old greenhouses [15]. A survey conducted in six states in the US such as Alabama, Florida, New Jersey, North Carolina, Ohio and Virginia suggested that the levels of runoff NO_3—N varied from 0.5 to 33 mg/L for container nurseries using controlled-released fertilizers (CRFs) and 0.1–135 mg/L for those using CRFs supplemented with water soluble fertilizers (WSFs) [16]. Also a survey completed from 11 nurseries in southern California showed that media NO_3—N concentrations in runoff exceeded 10 mg/L in most nurseries [17]. NO_3—N in irrigation runoff in a foliage plant production nursery in southern Florida ranged from 41 to 386 mg/L depending on irrigation methods [18].

Nitrate N is also leached from container substrates during crop production. In a container production of *Ilex crenata* Thunb. 'Compacta', Fare et al. [19] reported that the percentage of applied N leached as NO_3—N ranged from 46% when 13-mm irrigation was applied in 3 cycles to 63% when 13-mm irrigation was applied in a single cycle. Broschat [20] investigated N leaching from a container substrate comprised of 50% pine bark, 40% sedge peat and 10% sand and reported 3710 mg of NO_3—N could be leached per container during a 6-month production of *Spathiphyllum* Schott. This could be translated to the annual loss of 666 kg of NO_3—N per hectare. Container production of poinsettia (*Euphorbia pulcherrima* Willd. ex Klotzsch), a potted floriculture crop, fertilized with a solution containing 210 mg/L of N showed that 40 and 60% of applied N was leached from containers when fertigated with leaching fractions of 0.2 and 0.4, respectively (leaching fraction is defined as the volume of leachate divided by the irrigation solution applied) [21]. Production of container azalea (*Rhododendron* L. 'Karen') with a weekly application of N at 250 mg/L could result in the loss of N at 924 kg/ha [22]. Container production of a bedding plant *Impatiens walleriana* Hook. f. by overhead irrigation resulted in 25.6% of the total applied water leaching out of the container and 34% fell between containers, and weekly N concentrations ranged from 137 to 153 mg/L in leachate and 165–256 in runoff water during a 6-week production [23]. In Spain, NO_3—N in leachates of container-grown *Aloe vera* L., *Kalanchoe blossfeldiana* Poelln. and *Gazania splendens* Lem. ranged from 15 to 90 mg/L when plants were watered in 45% of the container capacity using nutrient solutions containing 372 mg/L NO_3—N and different concentrations of sodium.

Nitrate N resulted from leaching and runoff could enter rivers, lakes and estuaries contributing to water eutrophication. N concentrations greater than 0.4 mg/L have been shown to accelerate eutrophication, causing algal blooms [24]. NO_3—N contamination of groundwater is a major human health concern, particularly to infants when nitrate is transformed to nitrite in the digestive system [25, 26]. The nitrite can oxidize the iron in hemoglobin of red blood cells, resulting in the formation of methemoglobin. Because methemoglobin lacks the ability to bind (or release) oxygen, blood will be unable to carry sufficient oxygen to the individual body cells, causing the veins and skin to appear blue. This is a condition known as methemoglobinemia (sometimes referred to as "blue baby syndrome") [27]. Most humans over 1 year of age have the ability to rapidly convert methemoglobin back to oxyhemoglobin. Thus, the total amount of methemoglobin within red blood cells remains low despite relatively high levels of nitrate/nitrite uptake. In infants under 6 months of age, however, the enzyme systems responsible for reducing methemoglobin to oxyhemoglobin are incompletely developed and methemoglobinemia can occur. This also may happen in older individuals who have genetically impaired enzyme systems for metabolizing methemoglobin. Furthermore, prolonged nitrate and nitrite ingestion could increase risks of certain cancers [28].

The US Public Health Service adopted drinking water standards and set the recommended limit for NO_3—N at 10 mg/L in 1962 [29]. This drinking water standard was established to protect the health of infants, children, pregnant women, the elderly and immune-compromised individuals. The potential health hazard for others depends on the individual's reaction to NO_3—N and the total ingestion of NO_3—N and nitrites from all sources. From 1970 to 1992, the US Geological Survey found that 9% of the private wells that were tested exceed the recommended limit of 10 mg/L NO_3—N [30]. The US Environmental Protection Agency (USEPA) [31] has since adopted the 10 mg/L standard as the maximum contaminant level (MCL) for NO_3—N and 1 mg/L for nitrite-N for regulated public water systems. Subsequent reviews of this standard have not resulted in any changes.

Applied N can also be evolved as ammonia (NH_3) or nitrous oxide (N_2O) gases. It was estimated the 10% of manufactured N fertilizers could be volatilized as NH_3 gas [32] and 1% of N applied in inorganic forms was lost to the atmosphere as N_2O [33]. The volatilization of both NH_3 and N_2O are serious environmental concern as NH_3 contributes to photochemical smog [34] and N_2O is a potent greenhouse gas with a global warming potential of 310 times greater than carbon dioxide [35].

3. Methods for reducing N loss

Different strategies and methods have been proposed and used for reducing NO_3—N leaching and runoff during the production of container-grown plants. Chen et al. [36] suggested that approaches to NO_3—N leaching and runoff should take plant species, fertilizer application rates, container substrate and irrigation methods into consideration for developing best management practices (BMPs), which include (1) understanding plant species requirement for N and application of N based on plant need; (2) improving physical and chemical properties of container substrates and increasing their holding capacities for water and nutrients, particularly NO_3—N; (3) using controlled-release fertilizers to reduce NO_3—N leaching; and (4) irrigation system improvement by using either drip irrigation or subirrigation to reduce leaching and runoff.

The rationales for the solutions in Chen et al. [36] were as follows: (1) plants are generally inefficient in N utilization. It has been well documented that crops directly utilize less than half (rarely more than 40%) of applied N [37]. Moreover, overall N-use efficiency (NUE) declined with increasing N-fertilizer application [38]. However, recommended fertilizer rates for container-grown plants are often much higher than actual plant needs. As shown by Chen et al. [36], N rates for some container-grown crops ranged from 1067 to 2354 kg per hectare per year, which is 10–15 times higher than those recommended for many agronomic field crops. Such high recommendation rates, along with extensive irrigation further enhance N leaching and runoff. In addition, different plant species and even their different cultivars differ in N requirement. Thus, a nursery operation should have different fertilizer programs suited to each species or a group of species [36, 39]. (2) Since the commercialization of container substrates after the World War II, substrate components have been predominantly pine bark, peat, vermiculite and perlite. Components newly introduced are coconut coir and polymer gel [36]. Accumulated research evidence indicates that specific zeolites and biochars have an added adsorption capacity for nutrient elements, including N [36, 40, 41]. Incorporating selected zeolites and/or engineered biochars into substrate formation should improve nutrient holding capacity and reduce nutrient leaching. (3) N is the most abundant element in most fertilizer formulation. This is due to the fact that N is the most important nutrient to plant growth and development and a plant generally absorbs more N than other element. Common N compounds in fertilizer formulations include ammonium (NH_4^+), nitrate (NO_3^-) and urea [$CO\text{-}(NH_2)_2$]. Plants can directly absorb NH_4^+ and NO_3^-, but not urea. Urea in soil is hydrolyzed into NH_4^+ by microorganisms. NH_4^+ can also be nitrified by soil bacteria to NO_3^-. Between NH_4^+ and NO_3^-, most plant species prefer NO_3^- over NH_4^+ although a few plant species prefer NH_4^+. Additionally, as an anion, NO_3^- does not bind readily to the predominantly negatively charged soil and substrate colloids. Thus, NO_3^- is highly mobile in soil or substrate. To reduce the mobility of NO_3^-, encapsulated N fertilizers should be a better choice, and this is why CRFs have been developed [42]. (4) As water and fertilizer are interrelated in container plant production, one way to avoid N runoff or leaching into groundwater is to use zero runoff subirrigation [36]. Growers in Florida adopting either ebb-and-flow or capillary mat irrigation reported 20% reduction of fertilizer use and 75% reduction of water consumption in containerized plant production. Another irrigation method, which can achieve minimal runoff and less salt buildup in substrates, is to use surface irrigation systems, but to also capture, retain and recycle the runoff and stormwater within the boundaries of the production facility [43]. This is exemplified by whole greenhouse/nursery recycling system, called the total nursery recycling system. This recycling system includes (1) stormwater and/or irrigation runoff collection, (2) sedimentation, flocculation, filtration and disinfection, if necessary and (3) irrigation. Skimina [44] tested more than 100 species of landscape ornamental plants using this system and found that the range of plant growth response was 73–171% relative to control plants. However, few nurseries have used this total nursery recycle system for the production of greenhouse container plants. Growers were concerned about the feasibility and reliability of the water sources for the production of high-quality plants. As a result, the use of CRFs is considered to be a more convenient method for container plant production, while potentially reducing N leaching and runoff.

4. Controlled-release fertilizers and their applications

Controlled-release fertilizers are granules that are purposely designed to release nutrients in a controlled, delayed manner in synchrony with plant requirements for nutrients. CRFs belong to enhanced-efficiency fertilizers (EEFs), which is defined as "fertilizer products with characteristics that allow increased plant uptake and reduce the potential of nutrient losses to the environment (e.g., gaseous losses, leaching or runoff) when compared to an appropriate reference product" [45]. EEFs include CRFs, slow-release fertilizers (SRFs), stabilized N fertilizers, nitrification inhibitors and urease inhibitors. The terms, CRFs and SRFs, are generally considered analogous. However, Trenkel [42] and Shaviv [46] clearly defined their differences. In SRFs, the pattern of nutrient release is generally unpredictable and remains subject to change by soil type and climatic conditions. In contrary, the pattern, quantity and time of release can be predicted, within limits, for CRFs. This review, as indicated by the title, is intended to focus on CRFs only.

4.1. Common CRFs used in container plant production

Table 1 lists the leading producers and/or suppliers of CRFs including Agrium Inc., Calgary, Alberta, Canada; Chisso Asahi Fertilizer Co., Tokyo, Japan; Everris NA, Inc., a subsidiary of Israel Chemicals Ltds; Haifa Group, Haifa, Israel; Shandong Kingenta, Shandong, China; and J.R. Simplot, Boise, Idaho, US. CRFs produced by Agrium includes those with trade names: ESN, Polyon, Duration and XCU in which urea is coated by polymer. Popular CRFs include Nutricote and Meister are manufactured by Chisso Asahi Fertilizer, and urea is coated by resin. Everris, Inc. produces Agrocote, Osmocote and Poly-S where urea is coated by sulfur/polymer and resin, resin and sulfur and polymer, respectively. Urea in Multicote produced by Haifa Group is coated by resin, and Florikote produced by J.R. Simplot is coated by polymer [47].

Urea is a major N source for formulation of CRFs. Urea is actually the most widely used fertilizer globally because of its high N content (46%). Urea has the lowest transportation costs per unit of N and ease of application [32, 33]. Additionally, urea is highly soluble in water and has much lower risk of causing fertilizer burn to crops. Other N sources used in the formulation of CRFs include ammonium nitrate, ammonium phosphate and potassium nitrate. Sulfur was initially used as a material for coating urea. The Tennessee Valley Authority developed the production process for sulfur-coated urea more than 50 years ago [48] in which preheated urea granules were coated with molten sulfur and wax. The sulfur coating is an impermeable layer which can be slowly degraded through microbial activities and soil chemical and physical processes. The uniformity in coating coverage and thickness of coating determine the speed and effectiveness of urea release. Incompletely coated or cracked prills are immediately amenable to dissolution in soil water and hydrolysis by urease. However, due to its amorphous nature, sulfur alone cannot be used to produce well controlled-release urea. Subsequently, many other materials, such as binders, plasticizers and sealants were evaluated for reducing the immediate burst effect. Some tested materials reduced the burst effect but increased the cost and complexity [48]. As a result, sulfur alone has not been used as a coating agent. If used, it is in combination with some polymers. Polymer coating is a more sophisticated technology, and it consists of a core of soluble nutrients surrounded by a polymer coating. Each coated

Trade name	Manufacturer	Type of CRFs	Coating materials	Selected commercial products
Agrocote®	Everris, Inc.	Polymer/ resin-coated	Coated with polymer/sulfur and resin coatings	Agrocote® 19-6-12, Agrocote® 39-0-0 + 11% S
Duration®	Agrium, Inc.	Polymer-coated	Clay-coated PCU or micro-thin polymer membrane	Duration®CR, Duration® 44-0-0, Duration® 19-6-13
ESN®	Agrium, Inc.	Polymer-coated urea	Urea is coated with flexible micro-thin polymer	ESN® 44-0-0 (Environmentally smart nitrogen)
Florikote	J.R. Simplot	Polymer-coated	Coated with dual layer technology	Florikote® 40-0-0), Florikote® 12-0-40, Florikote® 19-6-13,
Meister®	Chisso-Asahi Fertilizer Co.	Resin-coated	Granular urea coated with a polymer composition of natural products, resin and additives	Meister® 15-5-15, Meister® 19-5-14
Multicote®	Haifa Group	Resin-coated	Nutrients encapsulated in a polymeric shell	Multicote® Agri 6 22-8-13, Multicote® Agri 6 34-0-7, Multicote® Agri 8 34-0-7
Nutricote®	Chisso-Asahi Fertilizer Co.	Polymer-coated NPK	Polymer coating with a special chemical release agent	Nutricote® NPK 20-7-10
Osmocote®	Everris, Inc.	Organic resin-coated	Granule contains NPK coated with organic resin	Osmocote® Exact, Osmocote® Exact Mini, Osmocote® Pro, Osmocote® Start
Polyon®	Agrium, Inc.	Polymer-coated	Coated with patented "Reactive Layers Coating" (ultra-thin ployurethane coating)	Polyon® 41-0-0, Polyon® NPK 20-6-13
Poly-S®	Everris, Inc.	Polymer-/sulfur-coated urea	Urea coated with sulfur followed by polymer	Poly-S® 37-0-0
TriKote®	Agrium, Inc.	Polymer-/sulfur-coated urea	Urea coated with polymer and sulfur	Trikote® 42-0-0

Table 1. Common controlled-release fertilizers (CRFs) used for production of container-grown plants, vegetables and turfgrass.

particle is known as a prill and nutrient release is controlled by the chemical composition and thickness of the polymer coating. Polymers could be thermosetting, thermoplastic or biodegradable. Some of the common thermoset polymers include urethane resin, epoxy resin, alkyd resin, unsaturated polyester resin, phenol resin, urea resin, melamine resin, phenol resin and silicon resin [49]. Among them, urethane resin is very commonly used [50]. Polyacrylamide is known to reduce soil erosion, and more studies should be conducted for its use in CRFs [46, 51]. Thermoplastic resins are not very commonly used because they are either not soluble in a solvent or make a very viscous solution which is not suitable for spraying; however, polyolefin is used for coating the fertilizer granules. Biodegradable polymers are naturally available and are known to be environmentally friendly because they decompose in bioactive environments and degrade by the enzymatic action of microorganisms, such as bacteria, fungi and algae and their polymer chains may also be broken down by nonenzymatic processes, such as chemical hydrolysis. Commercially, polymers used for coating urea include alkyd resin (Osmocote), polyurethane (Polyon, Multicote and Plantacote) and thermoplastic polymers.

4.2. N release patterns from CRFs

Different models have been proposed for explaining nutrient release patterns of CRFs [45, 52, 53]. It is generally agreed that nutrient release is governed by diffusion mechanisms. Shaviv [46] and Liu [54] proposed a multi-stage diffusion model. According to this model, after application of a coated fertilizer, irrigation water penetrates the coating to condense on the solid fertilizer core followed by partial nutrient dissolution. As osmotic pressure builds within the containment, the granule swells and causes the occurrence of two processes. One could be "catastrophic release". When osmotic pressure surpasses threshold membrane resistance, the coating bursts and the entire core are spontaneously released. This is also referred to as the "failure mechanism". In the second, if the membrane withstands the developing pressure, core fertilizer is thought to be released slowly via diffusion for which the driving force may be a concentration or pressure gradient, or combination thereof called the "diffusion mechanism". The failure mechanism is generally observed in frail coatings (e.g. sulfur or modified sulfur), while polymer coatings (e.g. polyolefin) are expected to exhibit the diffusion release mechanism [48]. Nutrient release from CRFs is generally classified into linear and sigmoidal patterns [42, 55]. In most cases, the energy of activation of the release, EA_{rel}, is calculated on the basis of estimates of the rate of the release (percentage release per day) during the linear period obtained from the release curves [52]. Nutrient release profiles are established in both laboratory and field tests. Laboratory tests include extraction of nutrients at 25, 40 and 100 °C. Field tests include the placement of net bags in the ploughed layer or soil in the actual production soil [42]. Shaviv [56] reported that nutrient release consists of three stages: the initial stage or lag period during which little release is observed; the constant release stage characterized with an increasing release; and the last or mature stage where nutrient release is gradually reduced.

Nitrogen release profiles from CRFs have been studied during container plant production. CRFs are either top dressed (granules are placed on the surface of container substrate) or incorporated (granules are mixed with container substrate before being used for potting). Plants are watered in a specific leaching fraction. Leachates are captured and collected weekly. NO_3–N and NH_4–N in each collected leachate are analyzed. This method is not designed to determine the amount of N released from a CRF over a period of time since N leaching, volatilization and absorption by plants occur simultaneously. It is intended to use the leached N as an indicator for analyzing N release patterns. Leached N can be plotted based on the cumulative N leached (the percentage of N leached in reference of total N applied) at a specific production time or period [57, 58] or simply plotted as concentration of N per container against time (days or weeks) sampled [20, 59]. Depending on the types and formulation of CRFs, container substrate components, production temperature and irrigation volume and frequency, different N release profiles have been reported. Based on the cumulative N leached, the release curves can be generalized to two types: linear [57, 60] and sigmoidal [58] curves. Regardless of N sources in CRFs, NO_3–N is the main N leached, accounting for 80–90%, suggesting that nitrification is active in container substrates [59]. Temperature is a force driving N release from CRFs. Cumulative N leached from both sand and bark substrates incorporated with an Osmocote fertilizer in Florida was much greater than in Ohio [58]. The methods of CRF application affect

N release or loss. More N leached from substrates incorporated with CRFs than those topdressed [59]. Furthermore, substrate moisture is a key factor influencing nutrient release from CRFs.

4.3. CRF application reduces N leaching and runoff in container plant production

Due to their controlled-release characteristics, research has been conducted since the 1960s on the feasibility of the use of CRFs for container plant production [61, 62]. With the increasing availability of CRF types and awareness of N leaching and runoff in the 1980s, research has shifted attention towards N release patterns and N leaching and runoff. **Table 2** presents some representative studies conducted in container-grown ornamental plants, turfgrass, citrus and field crops such as potato. At least six conclusions can be drawn from these studies: (1) the use of CRFs reduces N leaching and/or runoff. Depending on fertilizer types, plant species, application methods and environmental conditions, N in leachates or runoff resulting from CRF application could be approximately 50% less than WSF application. Mello et al. [63] showed that polymer-coated urea reduced N leaching by 64.5% compared to conventional urea in container production of *Lantan camara* L. Broschat [20] showed that 48 and 54% of applied N were leached from a liquid WSF and a granular WSF, respectively, in container production of *Spathiphyllum*, while N leached from two CRFs were 29 and 35%, respectively. N concentrations in runoff derived from container greenhouse production facilities was 43.1 mg/L compared to 4.4 mg/L after the same facilities switched from WSF application to the use of CRFs [64]. (2) CRF application also reduces N leaching in field crop production. NO_3—N in soil water collected by lysimeters 30 cm below potato production bed ranged from 7 to 45.1 mg/L from 39 to 95 days after planting compared to 15.6–172 mg/L fertilized with a WSF [65]. (3) CRFs reduce N_2O emission. Application of urea in turfgrass production resulted in 127–476% more N_2O emission into the atmosphere compared to 45–73% emission by using a CRF [66]. (4) Plant growth or yield resulting from CRF application are equal to or better than those produced by WSF including ornamental plants [16, 20], field crops [65, 67] and turfgrass [66]. (5) CRFs vary in N release and thus N leaching. N concentrations in leachates varied from 60 to 275 mg/L in container production of *Vibrunum* [16] and from 50 to 400 mg/L in other container ornamental plant production [68] due in part to the application of different CRFs. (6) CRF application may improve the rhizosphere microbial community. A study conducted in Japan showed that application of urea-formaldehyde fertilizers to onion bulbs and main roots of sugar beet changed the diversity of the microbial community and the abundances of certain bacterial species [69].

Furthermore, the use of CRFs has been shown to increase nutrient use efficiency and decrease fertilizer application. Trenkel [42] suggested CRFs can potentially decrease fertilizer use by 20–30% of the recommended rate of a conventional fertilizer while obtaining the same yield. In several field trials in Florida, young or non-bearing citrus trees fertilized with CRFs at a 50% of the recommended rate performed equally well compared to 100% of the recommended rate with WSF [70]. The same magnitude of reduction happened in potato production in Florida [71]. Applying CRFs generally reduces salt accumulation, thus minimizing the possibility of leaf burning. The use of CRFs reduces labor costs. Depending on plant species, one application of appropriate amount of CRFs will ensure plant growth until marketable size, while WSF fertilizers have to be applied as fertigation weekly, and sometimes daily.

Plant species	Growing substrate	Fertilizer	N leached or N conc. in leachates/runoff	Plant growth or comments	References
Spathiphyllum spp. Schott	Pine bark/peat/sand	Liquid WSF	48% of applied N leached	No plant dry weigh differences among fertilizer treatments	Broschat [20]
		Dry granular WSF	54% of applied N leached		
		Lightly-coated CRF	29% of applied N leached		
		Heavily-coat CRF	35% of applied N leached		
Lantana camara L.	Krome soil	Urea	N leached from containers fertilized with polymer-coated urea was 64.5% lower than those fertilized with urea	More flowers were produced by plants fertilized with polymer-coated urea than urea	Mello et al. [63]
		Polymer-coated urea			
A leaching column study without plants	Florida sandy soil	Ammonium nitrate	100% of applied N leached	Much less N was leached from Meister than isobutylidene coated urea, and all N was leached from ammonium nitrate.	Wang and Alva [79]
		Isobutylidene diurea	32% of applied N leached		
		Meister polyolefin resin-coated urea	12% of applied N leached		
A leaching column study without plants	Fine sandy soil	Urea	28% of applied N leached	Meister and Osmocote leached much less N than urea	Paramasivam and Alva [80]
		Poly-S	12% of applied N leached		
		Meister polyolefin resin-coated urea	6% of applied N leached		
		Osmocote	5% of applied N leached		
Different foliage and flowering crops	Peat/pine bark/sand	Polymer-coated urea (41-0-0)	23.1 mg/L	Polymer-coat urea provided stable and long-last release of NO_3^-—N and NH_4^+ than the other two. Trikote released more NH_4^+ than the other two. Plant growth was not significantly affect by treatments.	Blythe et al. [81]
		Trikote (42-0-0)	64.9 mg/L		
		Regalite Nitroform (38-0-0)	27.6 mg/L		
Container-grown *Viburnum odoratissimum* Ker-Gawl	Pine bark/peat/sand	Nutricote	275 mg/L	The highest concentrations of NO_3^-N leached from CRFs during a 4.5-month production period. Plant growth indices were not significantly affected by CRFs	Yeager and Cashion [16]
		Osmocote	220 mg/L		
		Prokote Plus	125 mg/L		
		Woodace	60 mg/L		

Plant species	Growing substrate	Fertilizer	N leached or N conc. in leachates/runoff	Plant growth or comments	References
A greenhouse leaching study without plants	Peat/vermiculite/sand	Nutricote 18-6-8	32% of applied N leached	Osmocote 18-6-12, Nutricote, and Woodave exhibited less response to temperature increase and thus less N leaching	Cabrera [59]
		Osmocote 18-6-12	36% of applied N leached		
		Osmocote 18-6-12 FS	51% of applied N leached		
		Osmocote 24-4-8 HN	49% of applied N leached		
		Polyon 25-4-12	45% of applied N leached		
		Prokote Plus 20-3-10	50% of applied N leached		
		Woodave 20-4-11	30% of applied N leached		
Container ornamental plants	Peat/pine bark/sand	Osmocote	50 mg/L	Osmocote steadily release of N	Merhaut et al. [68]
		Polyon	200 mg/L	N release reached a peak on week 9 then stabilized	
		Multicote	400 mg/L	N release reached a peak on week 8 then stabilized	
		Nutricote	400 mg/L	N release reached a peak on week 9 then stabilized	
Potato	Loamy sand	Polymer-coated urea	21.3 kg NO_3–N/ha	Apparent fertilizer N recovery with PCU (65% averaged over four rates) tended to be higher than split-applied soluble N (55%) at equivalent rates	Wilson et al. [67]
		Soluble N	26.9 kg NO_3–N/ha		
Foliage plants	Canadian peat/pine bark/lava rock	WSF	43.1 mg/L in runoff	Plant growth was not affected by switching from a WSF to a CRF	Wilson and Albabo [64]
		CRF	4.4 mg/L in runoff		
Turfgrass	A Timpanogos loam soil	Polymer-coated urea	1.25 mg N_2O–N/m²/h	Polymer-coated urea emitted significantly low amount of N_2O–N	Lemonte et al. [66]
		Urea	2.22 mg N_2O–N/m²/h		

Table 2. Nitrogen lost in leachates, runoff water or emitted into the atmosphere when controlled-release fertilizers (CRFs) only or CRFs with water soluble fertilizers (WSFs) used in crop production or leaching experiments.

4.4. Problems associated with the use of CRFs in container plant production

Several problems are associated with the use of CRFs in production of container-grown plants. Some are due to CRF design and formulation: (1) CRFs cost considerably more to manufacture than conventional fertilizers, thus they are more expensive. For example, one ton of a CRF (44% N) could be $650 compared to one ton of urea (46% N) at $481 [72]. (2) CRFs may not release nutrients based on plant requirements. This could be due to several factors: the formulation of nutrient elements, the permeability and durability of coating materials, plant species and growth stage difference, and inappropriate placement of CRFs, substrate moisture levels and microbial effects as well as production environmental conditions. The N release pattern of CRFs in laboratory tests is generally represented by a sigmoidal curve (**Figure 1**). Such release pattern is appropriate for field-grown crops, such as corn, wheat and tomato, as the lag phase is appropriate for seedling growth or allow transplants to get recovered and established from transplanting shock; log phase is designed for rapidly vegetative growth and the transition from vegetative growth to reproductive growth; and the stationery phase would allow nutrients absorbed or stored in vegetative organs to translocate to reproductive organs. The sigmoidal curve, however, may not be an ideal pattern for producing container-grown plants. Container plants are initiated with rooted cuttings or liners which already have well established root systems. Once the liners are planted in containers, they grow in an accelerated speed and require a steady supply of nutrient without lag phase. Thus, we propose here that CRFs for container-grown plants should have a nutrient release pattern, called "the expected curve for container plants" presented in **Figure 1**, not a sigmoidal curve. Many CRFs were predominantly developed based on the sigmoidal release curve, thus, they may not be ideally suitable for producing container-grown plants. (3) Thus far,

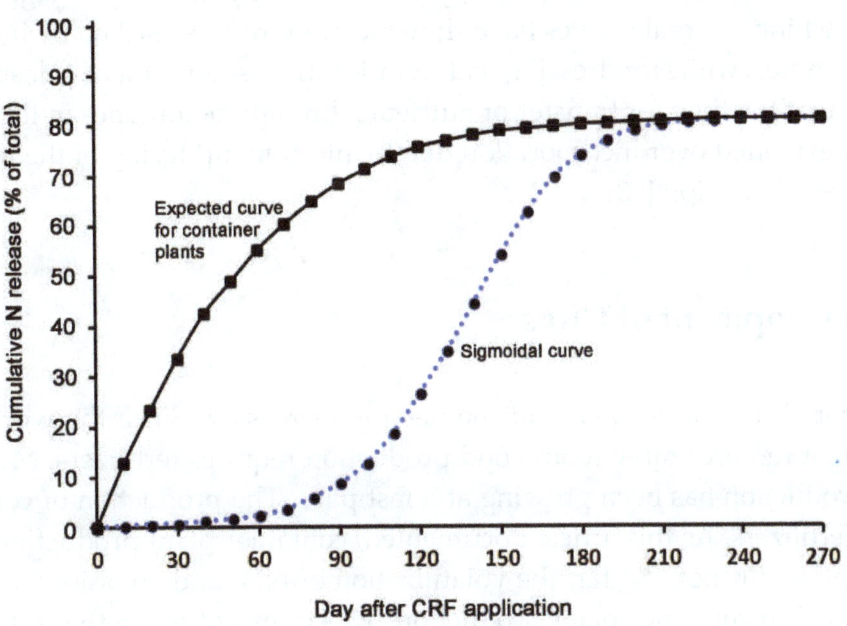

Figure 1. A proposed nutrient release curve versus the commonly preferred sigmoidal curve used for developing controlled-release fertilizers. Controlled-release fertilizer with the proposed curve could be more suitable for production of container-grown plants than those with a sigmoidal curve.

nutrient formulations of few CRFs are developed according to specific groups of plant species in nutrient requirements. Some species have low nutrient requirements. For example, ornamental foliage plants largely originate from the rainforest floor, and they inherently require low light levels and low nutrient supply for slow growth. This group of plants should be fertilized by CRFs that have complete nutrient elements with a rather slower release pattern. CRFs designed for use in subtropics and tropics should be different from those to be used in temperate regions. As shown by Birrenkott et al. [58], the same CRF for growing the same crop released different amount of N in Florida and Ohio.

Other problems with the use of CRFs are related to inappropriate application. The first is the misuse of CRFs. A CRF that is supposed to be used in the Southern USA, but used in the Northern USA, which may cause reduced release of required nutrients; as a result plant growth will be slow. If a CRF designed for container-woody ornamental plants is used for production of annual bedding plants, plant growth may slow down due to limited release of nutrients. The second problem is to apply either too little or too much CRFs. The use of an extra amount is the most common problem in container plant production. This practice not only wastes fertilizers and increases production costs, but also causes N leaching and runoff after excessive irrigation. A large number of plant species are produced in containers, but few species have been studied for N requirements [39]. Those studied were based on a particular substrate in a specific environmental condition. In reality, however, a wide range of substrates have been used in container plant production, and different substrates have different physical and chemical properties. Thus, the established N requirements may not be well suitable for plants to be produced in a different substrate. However, such information does provide reference guides for N application. Nevertheless, the use of extra amount practice must be changed, otherwise, even with the best CRFs available, N leaching and runoff could still occur in container plant production. Third, the methods of placing CRFs significantly affect N release or leaching. Several studies have shown that more N is leached by incorporation of CRFs with substrates, while topdressing had significantly less amount of N leaching [59]. The explanation is that the time for transfer of nutrients through membranes in topdressed CRFs is presumably extended over incorporation due to intermittent drying of the upper growing substrate between irrigation [73].

5. Future development of CRFs

It is certain that CRFs are needed, and the need is increasing. Since the world population keeps growing, it requires more food. Food production requires fertilizers. Meanwhile, container plant production has been growing at a fast pace. The production of container plants also requires fertilizers. As this article documented, container plant production is associated with N leaching and runoff. So far, the volatilization of NH_3 and emission of N_2O have not been well studied in container plant production. This does not mean that the volatilization and emission are not a problem since fertilization is estimated to account for 78% of the total emission of NH_3 and N_2O at the global scale [35]. Therefore, manufacturers should not only pay attention to N leaching but also take emission problems into consideration in the development of CRFs. Future fertilizers must be environmentally friendly and have minimal loss to the air and leaching and/or runoff of N to ground and surface water systems.

The development of CRFs has evolved from a sulfur-coating technology to a polymer-coated technology. With the advance of nanotechnology, future CRFs should integrate nanotechnology components for improving controlled-release characteristics [74]. The future CRFs should be biodegradable; materials used for producing CRFs should be capable of decomposing naturally in most common environmental conditions. Nutrient composition and formulations should be developed based on (1) different groups of plants: annual, perennial and evergreen; (2) the purpose of plant production: growth for fruit, grain or biomass increase (ornamental plants) and/or (3) their inherent needs for nutrients: low, medium and high requirements for major nutrient elements, particularly N. New coating materials that have better permeability and duration as well as biodegradability should be used for coating the nutrient elements. Depending on plant groups and production regions, appropriate coating materials should be used to ensure that nutrients are release largely based on plant requirements. Some natural polymers should be considered including chitosan, xanthan gum, carrageenan, pectin and modified clays [49]. Polymer-clay superabsorbent composites have been reported to be promising as their production costs are low with high water absorbency [75]. Additionally, future CRFs should consider the incorporation of beneficial microbes, such as plant growth promoting bacteria [76] and mycorrhizal fungi [77, 78] to maximize nutrient use efficiency and minimize negative impact on the environment.

6. Conclusion

There is an increasing trend for producing plants in containers worldwide. Container plant production, however, poses mounting concern over N leaching and/or runoff. This is due to the fact that plants are grown in confined substrates that are highly permeable and have low water and nutrient holding capacities, and a large amount of N and water are required for sustaining plant growth. In addition to N leaching and/or runoff, applied N may be volatilized as NH_3 and emitted as N_2O into the atmosphere, contributing to climate changes. This article documents that the use of CRFs can reduce N leaching and runoff and raises the question about NH_3 volatilization and N_2O emission in container plant production. It is firmly believed that the use of CRFs is an effective way of reducing N leaching and runoff and possibly NH_3 volatilization and N_2O emission. With the increase need for food and ornamental plants, the need for fertilizers, particularly CRFs will continuously increase. New environment friendly CRFs should be developed and used for crop and container plant production. On the other hand, since the amount of N lost is a function of fertilizer source, timing, soil infiltration and percolation rate, micropore flow, root density, soil moisture, and precipitation/irrigation rate and intensity, CRFs alone cannot resolve N loss problem. The application of CRFs along with integrated production practices should be carried out for minimizing N loss. Integration includes the application of CRFs based on plant species types and production purpose, irrigation of substrate according to plant need and appropriate methods of applying CRFs to the substrate.

Author details

Jianjun Chen[1]* and Xiangying Wei[1,2]

*Address all correspondence to: jjchen@ufl.edu

1 University of Florida, Institute of Food and Agricultural Sciences, Mid-Florida Research and Education Center, Apopka, FL, USA

2 Fujian University Key Laboratory of Plant-Microbe Interaction, College of Life Sciences, Fujian Agriculture and Forestry University, Fuzhou, China

References

[1] Jerardo A. Floriculture and Nursery Crops Yearbook: A Report from Economic Research Service. Washington, DC: US Department of Agriculture; 2007 http://www.ers.usda.gov/Publications/Flo/2007/09Sep/FLO2007.pdf

[2] Majsztrik JC, Ristvey AG, Lea-Cox JD. Water and nutrient management in the production of container-grown ornamentals. Horticultural Reviews. 2011;**38**:253

[3] USDA (U.S. Department of Agriculture). Nursery crops 2006 summary. In National Agricultural Statistics Service, USDA, Washington, DC; 2007. p. 63

[4] Ruter JM. Growth and landscape performance of three landscape plants produced in conventional and pot-in-pot production systems. Journal of Environmental Horticulture. 1993;**11**:124-127

[5] Harris JR, Gilman EF. Production method affects growth and post-transplant establishment of 'East Palatka' holly. Journal of the American Society for Horticultural Science. 1993;**118**:194-200

[6] Gilman EF, Beeson RC Jr. Nursery production method affects root growth. Journal of Environmental Horticulture. 1996;**14**:88-91

[7] Judd LA, Jackson BE, Fonteno WC. Advancements in root growth measurement technologies and observation capabilities for container-grown plants. Plants. 2015;**4**:369-392

[8] Majsztrik JC, Fernandez RT, Fisher PR, Hitchcock DR, Lea-Cox J, Owen JS, et al. Water use and treatment in container-grown specialty crop production: A review. Water Air Soil Pollution. 2017;**228**:151

[9] USDA (U.S. Department of Agriculture). Census of agriculture, 2007 summary and state data. USDA, Washington, DC; 2009. Available online: http://www.agcensus.usda.gov/Publications/2007/Full_Report/usv1.pdf [Accessed 3 December, 2017]

[10] Parrella MP, Wagner A, Fujino DW. The floriculture and nursery industry's struggle with invasive species. American Entomologist. 2015;**61**:39-50

[11] USDA (U.S. Department of Agriculture). Floriculture crops, 2013 summary. National Agricultural Statistics Service. USDA, Washington, DC; 2014

[12] Carman H. Economic Aspects of the California Nursery and Floral Industry, 2001-2009. Giannini Foundation of Agricultural Economics Information Series Report; 2011:11-1

[13] USDA (U.S. Department of Agriculture), National Agricultural Statistics Service (USDA-NASS). Census of Horticultural Specialties 2014, vol. 3, Special studies, part 3, AC-12-SS-3, USDA, Washington, DC; 2015

[14] Molitor HD. The European perspective with emphasis on subirrigation and recalculation of water and nutrients. Acta Horticulturae. 1990;**272**:165-173

[15] McAvoy R, Brand M, Corbett E, Bartok J Jr, Botacchi A. Effect of leachate fraction on nitrate loading to the soil profile underlying a greenhouse crop. Journal of Environmental Horticulture. 1992;**10**:167-171

[16] Yeager T, Cashion G. Controlled-release fertilizers affect nitrate nitrogen runoff from container plants. HortTechnology. 1993;**3**:174-177

[17] Mangiafico SS, Gan J, Wu L, Lu J, Newman JP, Faber B, et al. Detention and recycling basins for managing nutrient and pesticide runoff from nurseries. HortScience. 2008;**43**:393-398

[18] Wilson C, Albano J, Mozdzen M, Riiska C. Irrigation water and nitrate-nitrogen loss characterization in Southern Florida nurseries: Cumulative volumes, runoff rates, nitrate-nitrogen concentrations and loadings, and implications for management. HortTechnology. 2010;**20**:325-330

[19] Fare DC, Gilliam CH, Keever GJ, Olive JW. Cyclic irrigation reduces container leachate nitrate-nitrogen concentration. HortScience. 1994;**29**:1514-1517

[20] Broschat TK. Nitrate, phosphate, and potassium leaching from container-grown plants fertilized by several methods. HortScience. 1995;**30**:74-77

[21] Ku CS, Hershey DR. Growth response, nutrient leaching, and mass balance for potted poinsettia. I. Nitrogen. Journal of the American Society for Horticultural Science. 1997;**122**:452-458

[22] Ristvey AG, Lea-Cox JD, Ross DS. Nitrogen and phosphorus uptake efficiency and partitioning of container-grown azalea during spring growth. Journal of the American Society for Horticultural Science. 2007;**132**:563-571

[23] Andiru G, Pasian C, Frantz J. Effects of controlled-release fertilizer placement on nutrient leaching and growth of bedding impatiens. Journal of Environmental Horticulture. 2015;**33**:58-65

[24] Taylor MD, White SA, Chandler SL, Klaine SJ, Whitwell T. Nutrient management of nursery runoff water using constructed wetland systems. HortTechnology. 2006;**16**:610-614

[25] Johnson CJ, Bonrud PA, Dosch TL, Kilness AW, Senger KA, et al. Fatal outcome of methemoglobinemia in an infant. Journal of American Medical Association. 1987;**257**:2796-2797

[26] Spalding RF, Exner ME. Occurrence of nitrate in groundwater—A review. Journal of Environmental Quality. 1993;**22**:392-402

[27] Bryan NS, Loscalzo J. Nitrite and Nitrate in Human Health and Disease. New York: Springer, Science+Business Media, LLC; 2011

[28] Pennino MJ, Compton JE, Leibowitz SG. Trends in drinking water nitrate violations across the United States. Environmental Science & Technology. 2017;**51**:13450-13460

[29] Jasa PJ, Skipton SO, Varner DL, DH. Drinking water: Nitrate-nitrogen. NebGuide, University of Nebrask-Lincoln Extention, Institute of Agriculture and Natural Resources; 2006. Revised April 2006

[30] Oram B. Nitrates and Nitrites in Drinking Water and Surfacewaters. Dallas, PA, USA: Water Research Watershed Center; 2014 http://www.water-research.net/index.php/nitrate

[31] USEPA. National primary drinking water regulation. Code of Federal Regulation. Title 40. Chapter 1. Part 141.2002

[32] Schlesinger WH, Hartley AE. A global budget for atmospheric NH_3. Biogeochemistry. 1992;**15**:191-211

[33] Greenhouse Gas Working Group. Agriculture's role in greenhouse gas emissions and capture. Greenhouse Gas Working Group Rep. ASA, CSSA, and SSSA, Madison, WI; 2010

[34] Erisman J, Schaap M. The need for ammonia abatement with respect to secondary PM reductions in Europe. Environmental Pollution. 2004;**129**:159-163

[35] USEPA (U.S. Environmental Protection Agency). Inventory of U.S. greenhouse gas emissions and sinks: 1990-2005. EPA 430-R-07-002. USEPA, Washington, DC; 2007

[36] Chen J, Huang Y, Caldwell RD. Best management practices for minimizing nitrate leaching from container-grown nurseries. The Scientific World Journal. 2001;**1**:96-102

[37] Martínez-Alcántara B, Quiñones A, Legaz F, Primo-Millo E. Nitrogen-use efficiency of young citrus trees as influenced by the timing of fertilizer application. Journal of Plant Nutrition and Soil Science. 2012;**175**:282-292

[38] Dobermann AR. Nitrogen use efficiency—State of the art. University of Nebraska–Lincoln, Agronomy and Horticulture faculty publication. 2005;**316**:16. http://digitalcommons.unl.edu/agronomyfacpub/316

[39] Agro E, Zheng Y. Controlled-release fertilizer application rates for container nursery crop production in Southwestern Ontario, Canada. HortScience. 2014;**49**:1414-1423

[40] Yao Y, Gao B, Chen J, Yang L. Engineered biochar reclaiming phosphate from aqueous solutions: Mechanisms and potential application as a slow-release fertilizer. Environmental Science and Technology. 2013;**47**:8700-8708

[41] Yao Y, Gao B, Chen J, Zhang M, Inyang M, Li Y, et al. Engineered carbon (biochar) prepared by direct pyrolysis of Mg-accumulated tomato tissues: Characterization and phosphate removal potential. Bioresource Technology. 2013;**138**:8-13

[42] Trenkel ME. Slow-and controlled-release and stabilized fertilizers: An option for enhancing nutrient use efficiency in agriculture. IFA, International fertilizer industry association; 2010

[43] Chen J, Beeson RC, Yeager TH, Stamps RH, Felter LA. Evaluation of captured rainwater and irrigation runoff for greenhouse foliage and bedding plant production. HortScience. 2003;**38**:228-233

[44] Skimina CA. Recycling water, nutrients, and waste in the nursery industry. HortScience. 1992;**27**:968-968

[45] AAPFCO. Official publication number 65. Association of American Plant Food Control Officials. http://www.aapfco.org/publications.html <Accessed 8 November, 2017>. 2013

[46] Shaviv A. Controlled Release Fertilizers. IFA International Workshop on Enhanced-Efficiency Fertilizers, Frankfurt. International Fertilizer Industry Association, Paris, France; 2005

[47] Ozores-Hampton M. Description of enhanced-efficiency fertilizers for use in vegetable production. UF/IFAS Extension, document HS1247. University of Florida, Immokalee, FL, U.S.A. http://edis.ifas.ufl.edu/pdffiles/HS/HS124700.pdf <Accessed November 12, 2017>

[48] Azeem B, KuShaari K, Man ZB, Basit A, Thanh TH. Review on materials and methods to produce controlled release coated urea fertilizer. Journal of Controlled Release. 2014;**181**:11-21

[49] Sempeho SI, Kim HT, Mubofu E, Hilonga A. Meticulous overview on the controlled release fertilizers. Advances in Chemistry. 2014;**2014**:16. Article ID 363071. http://dx.doi.org/10.1155/2014/363071

[50] Ukessays, Controlled release fertilizers and nanotechnology traces biology essay; 2015. https://www.ukessays.com/essays/biology/controlled-release-fertilizers-and-nanotechnology-traces-biology-essay.php <Accessed November 12, 2017>

[51] Subbarao CV, Kartheek G, Sirisha D. Slow release of potash fertilizer through polymer coating. International Journal of Applied Science and Engineering. 2013;**11**:25-30

[52] Du C, Zhou J, Shaviv A. Release characteristics of nutrients from polymer-coated compound controlled release fertilizers. Journal of Polymers and the Environment. 2006;**14**:223-230

[53] Guo M, Liu M, Liang R, Niu A. Granular urea-formaldehyde slow-release fertilizer with superabsorbent and moisture preservation. Journal of Applied Polymer Science. 2006;**99**:3230-3235

[54] Liu L, Kost J, Fishman ML, Hicks KB. A review: Controlled release systems for agricultural and food applications. ACS Symposium Series New Delivery Systems for Controlled Drug Release from Naturally Occurring Materials. 2008;**992**:265-281

[55] Zhang M, Yang Y, Song F, Shi Y. Study and industrialized development of coated controlled release fertilizers. Journal of Chemical Fertilizer Industry. 2005;**32**:7-12

[56] Shaviv A. Advances in controlled-release fertilizers. Advances in Agronomy. 2001;**71**:1-49

[57] Colangelo DJ, Brand MH. Nitrate leaching beneath a containerized nursery crop receiving trickle or overhead irrigation. Journal of Environmental Quality. 2001;**30**:1564-1574

[58] Birrenkott BA, Craig JL, McVey GR. A leach collection system to track the release of nitrogen from controlled-release fertilizers in container ornamentals. HortScience. 2005;**40**:1887-1891

[59] Cabrera RI. Comparative evaluation of nitrogen release patterns from controlled-release fertilizers by nitrogen leaching analysis. HortScience. 1997;**32**:669-673

[60] Mikkelsen RL. Using hydrophilic polymers to control nutrient release. Nutrient Cycling in Agroecosystems. 1994;**38**:53-59

[61] Maynard D, Lorenz O. Controlled-release fertilizers for horticultura1 crops. Horticultural Reviews. 1979;**1**:79-140

[62] Sharma GC. Controlled-release fertilizers and horticultural applications. Scientia Horticulturae. 1979;**11**:107-129

[63] Mello SC, Li YC, Migliaccio KW, Linares EP, Colee J, Angelotti-Mendonça J. Effects of polymer coated urea and irrigation rates on lantana growth and nitrogen leaching. Soil Science Society of America Journal—Soil Biology and Biochemistry. 2017;**81**:546-555

[64] Wilson PC, Albano JP. Impact of fertigation versus controlled-release fertilizer formulations on nitrate concentrations in nursery drainage water. HortTechnology. 2011;**21**:176-180

[65] Pack JE, Hutchinson CM, Simonne EH. Evaluation of controlled-release fertilizers for northeast Florida chip potato production. Journal of Plant Nutrition. 2006;**29**:1301-1313

[66] LeMonte JJ, Jolley VD, Summerhays JS, Terry RE, Hopkins BG. Polymer coated urea in turfgrass maintains vigor and mitigates nitrogen's environmental impacts. PLoS One. 2016;**11**:e0146761

[67] Wilson ML, Rosen CJ, Moncrief JF. Effects of polymer-coated urea on nitrate leaching and nitrogen uptake by potato. Journal of Environmental Quality. 2010;**39**:492-499

[68] Merhaut DJ, Blythe EK, Newman JP, Albano JP. Nutrient release from controlled-release fertilizers in acid substrate in a greenhouse environment: I. Leachate electrical conductivity, pH, and nitrogen, phosphorus, and potassium concentrations. HortScience. 2006;**41**:780-787

[69] Ikeda S, Suzuki K, Kawahara M, Noshiro M, Takahashi N. An assessment of urea-formaldehyde fertilizer on the diversity of bacterial communities in onion and sugar beet. Microbes and Environments. 2014;**29**:231-234

[70] Obreza TA, Sartain JB. Improving nitrogen and phosphorus fertilizer use efficiency for Florida's horticultural crops. HortTechnology. 2010;**20**:23-33

[71] Hutchinson C, Simonne E, Solano P, Meldrum J, Livingston-Way P. Testing of controlled release fertilizer programs for seep irrigated Irish potato production. Journal of Plant Nutrition. 2002;**26**:1709-1723

[72] Ruark M. Advantages and disadvantages of controlled-release fertilizers. Presentation to Wisconsin Fresh Fruit and Vegetable Conference, 1/17/2012. Dept. of Soil Science, University of Wisconsin, Madison; 2012. http://www.soils.wisc.edu/extension/materials/Overview_of_fertilizer_technologies_2012_WIFFVC.pdf <Accessed November 13, 2017>

[73] Barron HM. The use of slow-release fertilizers. Combined Proceedings of International Plant Propagators' Society. 1974;**24**:221-229

[74] Zhang M, Gao B, Chen J, Li Y, Creamer AE, Chen H. Slow-release fertilizer encapsulated by graphene oxide films. Chemical Engineering Journal. 2014;**255**:107-113

[75] Ekebafe L, Ogbeifun D, Okieimen F. Polymer applications in agriculture. Biokemistri. 2011;**23**:81-89

[76] Vessey JK. Plant growth promoting rhizobacteria as biofertilizers. Plant and Soil. 2003; **255**:571-586

[77] Wei X, Chen J, Zhang C, Pan D. A new *Oidiodendron maius* strain isolated from *Rhododendron fortunei* and its effects on nitrogen uptake and plant growth. Frontiers in Microbiology. 2016;**7**:1327

[78] Wei X, Chen J, Zhang C, Pan D. Differential gene expression in *Rhododendron fortunei* roots colonized by an ericoid mycorrhizal fungus and increased nitrogen absorption and plant growth. Frontiers in Plant Science. 2016;**7**:1594

[79] Wang FL, Alva AK. Leaching of nitrogen from slow-release urea sources in sandy soils. Soil Science Society of America Journal. 1996;**60**:1454-1458

[80] Paramasivam S, Alva AK. Leaching of nitrogen forms from controlled-release nitrogen fertilizers. Communications in Soil Science and Plant Analysis. 1997;**28**:1663-1674

[81] Blythe EK, Mayfield JL, Wilson BC, Vinson EL, Sibley JL. Comparison of three controlled-release nitrogen fertilizers in greenhouse crop production. Plant Nutrition. 2002;**25**: 1049-1061

3

Field Scale Simulation of Nitrogen Dynamics using LEACHN and OVERSEER® Models

Babar Mahmood

Abstract

Computer models have been used extensively to study the dynamics of nitrogen (N) at effluent-irrigated land treatment systems (LTS). Nitrogen version of leaching estimation and chemistry (LEACHN) model and OVERSEER® are two such models that have the ability to simulate N movement through the soil-water-plant system. This chapter covers brief description of two models, that is, LEACHN and OVERSEER® that were used in this study. This is the third phase of previously conducted studies, and the focus of this third phase was (i) to use the LEACHN model (as optimised based on best N transformation rate constants in a previous study) to simulate N dynamics (under different irrigation scenarios, that is, natural rainfall only, rainfall and irrigation with no N, and rainfall with irrigation containing N) for the medium effluent irrigation treatment plot at an existing land disposal site and (ii) to use another model (i.e., OVERSEER®) to simulate N movement at the same land disposal site and compare its prediction with LEACHN model's predictions (for the low effluent irrigation treatment at the site). This study showed that the LEACHN model has the ability to simulate the fate and transport of N (under different irrigation scenarios) at field scale level. Also, OVERSEER® model could be used to simulate N dynamics at an effluent-irrigated land disposal site. The amount of N leached as predicted by OVERSEER® was reasonably close to LEACHN model predictions.

Keywords: LEACHN model, OVERSEER® model, N dynamics at field scale level, nitrate-N leaching

1. Introduction

As we know that it is impractical to directly measure the nutrient losses to the environment and therefore simulation models could be the best alternative to assess the potential loss of nutrients to the local environment. Over the past decade, the development and use of simulation models for predicting nutrient and pesticide behaviour in the root zone of agricultural land systems, and in the underlying unsaturated zone, have received considerable attention. The models currently available for predicting the fate and transpiration of pesticides and nutrients in soils and groundwater have been critically reviewed and used, among others [1–7].

Cichota and Snow [1] presented an overview of some models used for nutrient loss estimation at farm scale (including OVERSEER®, NPLAS—nitrogen and phosphorus load assessment system, SPASMO—soil-plant-atmosphere system model, EcoMod, LUCI—land use change and intensification, and APSIM—agricultural production systems simulator), and at large scales, that is, catchment level (including NLE—nitrogen leaching estimation, SPARROW—spatial referenced regression on watershed attributes, ROTAN—rotorua and taupo nitrogen, CLUES—catchment land use and environmental sustainability, and AquiferSim).

They [1] also reviewed some soil process models such as groundwater loading effects of agriculture systems (GLEAMS), Leaching Estimation and Chemistry Mode (LEACHM) and HYDRUS. In their review, they said that most of the models have shown their usefulness to estimate nutrient losses in order to prevent environmental impacts at a small or larger scale levels. However, there is a *lack* of information about how these models work and what is their main focus, and therefore, it is important to know the main purpose, strengths and weaknesses of a model (before it is used to perform any kind of simulation) so that the most appropriate model could be selected.

It is well known that the simulation models are useful to understand the interaction between the transformation and transport of N in the field. The growing concern about the environmental impact of effluent-irrigated systems has increased the desire to predict the transport and transformation of N in the soil-plant system more accurately. Several models simulating N transformations and transport in the soil-plant system have been developed and tested over time [2–4, 8–13]. Evaluating N transformation and transport models under field conditions is a complex research challenge. A major difficulty in evaluating these models under field conditions results from the strong interaction between physical and biological factors, plant uptake and N cycling processes.

In recent studies [14, 24], the nitrogen version of leaching estimation and chemistry (LEACHN) model was optimised and an in-depth analysis of changing the N transformation rate constants (i.e., mineralisation—K_{min} – day^{-1}, nitrification—K_{nit} – day^{-1} and denitrification—K_{den} – day^{-1}) and bulk density on LEACHN model predictions was undertaken. The in-depth analysis (based on N transformation rate constants) (i.e., second phase of the study) provided an in-depth understanding of movement and distribution of effluent N down the profile at the studied land disposal site. The testing of the parameterised/optimised LEACHN model (as part of second phase of the study) showed that this model had the ability to predict the timing and amount of leachate nitrate-N (NO_3-N) concentrations at an effluent disposal site. The work reported here (i.e., *third* phase) is the continuation of a previous study [14]. The key research *questions* were: (i) Is LEACHN model able to simulate N dynamics (under different irrigation scenarios) at a land disposal site and (ii) Is it possible to use OVERSEER® model to simulate N movement at a land disposal site? Therefore, the *main* focus of this chapter was to use LEACHN model (as optimised using the best N transformation rate constant values in the previous study [14]) to refine our understanding of the predicted fate of N at the existing effluent-irrigated land treatment site under different irrigation scenarios. Also, it was decided to use another model (i.e., OVERSEER®) to simulate N dynamics for the same site and then compare its predictions with LEACHN model's predictions for one of the effluent irrigation treatments (i.e., the low effluent irrigation treatment for pasture plot was chosen). The details of optimisation of LEACHN model are not given here, and can be found in [14] as recently published. Thus, the specific objectives of this study were to (i) undertake irrigation scenario analysis (i.e., rainfall alone with no effluent irrigation, rainfall and irrigation of effluent containing no N, and rainfall and irrigation with effluent containing N) using the optimised model in order to understand the predicted fate of N added in effluent at a land disposal site and (ii) explore the use of OVERSEER® model at land treatment systems (LTS), and then compare its predictions of the amount of N leached and average NO_3-N concentrations with optimised LEACHN model predictions. As we know that the movement of N through the soil-water matrix is a very complex process. Therefore, in terms of novelty, the work presented here in this chapter refines our knowledge and understanding of N dynamics movement at a land treatment facility. This work explores and compares the abilities of two different models (i.e., LEACHN—a process-based models, and OVERSEER®—a farm management tool), which has not been undertaken earlier. Further, the testing of OVERSEER® model for a LTS was challenging and it was done in an innovative way as this model is not designed to predict N leaching at a LTS.

2. Material and methods

As noted above, the work described in this study (i.e., *third* phase) covers two main areas.

1. To use the optimised LEACHN model to gain a greater understanding of the dynamics of N within an existing effluent-irrigated land disposal site in Carterton District Council (CDC) of Wellington region, New Zealand.
2. To use the OVERSEER® model to simulate N movement at the CDC land disposal site.

2.1. Model descriptions

2.1.1. LEACHN

LEACHN is the nitrogen version of the LEACHM model, which has evolved from modelling efforts in the last 20 years and has successfully been used by several workers to describe nitrate and pesticide movement in the field [4, 14–19, 24]. It is a research model that can be used for management purposes. There are five modules of LEACHM. One of these modules, LEACHN, describes nitrogen transport and transformation and that was optimised in the second phase [14] of the study. The details of other modules and the soil hydrological properties, interactions, transport and transformation of N in LEACHN are *not* provided here, as it is covered in more detail in the second phase of this study [14, 24].

2.1.2. *OVERSEER®*

OVERSEER®is a decision support system (DSS) farm model used by farmers, advisors and policy makers. It estimates the nutrient budget for a farm taking into account all inputs and outputs (including internal cycling of nutrients around the farm) to and from the farm. This model is widely used in New Zealand as a decision support tool by consultants and regional councils. This model was based on the knowledge obtained primarily from New Zealand and in consultation with farmers and consultants, and therefore, it is well suited for handling management practices and environmental conditions specific to New Zealand. Initially, this model was used to examine the impacts of land use and management practices on nutrient losses as a way of estimating fertiliser requirements, but recently the model has been used to monitor farm nutrient losses as a tool for applying new environmental policies [20].

OVERSEER®has proved to be a very useful tool to examine nutrient use and flows within a farm (i.e., as products, fertiliser, effluent, supplements or transfer by animals) and to assess nutrient use efficiency and environmental impacts at the farm scale. It is an empirical model (i.e., based on observations or experiments) and empirical relationships, internal databases and readily available data from existing farms are used by the model to calculate nutrient budgets at a farm scale level [20, 21]. The overall model contains separate sub-models dealing with pastoral, cropping and horticulture enterprises. Of these various sub-models, the pastoral model is probably the most *robust* because of the large quantity of trial data that is available in New Zealand on grazed pasture systems.

The model is designed so that the input data required is meaningful to farmers and is easily available. This is in contrast to LEACHN, which is a data hungry model. For reliable performance, OVERSEER® requires that reasonable input data are given [21]. This means that the amount of fertiliser required to support the given level of production needs to be known. The model assumes that the system is in equilibrium (i.e., model does not account for transition during a change from one practice to another). The OVERSEER®model is designed to predict long-term average behaviour of the system (as compared to LEACHN that can predict on a daily, weekly and monthly basis), and therefore, it is not suitable for extreme case scenarios or a system in transition, or for estimating nutrient losses from particular years.

In OVERSEER®,a *dairy* farm can be divided into blocks that can include effluent disposal and non-effluent disposal blocks. This provided the opportunity to assess whether the OVERSEER® model could be used to simulate N leaching losses at a LTS by assuming that it was operating as an effluent-irrigated block on a dairy farm. If the OVERSEER® model could be *used* at a LTS, this would have considerable advantages as many regional councils already understand and accept the principals and predictions from the OVERSEER®model.

2.2. Experimental site

A brief detail of experimental site is provided here. Further details can be found in Refs. [14, 15]. A diagram and the layout map of the existing CDC land disposal site are shown in **Figures 1** and **2**. The site was located close to the Carterton's wastewater land treatment facility. An irrigation system was designed to apply effluent at three rates, that is, 30 (Low),

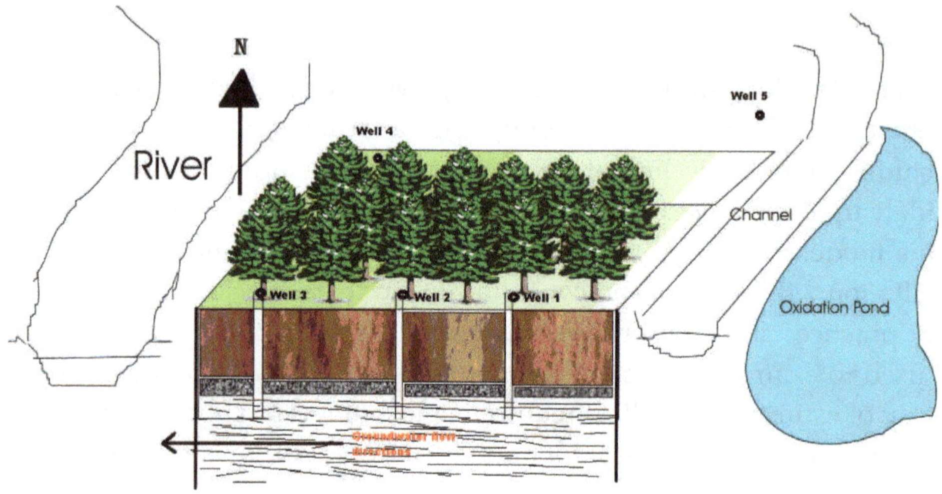

Figure 1. The CDC effluent-irrigated land disposal site.

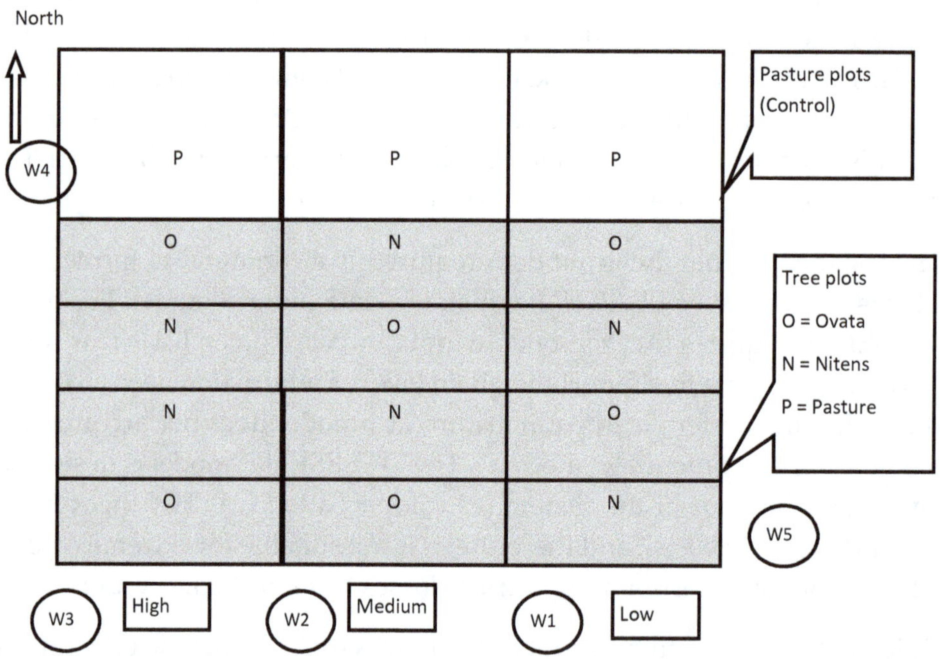

Figure 2. A layout map of the effluent-irrigated land treatment site planted with trees and pasture, showing low, medium and high irrigations areas, and the locations of monitoring wells (sourced from Ref. [14], and as modified from Refs. [15, 24]).

45 (Med) and 100 (High) mm per week. The site was planted with pasture and short rotation trees, that is, *Eucalyptus ovata* and *Eucalyptus nitens* (refer to **Figures 1** and **2**). The total study area was 540 m^2. There were three blocks (i.e., 180 m^2 per block) and each block has three plots (i.e., one control plot of pasture and two plots of randomly planted trees; **Figure 2**). The location of monitoring wells is also shown in **Figures 1** and **2**.

The experiment was conducted over the time period of 38 weeks (i.e., December 1997–August 1998). Soil moisture content (SMC) measurements were made using the time domain reflectometry (TDR) machine (prior to irrigation and then fortnightly) at 0–150, 0–300, and 0–450 mm soil depths in all pasture and tree plots. Soil-water samples were collected using ceramic cups installed at 150, 300, and 450 mm depths before the first irrigation and then bi-weekly. The groundwater samples were also collected before starting effluent irrigation and then bi-weekly from the monitoring wells. All soil and water samples were analysed for NO_3-N and ammonium-N (NH_4-N) using the standard methods of water analysis [22]. Please *refer* to [14, 24] for more experimental details.

2.3. Use of the optimised LEACHN model to explore the effect of effluent irrigation on the dynamics of water and N at the CDC LTS

It was assumed that the values for K_{min}, K_{nit} and K_{den} (i.e., 0.02, 0.02 and 0.0000017 per day, respectively) *used* in the optimised LEACHN model [14] were adequate to describe the NO_3-N concentrations at the CDC's land disposal site. The optimised model was then used to investigate in more detail the effect of adding effluent to the soil at the land disposal site. The *medium* irrigation treatment of the tree plot (refer to **Figure 2**) was only used for this analysis, and this treatment was compared first with the model predictions of what would have occurred at the site under natural rainfall with no added effluent and second with the model predictions of what would have occurred if the medium irrigation strategy had been followed, but with pure water, rather than N-containing effluent. The chosen time period was 260 days (almost 38 weeks) to simulate N dynamics for the medium effluent irrigation treatment pasture plot at the land disposal site.

2.4. The ability of the OVERSEER® nutrient budget model to predict N dynamics at the CDC LTS

The OVERSEER® model was used to compare the amount of N applied and leached for the low application rate of effluent irrigation treatment on pasture plot (refer to **Figure 2**) at the CDC LTS. This part was interesting, innovative and bit *challenging* as the OVERSEER® model was not originally designed to predict N leaching from a LTS. But, as mentioned earlier that the OVERSEER® model does have the ability to simulate N leaching from effluent disposal areas on *dairy* farms. In doing this, the OVERSEER® model (version 6 that was used for the study) does not have a facility to enter the application rate of effluent N directly into the model. Instead, the OVERSEER® model calculates the application rate of effluent N from the number of dairy cows in the herd on the dairy farm and the area of the effluent disposal area. For a given number of cows, if the size of the effluent area is large, the application rate of effluent N per hectare will be low, and if the area of the effluent area is very small, then the application rate will be correspondingly high.

This feature of OVERSEER® provided the opportunity to use the model to simulate N leaching on an LTS (for the low effluent irrigation pasture plot) such as that at Carterton. This was done in the OVERSEER® model by setting up the CDC LTS as the effluent disposal block of a *notional* dairy farm. The notional "size" of the effluent block was adjusted so that the application rate of effluent organic N corresponded with the application rate of organic N in the low irrigation treatment on the pasture plot at the CDC LTS. The reason for matching the application rates of organic N in this way is explained below.

Although by manipulating the size of the notional effluent area, it was possible in OVERSEER® to adjust the rate of N application on the effluent block to a value similar to that was applied at the land disposal site for the low effluent irrigation pasture plot, there are differences between farm dairy effluent (FDE) and municipal sewage effluent in the concentration of total N (which affects the amount of water added per kg of N) and the ratio of organic to inorganic N. Fortunately, the OVERSEER® model can account for these differences through its capacity to apply irrigation water containing inorganic N, as well as applying FDE.

FDE is more concentrated than the municipal sewage effluent at the CDC LTS. Its N concentration, on average, varies between 200 and 500 mg/L [23]. Also, it has been reported that organic N is the main component of total N in FDE (up to approximately 80% [23]). In this analysis, it was assumed that OVERSEER® would use an average total N concentration for FDE of 250 mg/L, of which 80% would be in organic form.

As noted above, the OVERSEER® model (version 6) has the capacity to specify an amount of irrigation on a monthly basis, and also to specify the nutrient concentration in that irrigation water. If a N concentration in the irrigation water is specified, the OVERSEER® model assumes that this N is in the inorganic form. By using this feature of the OVERSEER® model, it was possible to mimic the N application regime in the *low* effluent treatment at the CDC LTS exactly, in terms of the depth of water applied, the total N applied and the ratio of organic to inorganic N. This was done as follows:

1. The quantities of water, organic N and inorganic N applied in the low effluent treatment at the CDC LTS were calculated.
2. The application rate of FDE needed to apply the same amount of organic N as at the CDC LTS was then calculated (assuming that 80% of total N in FDE was organic N) and the "size" of the notional effluent area was adjusted so that quantity of organic N was being applied.
3. This application rate of FDE was then supplying the correct amount of organic N, but the quantities of water and inorganic N applied were much less than at the CDC LTS. These shortfalls were then added in the OVERSEER® model through irrigation.
4. The amount of additional irrigation water required to match the total water applied in the low effluent treatment at the CDC LTS was then calculated.
5. The inorganic N concentration in this irrigation water needed to ensure that the same quantities of inorganic N were applied as in the low effluent treatment at the CDC LTS was then calculated.

When setting up the notional effluent block in the OVERSEER® model, it was specified in the model that all the pasture grown on the effluent-irrigated block was taken off as supplements in order to mimic the "cut & carry" system at the CDC LTS.

The behaviour of N predicted by OVERSEER® for this farm scenario was then compared with the predictions of the LEACHN model. The annual rainfall data for the Carterton land treatment site were used in the scenarios.

3. Results and discussion

3.1. The ability of the LEACHN model to predict N dynamics at the CDC LTS

This study used the optimised LEACHN model to investigate the fate of the added water and N in the effluent at the CTC LTS (**Table 1**). The model predicted that under natural rainfall and with no effluent irrigation (Scenario S1 in **Table 1**), there would be very little accumulated leachate from day 1 to 196. However, there was some leaching from day 210 onwards (a total of 177 mm during winter). The predicted leachate NO_3-N concentrations were very low (<1 mg/L) and the cumulative mass of NO_3-N leached for this scenario was <1 kg/ha (**Figure 3**) for the simulation period (260 days). The reason for these low predicted NO_3-N concentrations in the drainage water is discussed later in this section. The predicted net N mineralised in scenario S1 was 6 kg/ha.

When LEACHN was used to simulate the addition of irrigation water containing no N at the medium effluent application rate (Scenario S2 in **Figure 3**), the irrigation caused drainage from day 28 onward (i.e., during summer). The accumulated drainage for scenario S2 was 1080 mm as compared to 177 mm for S1. The amount of water added (as rainfall and irrigation water) in scenario S2 was 2256 mm as compared to 681 mm added in scenario S1 as rainfall only. In Scenario S2, the predicted leachate NO_3-N concentrations were high (up to 32.0 mg/L) at the start of the drainage period but then dropped away to much lower concentrations (1 mg/L) towards the end of the simulation period (**Figure 3**). The cumulative mass of NO_3-N leached was 20 kg/ha (refer to **Tables 1** and **2**; **Figures 3** and **4**), and net N mineralised was 9 kg/ha (i.e., 3 kg/ha more than when no irrigation water was applied).

Irrigation scenarios	**Cumulative mass of NO_3-N leached**	**Cumulative drainage**	**Net N mineralised**
	(kg/ha)	**(mm)**	**(kg/ha)**
S1—Just rainfall and no effluent	<1	177	6
S2—Rainfall and effluent with no N	20	1080	9
S3—Rainfall and effluent with N	92	1080	9

Table 1. The amount drainage, NO_3-N leached, and net N mineralised in response to three different irrigation scenarios, as calculated by LEACHN.

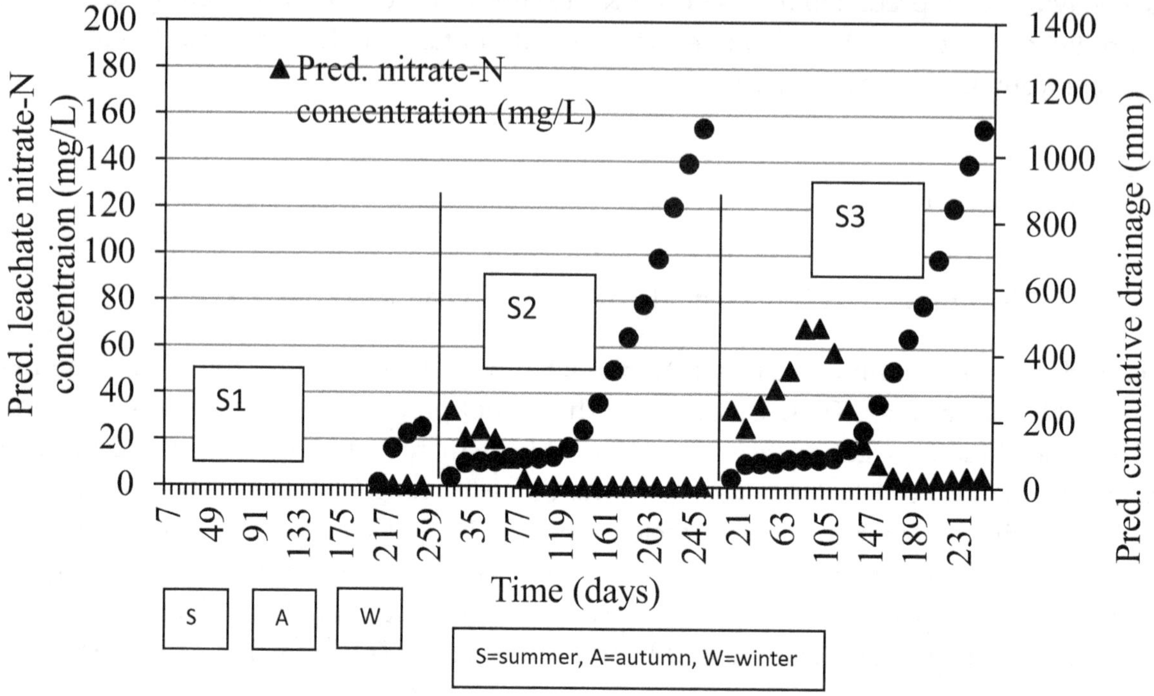

Figure 3. The predicted leachate NO_3-N concentrations and cumulative drainage volume for three irrigation scenarios (S1—rainfall alone and no effluent irrigation, S2—rainfall and irrigation with effluent containing no N, and S3—rainfall and irrigation with effluent that contains N) using the LEACHN model.

When the normal N-containing effluent was added (Scenario S3 in **Figure 3**) LEACHN predicted that the cumulative drainage would be 1080 mm (the same as in scenario S2). The NO_3-N concentrations at the commencement of drainage were similar to those at the start of scenario S2 (32 mg/L) but then increased to a maximum of 68.6 mg/L after about 90 days. From then on, the NO_3-N concentrations dropped steadily to a minimum value of 2.75 mg/L. The net N mineralised was the same as in scenario 2 (i.e., 9 kg/ha) but the cumulative mass of NO_3-N leached was greater (i.e., 92 kg/ha—**Tables 1** and **2**; **Figure 4**) than that in scenario 2 (20 kg/ha; **Figure 4**). The pattern of NO_3-N concentrations over time was reflected in the cumulative loads of NO_3-N which were similar to those in scenario S2 until about day 115. After this date, however, the cumulative load of leached NO_3-N in scenario S3 increased rapidly—presumably reflecting the arrival of the first of the effluent N at the drainage depth, that is, 500 mm.

Although LEACHN predicted that more NO_3-N would be leached in scenario S3 (92 kg/ha) than in scenario S2 (20 kg/ha), the difference (i.e., 72 kg/ha) was much less than the amount of N added in the effluent (184 kg/ha; **Table 2**). The net mineralisation was predicted to be the same in both scenarios and so the remainder of the added N was divided between increased volatilisation and plant uptake, and a greater amount of inorganic N remaining in the soil profile (**Table 2**).

The predicted differences in plant N uptake and soil mineral N concentrations are interesting. It is apparent from **Tables 2** and **3**, and **Figure 5** that the LEACHN model predicts that the quantities of inorganic N in the profile will decrease during the period of the simulation—particularly in scenarios S1 and S2. This is a result of plant uptake being greater than the

	Units	Irrigation scenarios		
		Rainfall alone (S1)	Rainfall and irrigation without N (S2)	Rainfall and effluent irrigation with N (S3)
Total N applied	(kg/ha)	0	0	184
Humus-N_i	(kg/ha)	17,922	17,922	17,922
Humus-N_f	(kg/ha)	17,916	17,913	17,913
Initial NH_4-N	(kg/ha)	19	19	19
Initial NO_3-N	(kg/ha)	31	31	31
NH_4-N in the soil profile	(kg/ha)	0.4	0.6	19
NO_3-N in the soil profile	(kg/ha)	0.00	0.05	5
NH_4-N leached	(kg/ha)	0.2	1	3
NO_3-N leached	(kg/ha)	<1	20	92
Plant uptake of NH_4-N	(kg/ha)	8	8	25
Plant uptake of NO_3-N	(kg/ha)	47	30	66
Volatilised	(kg/ha)	<1	<1	34
Denitrified	(kg/ha)	0.00	0.05	0.22
Cumulative leachate	(mm)	177	1080	1080

Table 2. A quantitative N balance for day 252 for three irrigation scenarios, as calculated by LEACHN.

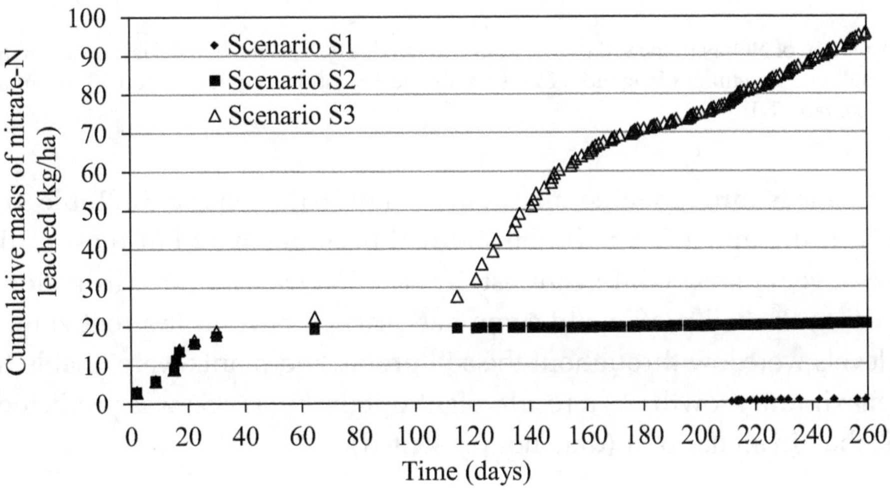

Figure 4. The cumulative mass of NO_3-N leached over the simulation time period of 260 days for the three irrigation scenarios.

Day no.	Soil depth	Soil NO_3-N for scenario S1	Soil NO_3-N for scenario S2	Soil NO_3-N for scenario S3
	(mm)	(mg/dm³)	(mg/dm³)	(mg/dm³)
0	100	6.63	6.63	6.63
	200	7.02	7.02	7.02
	300	5.31	5.31	5.31
	400	5.31	5.31	5.31
	500	7.03	7.03	7.03
56	100	1.44	0.12	4.16
	200	2.94	0.47	4.85
	300	4.97	1.50	6.56
	400	5.71	2.94	7.90
	500	6.61	4.36	9.09
126	100	0.00	0.00	2.90
	200	0.00	0.00	3.82
	300	0.02	0.00	5.75
	400	0.10	0.00	7.29
	500	0.18	0.00	8.11
252	100	0.00	0.00	0.68
	200	0.00	0.00	0.81
	300	0.00	0.00	1.10
	400	0.00	0.03	1.18
	500	0.00	0.05	1.21

Table 3. The distribution of NO_3-N in the soil profile (mg/dm^3) on day 0, 56, 126 and 252 for the three irrigation scenarios (i.e., S1—just rainfall and no effluent irrigation, S2—rainfall and irrigation without N, and S3—rainfall and irrigation with effluent that contains N).

supply of inorganic N through mineralisation. By day 56 in scenario S1 (**Table 3**), the amount of inorganic N in the upper layers of the soil had been depleted but most of the inorganic N in the lower layers remained. In contrast, by day 56 in scenario S2, irrigation had moved inorganic N down the soil profile and some had leached beyond the root zone. As a result, inorganic N levels were low throughout the soil profile and plants were unable to access sufficient N for maximum growth. As a result, plant uptake by day 56 was predicted to be lower in scenario S2 than in scenario S1 (data not presented).

By day 126, virtually all the inorganic N in the profile in scenarios S1 and S2 had been exhausted by a combination of plant uptake (S1 and S2) and leaching (S2). As a result, the predicted plant

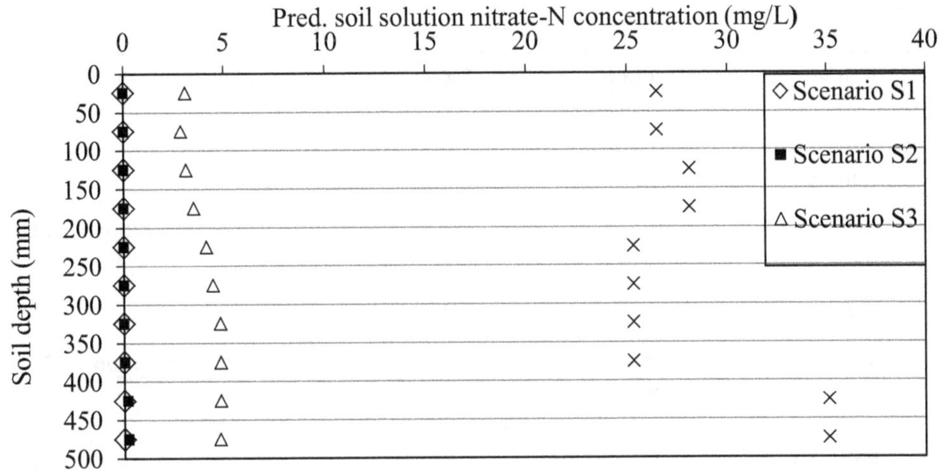

Figure 5. Predicted soil solution NO_3-N concentrations down the soil profile on the starting day and day 252 for the three irrigation scenarios (i.e., S1—just rainfall and no effluent irrigation, S2—rainfall and irrigation without N, and S3—rainfall and irrigation with effluent that contains N).

N uptake in scenarios S1 and S2 was considerably less than in scenario S3 in which effluent containing N was irrigated on to the soil. Such a plant response to added N is commonly observed in New Zealand agricultural systems.

3.2. The ability of the OVERSEER® nutrient budget model to predict N dynamics at the CDC LTS

The focus of this exercise was to explore if it is possible to use the OVERSEER® model (with some modifications to input parameters) to simulate N dynamics at an effluent-irrigated LTS (similar to CDC). As described in methodology, the amounts of water, organic N and inorganic N applied in the low irrigation treatment of the pasture plot at CDC land disposal site (refer to **Table 4**) were entered into the OVERSEER® model as a notional combination of FDE and irrigation water containing a low concentration of inorganic N.

The outputs from the OVERSEER® model were compared (**Table 5**) with the outputs from the LEACHN model optimised as described in the previous section. Both models predicted similar annual total leachate (1629 mm and 1682 mm for the LEACHN and OVERSEER®models, respectively). The inputs of inorganic N in the applied effluent (and irrigation) were as measured at the CDC LTS and were therefore the same for both models. Similarly, in both models, the plant uptake of N was set at 132 kg N/ha, which was measured at the CDC site. Given these similar inputs, the OVERSEER®model predicted a leaching loss of 69 kg N/ha, which was reasonably similar to the 85 kg N/ha predicted by the LEACHN model. The OVERSEER®model predicted very small emissions of N to the atmosphere (4 kg N/ha), whereas the LEACHN model predicted that 32 kg N/ha would be lost to the atmosphere. Most of the predicted emissions in the LEACHN model were as ammonia volatilisation.

Input	Unit	CDC low effluent irrigation treatment		OVERSEER® model	
Water	mm/year	Effluent application	1560	FDE application	67
				Irrigation	1493
Total water	mm/year		**1560**		**1560**
Organic N	kg/ha/year	Effluent application	131	FDE application	132
Total organic N	kg/ha/year		**131**		**132**
Inorganic N	kg/ha/year	Effluent application	180	FDE application	34
				Irrigation	146
Total inorganic N	kg/ha/year		**180**		**180**

Table 4. The amounts of water, inorganic and organic N that were applied as FDE and irrigation in the OVERSEER® model to create a notional effluent disposal site similar to the low irrigation treatment pasture plot at CDC.

Inputs	Units	LEACHN model	OVERSEER® model
Inorganic N applied	kg/ha	180	180
Organic N applied	kg/ha	(131)	132
N fixation	kg/ha	-	2
Total N inputs	kg/ha	311	313
Outputs			
N leached	kg/ha	85	69
Plant N uptake	kg/ha	132	133
Loss to atmosphere	kg/ha	32	4
Total N outputs	kg/ha	249	206
Change in soil N			
Soil inorganic N	kg/ha	−39	0
Soil organic N	kg/ha	(101)	107
Total change in soil N	kg/ha	(62)	107
Water			
Rainfall	mm		
Effluent and irrigation	mm	1560	1560
Cumulative leachate	mm	1629	1682
Average NO_3-N concentration	mg/L	5.2	4.1

Table 5. Comparison of the predictions from the LEACHN and OVERSEER® models in terms of N leached (kg/ha/year) and average NO_3-N concentrations (mg/L) when the amount of N applied is the same.

The OVERSEER® model predicted that there would be no change in soil inorganic N over the year but that the organic N would increase by 107 kg N/ha. The LEACHN model predicted a decrease in inorganic N, but prediction of changes in organic N by the LEACHN model is difficult. At the CDC LTS, the effluent applied contained both inorganic and organic N (**Table 4**) and the organic N was included in the inputs to the OVERSEER® model. It is, however, not easy to include inputs of organic N in the LEACHN model, and therefore only the inorganic N content of the CDC effluent was included in the LEACHN simulations described throughout the study.

In **Table 5**, a notional N balance was calculated for the LEACHN model by adding the input of organic N (number in brackets) and then calculating the final increase or decrease in organic N by difference. When this was done, the predicted increase in organic N was less than predicted by the OVERSEER® model—reflecting the greater losses by leaching and volatilisation.

It would therefore appear that the existing OVERSEER® model can be adapted to simulate a LTS and gives a similar prediction to a more detailed process-based model such as LEACHN. This could be of value because regional councils are increasingly recognising the use of OVERSEER® as a monitoring tool to demonstrate compliance with environmental regulations.

4. Summary and conclusions

This study was about the application of two models (i.e., OVERSEER® and LEACHN) to predict the fate of water and N dynamics at an effluent-irrigated land disposal facility, and therefore the following conclusions could be drawn from this piece of work.

1. The irrigation scenario analysis showed that the LEACHN model is able to build up a very detailed picture of what is happening to soil N in the soil profile in response to three irrigation scenarios, that is, no effluent application (i.e., just natural rainfall—S1), application of pure water (i.e., rainfall and effluent without N—S2) and effluent application with N composition (S3). This is very useful piece of information to understand the fate of water and N added to a land disposal site.

2. The irrigation scenario (using the optimised LEACHN model) showed that the accumulated leachate was almost nil from day 1 to 196 under natural rainfall (with no N effluent irrigation). There was a total of 177 mm leaching during winter from day 210 onwards and leachate NO_3-N concentration were quite low (<1 mg/L) during that time. The accumulated mass of NO_3-N leached in the first 177 mm of S1 and S2 was <1 and 20 kg/ha, respectively.

3. The predicted soil solution NO_3-N concentrations did not change greatly in the top 500 mm soil depth when more water (i.e., rainfall and irrigation with water containing no N) was added to the system and consequently the soil N was leached and therefore there was a lack of available soil N for plants uptake in irrigation scenario 2. The LEACHN model predicted that more nitrate would be in the soil profile and the soil solution NO_3-N concentration would be significantly increased (up to 5.0 mg/L) in irrigation scenario 3 (i.e., rainfall and irrigation with effluent containing N).

4. The comparison of two different models showed the amount of N leached, as predicted by OVERSEER® model, was 69 kg/ha/year which was reasonably close to that of predicted by LEACHN model (85 kg/ha/year). The comparison showed that, depending on the level of precision required, the OVERSEER® model could be used to simulate N dynamics at a LTS similar to CDC.

5. The strength of LEACHN lies in its ability to predict what would happen to soil N down the profile when irrigation with or without N is added at a LTS and when, where and how much NO_3-N is going to leach. It can be concluded that LEACHN model provides more accurate useful information on how much NO_3-N is going to leach and when this is likely to happen. At the same time, it provides useful insights into the movements and transformation of soil and effluent N down the profile, which could be very helpful for the operators/managers of effluent land disposal facilities. Therefore, a LEACHN model could be a useful decision-making tool to design effluent loading rates that minimise the risk of groundwater NO_3-N contamination at the effluent land disposal sites.

4.1. Recommendations

(a) In the simulations reported in this study, the focus in the LEACHN model was on the fate of inorganic N added in effluent. There is, however, the capacity within the LEACHN model to include the addition of plant residues, manure and urea. It would be useful in future studies to explore ways in which the ability of LEACHN to include different types of N input could be used to mimic more exactly the management regimes at LTSs.

(b) The study demonstrated that when using the LEACHN-based DSS to explore the consequences of different effluent application strategies care must be taken to ensure that realistic rainfall patterns are input into the model—rather than estimates of long-term averages.

4.2. Practicality

Overall, the findings of this chapter suggest that LEACHN model has the ability to simulate N dynamics (i.e., movement and transformation down the soil profile) at effluent-irrigated land disposal sites. In future, LEACHN-based DSS could be used by LTS managers/environmental engineers/scientists to monitor and manage effluent-irrigated land treatment sites in a way that reduces the risk of groundwater nitrate-N contamination at the sites.

Acknowledgements

I would like to acknowledge that a large portion of this chapter is based on one of the chapters of author's thesis [24]. Some modifications were made to fulfil the purpose of this chapter.

I would like to express my sincere gratitude and indebtedness to Professor Russ Tillman (from Massey University, Palmerston North Campus) for his guidance, perceptive questioning and support during this research work. The completion of this study could not have been possible without the friendship, inspiration and encouragement that I received from Professor Tillman and for these, I am particularly grateful. I would also like to thank Department of Civil Engineering, Unitec Institute of Technology for providing me with time and resources that I needed to complete this study.

I am grateful for the moral support and joyful company of my loving wife Saima Babar, my son Daniyal Mahmood, and my two lovely daughters Bismah Mahmood and Zoya Mahmood, who patiently shared in the sacrifices that were a necessary part of my research work. Finally, my deepest gratitude and thanks to my parents (late), brother and sisters whose love, patience and constant encouragement have been invaluable in my life.

Author details

Babar Mahmood

Address all correspondence to: bmahmood@unitec.ac.nz

Department of Civil Engineering, Unitec Institute of Technology, Auckland, New Zealand

References

[1] Cichota R, Snow VO. Estimating nutrient loss to waterways — an overview of models of relevance to New Zealand pastoral farms. New Zealand Journal of Agricultural Research. 2009;**52**:239-260

[2] Johnson IR, Chapman DF, Snow VO, Eckard RJ, Parsons AJ, Lambert MG, Cullen BR. DairyMod and EcoMod: Biophysical pasture simulation models for Australia and New Zealand. Australian Journal of Experimental Agriculture. 2008;**48**:621-631

[3] Kunjikutty SP, Praher SO, Patel RM, Barrington SF, Kim SH. Simulation of nitrogen transport in soil under municipal wastewater application using LEACHN. Journal of the American Water Resources Association. 2007;**43**:1097-1107

[4] Mahmood B. Analysis of a LEACHN-based management technique in predictive mode. Agricultural Water Management. 2005;**75**:25-37

[5] Nolan BT, Bayless ER, Green CT, Garg S, Voss FD, Lampe DC, Barbash JE, Capel PD, Barbara A. Evaluation of Unsaturated-Zone Solute-Transport Models for Studies of Agricultural Chemicals [Internet]. Reston, Virginia, USA: Geological Survey; 2005. Available from: https://pubs.usgs.gov/of/2005/1196/ofr20051196.pdf [Accessed: 18 March 2016]

[6] Donigian Jr AS, Rao PSC. Selection, application, and validation of environmental fate models. In: Proceedings of the International Symposium on Water Quality Modelling of Agricultural Non-Point Sources. Part II, USDA-ARS-81, US Government Printing Office, Washington, DC. 1990. pp. 557-600

[7] Wagenet RJ, Rao PSC. Modelling pesticide fate in soils. In: Pesticides in the Environment: Processes, Impacts, and Modelling. Madison, WI: Soil Science Society of America; 1990

[8] Bryant JR, Snow VO, Cichota R, Jolly BH. The effect of situational variability in climate and soil, choice of animal type and N fertilisation level on nitrogen leaching from pastoral farming systems around Lake Taupo, New Zealand. Agricultural Systems. 2011;**104**:271-280

[9] Vanclooster M, Viaene P, Diels J, Freyen J. A deterministic evaluation analysis applied to an integrated soil-crop model. Ecological Modelling. 1995;**81**:83-195

[10] Eckersten H, Jansson PE, Johnsson H. SOILN Model, User's Manual. 2nd ed. Division of Agricultural Hydrotechnics Communications. Vol. 94. Uppsala: Department of Soil Science, Swedish University of Agricultural Sciences; 1994. p. 4

[11] Jabro JD, Jemison JM Jr, Lengnick LL, Fox RH, Fritton DD. Field validation and comparison of LEACHM and NCSWAP models for predicting nitrate leaching. Transactions of American Society of Agricultural Engineers (ASAE). 1993;**36**:1651-1657

[12] Hutson JL, Wagenet RJ. LEACHM (Leaching Estimation and Chemistry Model): A process-based model of water and solute movement, transformations, plant uptake and chemical reactions in the unsaturated zone. Version 3. Department of Soil, Crop and Atmospheric Sciences, Research Series No. 92-93. Ithaca, New York: New York State College of Agriculture and Life Sciences, Cornell University; 1992

[13] Johnsson JL, Bergstrom L, Jansson PE, Paustian K. Simulated nitrogen dynamics and losses in a layered agriculture soil. Agriculture, Ecosystems and Environment. 1987;**18**:333-356

[14] Mahmood B, Tillman R. Sensitivity of LEACHN model's predictions to changes to nitrogen transformation rate constants and bulk density. Transactions of the American Society of Agricultural Engineers (ASAE). 2015;**58**(2):353-366. DOI: 10.130331/trans.58.10636

[15] Mahmood B, Russell JM, Wall GL. Field scale nitrate simulations. Transactions of the American Society of Agricultural Engineers (ASAE). 2002;**45**(6):1835-1842

[16] Sogbedji JM, van Es HM, Hutson JL. N fate and transport under variable cropping history and fertiliser rate on loamy sand and clay loam soils: I. Calibration of the LEACHMN model. Plant and Soil. 2001;**229**(1):57-70

[17] Ramos C, Carbonell EA. Nitrate leaching and soil moisture prediction with the LEACHM model. Fertiliser Research. 1991;**27**(7):171-180

[18] Wagenet RJ, Hutson JL. Predicting the fate of non-volatile pesticides in the saturated zone. Journal of Environmental Quality. 1986;**15**:315-322

[19] Wagenet RJ, Hutson JL, Biggar JW. Simulating the fate of a volatile pesticide in unsaturated soil: A case study with DBCP. Journal of Environmental Quality. 1989;**18**:78-84

[20] Wheeler DM, Ledgard SF, Monaghan RM, McDowell R, deKlein CAM. Overseer® nutrient budget model—what is it, what it does. In: Proceedings of the Workshop on Implementing Sustainable Nutrient Management Strategies in Agriculture. Occasional Report No. 19. ISSN 0112-9902. Palmerston North, New Zealand: Massey University; 2006. pp. 231-236

[21] Wheeler DM. OVERSEER® scenarios—"what if" analysis. In: Nutrient Management in a Rapidly Changing World. Occasional Report No. 22. Palmerston North, New Zealand: Fertiliser and Lime Research Centre, Massey University; 2009

[22] Gillian G. Methods of Soil, Plant, and Water Analysis. Forest Research Institute Bulletin No. 70. Rotorua, New Zealand: Forest Research Institute; 1984

[23] Longhurst RD, Roberts AHC, Connor MB. Farm dairy effluent: A review of published data on chemical and physical characteristics in New Zealand. New Zealand Journal of Agricultural Research. 2000;**43**:7-14

[24] Mahmood B. Decision supports tools to reduce the risk of groundwater nitrate contamination at land treatment systems [PhD thesis]. Palmerston North Campus, New Zealand: Institute of Engineering and Technology, Massey University; 2013

4

Prospects of N Fertilization in Medicinal Plants Cultivation

Felix Nchu, Yonela Matanzima and
Charles P. Laubscher

Abstract

High global demand for plant-derived medicines is threatening the existence of many wild indigenous plant species. However, the high demand of medicinal plants has also created huge business opportunities in commercial farming of medicinal plants. Large-scale production of secondary metabolites by plants and medicinal materials will be crucial in the medicinal plant industry. As commercial cultivation of medicinal plants gains traction among farmers, N fertilizers will be increasingly used to enhance plant growth and yield. Therefore, the implementation of better nitrogen use efficiency is critically important. Excessive use of N can lead to many problems; it is costly, it can cause environmental pollution and its high levels in plant tissues can be toxic to plants, herbivores and humans. This chapter discusses the potential risks, opportunities and setbacks associated with the use of N in cultivation of medicinal plants.

Keywords: nitrogen fertilizer, medicinal plants, toxicity, yield, secondary metabolite

1. Introduction

Exploitation of plant resources for the treatment of human and animal diseases has placed significant pressure on plant biodiversity. It has been reported that more than 3.5 billion people in the developing world rely on plants as components of their primary health care [1]. However, the use of medicinal plants is not only limited to the developing world, in fact demand for herbal medicine is also rising in many developed countries, for example, in Germany, it is estimated that 600–700 plant-based medicines are available and prescribed by 70% of physicians [2]. The demand for plant-derived medicines has created a large business in indigenous plants in South Africa, which is estimated to be worth R270 million annually [3, 4].

In South Africa alone, there are some 27 million indigenous medicinal consumers [5]. The well-known examples of plant species that are currently traded in South Africa include *Artemisia afra* (Asteraceae), *Melianthus comosus* (Melianthaceae), *Aloe ferox* (Asphodelaceae), *Aloe arborescens* (Asphodelaceae), *Salvia Africana-caerulea* (Lamiaceae) and *Helichrysum cymosum* (Asteraceae) [6].

Plant parts obtained from single or varied species are used to prepare medicinal products. Medicinal plant parts contain bioactive principles that are often referred to as secondary metabolites. Primary metabolites such as enzymes and proteins, lipids, chlorophyll and carbohydrates are fundamental to the life of the plant, while secondary metabolites (terpenoids, the alkaloids, the phenylpropanoid and some phenolic compounds) do not appear to be necessary to sustain life at a fundamental biochemical level. However, secondary metabolites play important defence and chemical ecological roles [7]. Medicinal properties can be obtained from the following plant parts: leaves, bulbs, essential oil, fatty acid, flowers, fruit, gum, stem, roots, rhizome, seed, tuber and wood. Plant secondary metabolites are thought to be responsible for the antimicrobial, antioxidant, anti-inflammatory and insecticidal activities of plant extracts [8]. These plant-derived extracts and compounds are exploited for the treatment of human and animal diseases. Large-scale production of secondary metabolites by plants is crucial in the medicinal plant industry. However, the production of secondary metabolites by plant depends on endogenous and exogenous factors [9]. Nitrogen is one of the most important nutrients needed by plants for growth. Information on the role of nitrogen in plant physiology is plentiful in literature. Nitrogen is involved in many physiological processes in plants including growth and photosynthesis. Consequently, nitrogenous fertilizers are among the most used fertilizers in the world. Nevertheless, excessive use of N can have negative economic and the environmental implications. Intensive N fertilization can lead to toxic N levels in plant tissues and herbivores. Thus, there are calls for implementation of better nitrogen use efficiency (NUE) [10].

Researchers have recognized the potential benefits of manipulating nutrient nitrogen supply for optimal plant growth and the need to minimize some of the setbacks associated with nitrogen fertilization. This has incentivized the quest for the development of precision fertilization and innovative plant cultivations methods. For examples, the use of sustainable, innovative and precision agronomic technologies such as hydroponics, aeroponics, aquaponics and organic farming can optimize the manufacturing of natural molecules of pharmaceutical and cosmetic significance. According to Masclaux-Daubresse et al. [10], increasing nitrogen use efficiency in the contexts of plant nutrition and limiting nitrogen fertilizer use is important. It is essential to preserve the environment, while promoting sustainable and productive agriculture. Therefore, knowledge on nitrogen availability and conservation in growth media, and nitrogen uptake, assimilation and translocation by plant are critically important to the development of efficient nitrogen fertilization strategies. This chapter discusses the potential risks, opportunities and setbacks associated with the use of N in cultivation of medicinal plants.

2. Demand for medicinal plants and rationale for commercial cultivation

Until recently, the most commercial farmers have been focused on improving quality and quantity of agricultural and horticultural crops over medicinal plants. Medicinal plants are used in traditional practices worldwide and their use has been increasing steadily. Medicinal plants constitute an important component of health care systems, globally. The trade of medicinal plants is estimated to be worth R270 million annually [3]. According to Sobiecke [11], globally, products that are derived from traditional medicine are estimated to be worth R2.9 billion per year. On the demand for medicinal plants is increasing worldwide and it is estimated up to 700,000 tonnes of plant material are consumed annually to the value of about 150 million US dollars [4]. The World Health Organization estimates that 21,000 species are used for medicinal purposes around the world and in India 150 species are used commercially [12]. In Zimbabwe, herbal medicine is the most affordable and easily accessible form of treatment in primary health care and up to 93 medicinal plant species are used in the south-central region of Zimbabwe [13]. In Pakistan, more than 500 species of plants are used in herbal medicine [14]. Street and Prinsloo [15] presented 10 highly used South African medicinal plants, such as *Agathosma betulina* (Rutaceae), *A. ferox* (Asphodelaceae), *Aspalathus linearis* (Fabaceae), *Harpagophytum procumbens* (Pedaliaceae), *Hypoxis hemerocallidea* (Hypoxidaceae), *Merwilla natalensis* (Hyacinthaceae), *Pelargonium sidoides* (Geraniaceae), *Siphonochilus aethiopicus* (Zingiberaceae) and *Sutherlandia frutescens* (Fabaceae) in a review paper. Although some critics have argued that traditionally it is not acceptable to use cultivated medicinal plants, a recent report on the perception of cultivation of medicinal plant species indicated that very high proportions (over 69%) of respondents are willing to buy and make use of cultivated medicinal plants [16]. This trend suggests that developing efficient and sustainable agro-technology should be one of the focal areas for research.

Cultivation of medicinal plants is gaining momentum among subsistence and commercial farmers [17]. Farming of medicinal plants has many advantages, for examples it can contribute to job creation and improvement of household earnings, and it can reduce over-exploitation and harvesting of some wild and endangered species. Similar to the cultivation of food crops, medicinal plant cultivation programmes should have specific goals, which include to increase medicinal plant yield and plant growth rate, increase and standardized quality and quality of secondary metabolites produced and reduce toxicity to humans. It is worth noting that commercial cultivation may inadvertently lead to environmental degradation and loss of genetic diversity as well as loss of incentives to conserve wild populations [18]. However, Wiersum et al. [4] argued that the impact of the cultivation of medicinal plant can be beneficial if it is done within the context of protecting and strengthening the cultural values of biodiversity and creating a positive attitude towards biodiversity conservation in general.

3. Nutrient nitrogen

Nitrogen is one of the most important nutrients needed by plants; it is an important element for the formation amino acids, it is essential for plant cell division, it is directly involved in photosynthesis, it is an important component of vitamins and it aids in the production of carbohydrates. Physiologically, N is mostly available to plants in the forms of ammonium and nitrate and preference for one of the two forms to be taken up by plants tend to be influenced by the plant species and soil conditions, including pH and soil temperatures [10, 19].

Nitrate uptake is followed by reduction to nitrite, which is then transported to the chloroplast wherein it is reduced to ammonium and is mostly assimilated in the plastid/chloroplast and finally undergoes nitrogen remobilization, whereby leaf proteins and especially photosynthetic proteins of plastids are extensively degraded during senescence, providing an enormous source of nitrogen that plants can tap to supplement the nutrition of growing organs such as new leaves and seeds [10]. Nitrogen is available to plants from varied sources and includes inorganic fertilizers (ammonium nitrate, ammonium sulphate, urea, calcium ammonium nitrate and diammonium phosphate and sodium nitrate), organic (compost, manure, seaweed, fish meal and fish emulsion and guano) sources. Although nitrogen occurs naturally in soils, generally, the quantity is quite low and varies geographically warranting external N inputs in the form of fertilizers.

Both organic and inorganic N fertilizers have advantages and disadvantages. Inorganic fertilizers provide readily available nitrogen; however, they are easily lost by leaching, denitrification, volatilization and run-off. Furthermore, inorganic fertilizers have been frequently linked to cases of environmental contamination, soil acidification and salinity. On the other hand, organic fertilizers release of N to plant tends to be slower and depends on the mineralization rates. Nevertheless, organic fertilizers improve the soil physical and chemical properties. Some of the setbacks associated with the use of organic or inorganic fertilizers are predominant in plant cultivation whereby the growth medium is soil. Inherent variations in biophysicochemical properties of soils make it difficult to accurately determine the effects of fertilization on plant growth, yield and quality of produce. Factors such as seasonal changes, development stages, levels of pathogens, geographical differences and nutrient status of the soil affect the amount of secondary metabolites plants produce [20, 21]. These factors can potentially influence the standardization of the quality of medicinal materials. Consequently, more precise plant cultivation techniques are increasingly being used in crop cultivation.

According to Jehnson [22] and Hayden [23], hydroponics technology is a technique of growing plants in a nutrient solution (water and fertilizers) with or without the use of artificial medium (e.g. sand, rockwool, vermiculite, gravel, peat moss, coir and sawdust) to provide a mechanism of support. The advantages of using hydroponics include high-density maximum crop yield, crop production can be achieved in areas where good soil for production is not available, plants can be grown during off-season and temperature can be manipulated [22, 24]. In hydroponics, N is supplied to plants in the form of dissolved salts, which is usually prepared in small and precise quantities, and different nutrient recipes and combinations can be used. Hydroponic technology can be used to manipulate production of plant secondary metabolites [25]. It can favour plant vigour, decrease poisonous levels of plant toxins, increase uniformity and probability of obtaining bioactive extracts [26]. Other related technologies such as aquaponics and aeroponics can also be used to cultivate some medicinal plant species; however, they are still to be fully explored. Aquaponics is the combination of hydroponics and aquaculture in an integrated system to raise fish and grow plants, simultaneously, while aeroponics is a liquid hydroponics system with no other supporting medium for the roots of the plants [22]. In aeroponics plants are grown in misty environment.

4. Physiological effect of nitrogen on medicinal plants

Fertilization programme in medicinal plants has two important objectives: high vegetative growth and high quantity and quality of secondary metabolites produced. Meeting these objectives could lead to high medicinal materials and increased medicinal value of a plant. Generally, N supply favour increased vegetative growth. Argyropoulou et al. [27] investigated the effect of nitrogen starvation on morphological, physiological and biochemical parameters of basil plants cultivated aeroponically. They observed that net photosynthesis rate, transpiration rate, the stomatal conductance and the concentration of total chlorophylls were strongly restricted by N deprivation rate and that total phenolic concentration significantly increased in N-starved plants indicating that biosynthesis of secondary plant metabolites is favoured in nitrogen-deficient plants. Periwinkle, a medicinal plant that is rich in terpenoid alkaloids, when exposed to mixture of nitrate and ammonium, produced the highest content of amino acids, proteins, total alkaloids, vincristine and vinblastine compared to each of the different N forms. It was also observed in the same study that increase in N level beyond 11 mM had an antagonistic effect on alkaloid content [28]. Previous studies have indicated that when plants have N deficiency they tend to have increased concentration of C-based secondary metabolites [29, 30]. Future studies that identify critical N levels for important medicinal plant species will guaranty both high production of medicinal material and quantity and quality of bioactive medicinal principles.

5. Nutrient nitrogen threshold

Nitrogen is a major constituent of enzymes, proteins, chlorophyll and is involved in many important biochemical processes in plants including photosynthesis. However, it has been shown in many studies that N effects on plant physiological processes like syntheses of amino acids and phenolics are dependent on tissue N concentration, plant species and other exogenous factors like water availability, temperature and light. Yañez-Mansilla et al. [31] hypothesized that there is an optimum N concentration threshold that ensure a high phenolic concentration and antioxidant capacity without detrimental effects on plant performance and proposed a threshold of 15 g N/kg DW as an optimum concentration for ensuring high antioxidant activity and quality in blueberry leaves, based on results obtained in their study. In order to meet requirements of new regulations in the coastal valleys of central California, USA, field trials were carried out by Bottoms [32] to identify commercial fields in which N application could be reduced or eliminated in order to improve nitrogen (N) fertilizer efficiency. Crop growth, N uptake and the value of soil and plant N diagnostic measures were evaluated in 24 iceberg and romaine lettuce plants and it was concluded that soil NO_3–N greater than 20 mg/kg was a reliable indicator that N application could be reduced or delayed. Many farmers, scientists, consumers and governments are becoming aware of the risks associated with excessive nitrogen fertilization and are seeking environmentally friendly and sustainable approaches of N fertilization. Medicinal crops farmers would have to take cognizance of the need to balance high yield, quality medicinal materials and minimum environmental toxicity. It is expected that indigenous plant species, especially those occurring in their natural habitats are adapted to their local conditions and may tend to have low critical levels for most of the nutrients. For example, medicinal plants occurring in the fynbos biome of South Africa are adapted to nutrient-poor and low pH soils. Therefore, exposing these species to high N concentration may have minimal effect on plant physiology and can even have detrimental effects on plant growth.

6. Economics of nitrogen fertilization

Many studies have demonstrated that plant yield increases with N fertilization. The quest by farmers for high yield and high profit margins has encouraged the implementations of inappropriate N fertilization programmes. Excessive and inadequate N supply to plants could induce deleterious effects in plants and the environment. With increasing N fertilization costs, it is important to determine optimum N fertilization rates in order to achieve economically viable N fertilization in crop production. In a study carried out in Viçosa, Minas Gerais State, Brazil that aimed at determining the economic optimum N fertilization rates under cold and ambient conditions of four potato cultivars, it was found that economic optimum N fertilization rates ranged from 147 to 201 kg/ha depending upon cultivar and relative prices of N and potato tubers [33]. Farquharson et al. [34] recognized the importance of environmental effects such N_2O emission of N fertilization in Australian wheat production and using an economic framework model, they predicted that the best fertilizer decision is reduced by about 4 kg N/ha (5%) when the Intergovernmental Panel on Climate Change (IPCC)-based environmental cost of N fertilizer is considered. Nyborg et al. [35] reported that economics of nitrogen fertilization of barley and rapeseed is influenced by nitrate-nitrogen level in the soil and suggested that soil testing to determine N_2O–N levels is essential for maximum economic returns from N fertilization. Based on the above-mentioned arguments, the use of precision N fertilization approach is encouraged, for example, in hydroponics it is possible to manipulate plants to produce higher yields of bioactive fractions [36].

7. Case study

Preliminary assessment of the effects of nutrient nitrogen on growth and antimicrobial activities of *H. cymosum* grown under greenhouse conditions.

7.1. Introduction

H. cymosum subsp. cymosum (Asteraceae) is an indigenous South African medicinal plant (**Figure 1**). It has high medicinal value and is heavily harvested from the wild. This species is distributed along the coastal areas of the Eastern and Western Cape Provinces. The soil of the coastal region of the Western Cape region is typically acidic and nutrient-poor and is derived from the weathering of granite [37]. The objective of this study was to assess the effect of N fertilization on growth, tissue nutrient content and antimicrobial activities of acetone leaf extracts of *H. cymosum* cultivated on field collected soil samples under greenhouse conditions.

Figure 1. Hydroponics cultivation of the medicinal plant species *H. cymosum* in a greenhouse.

7.2. Materials and methods

Soil was collected from a commercial vegetable farm located in Kuilsriver, Western Cape, South Africa and the soil subsamples analysed (physico-chemical analysis) [38]. The field collected soil was used to prepare 3 kg potted soil samples. Ammonium nitrate salt was dissolved in 500 ml of sterile distilled water and the solution was poured into the potted soils to obtain a final soil concentration of nitrogen that was 136 ppm. Potted soils were placed in rows on a steel table. In the control treatment, only 500 ml of sterile distilled was added and the baseline N concentration was 32 ppm. Six weeks old rooted cuttings of *H. cymosum* were transplanted individually into each pot. A total of 16 pots, grouped into two treatments with eight replicates per treatment were used. Parameters such as plant height, nutrient concentration of leaves and leaf numbers were assessed in order to determine the effects of nitrogen and potassium on growth of *H. cymosum* at the end of the experiment, 13 weeks post-treatment. Leaf tissue analysis was carried out [39, 40]. Fresh foliage harvested at 13 weeks post-treatment was air dried at room temperature for 4 weeks. Dried plant materials were cut into smaller pieces and ground using a Jankel and Kunkel Model A 10 mill into fine powder. Powdered leaf material (5 g) was extracted with 100 ml of acetone in a glass beaker with the aid of a vortex mixer for 15 min and the supernatant filtered using Whatman No.1 filter paper. The extracted material was left to dry overnight. The micro-dilution method previously described by Eloff [41] was employed with slight modifications to determine the minimum inhibitory concentration (MIC) for the extracts. *Fusarium oxysporum* fungal culture was introduced to all microplates (10^5 spores/ml). Mancozeb (60 mg/10 ml) was prepared using sterile distilled water as a positive control and a mixture of sterile distilled water and acetone was used as a negative control. Data were analysed using a one-way analysis of variance (ANOVA).

7.3. Results

There was no significant difference ($P > 0.05$) in plant height exposed to higher level of N (51.4 ± 4.9 cm) compared to those exposed to low level N (Control) (55.1 ± 5.1 cm) at 13 weeks post-treatment. Similarly, no significant difference ($P < 0.05$) was observed in the number of branches in plants exposed to the different N treatments. Comparatively, N-treated (1.9 ± 0.2 ppm) plants had a significantly high levels of tissue content N in the leaves (df 1,6; $F = 7.8$; $P = 0.03$) than those exposed to low nutrient N treatment (1.4 ± 0.1 ppm) at 13 weeks post-treatment (**Table 1**). MIC bioassay did not show a significant effect ($P > 0.05$) on antifungal activity following N treatment compared to control (0.187 mg/ml) (**Table 2**).

7.4. Discussion

Nitrogen-treated plants had higher N content in the leaves compared to low N-exposed plants suggesting that the treatment with an increased level of N could have induced high uptake of nitrogen. The plant growth was not significantly different in plants treated with 136 ppm of nitrogen compared to control plants (32 ppm). This result suggests that higher nitrogen supply may not always result in high vegetative growth. A plausible explanation could be that plants occurring naturally in nutrient-poor area may have low optimum nutrient requirement and may not warrant excessive N treatment. Also, high N fertilization of medicinal plants may not necessarily reduce bioactivity of their extracts.

Treatment	**N content ppm**
N	1.4 ± 0.1
Control	1.9 ± 0.2

Table 1. Tissue nutrient content (ppm) in aerial parts of *H. cymosum* following exposure to control and N treated field collected soil samples after 13 weeks post-treatment.

Acetone extracts	**Minimum inhibitory concentration (MIC mg/ml) of acetone extract of *Helichrysum cymosum* against *Fusarium oxysporum***	
	24 h	**48 h**
N	0.82 ± 0.01	0.187 ± 0
Control	0.93 ± 0	0.187 ± 0

Table 2. MIC antifungal activity of the acetone extract of *H. cymosum*.

8. Nitrogen toxicity

Excessive nitrate fertilization can induce high accumulation of nitrates in plant tissues to levels that are potentially toxic to humans and livestock. However, Qiu et al. [42] showed that

that genotypic variation in nitrate accumulation is associated with differences in water content for rape, Chinese cabbage and spinach. Vegetables account for over 70% of the total nitrogen intake of humans [43]. Increased concentration of nitrite and nitrates in diet are risk factors for many diseases in mammals [44]. Although nitrate intake from vegetables is receiving substantial attention, it is important that cultivated medicinal plants receive similar attention as the industry develops. Commercial cultivation of medicinal plants could lead to excessive N fertilization and high concentration of nitrites and nitrates in medicinal plant parts and subsequently in herbal decoctions and infusions. This can negate the beneficial effects of medicinal plants. Also, accumulation of unused nitrates in soils could have unfavourable effect on soil biological, physical and chemical properties. Furthermore, leached nitrates in water runoffs could lead to eutrophication of freshwater resources. Since plants have different N needs/ requirements, research on the N requirement of each plant in different growing conditions is important in order to achieve high yield, safe and good quality medicinal materials from plants.

9. Regulation of N fertilization

The development of fertilization policies in many countries is an indication of recognition of the risk that is associated with the use of fertilizers including nitrogenous fertilizers. One of the main challenges facing regulation of the use of fertilizer inputs include high variations of rate of N fertilization across regions and crops and the stage of economic development [45]. The increasing demand for efficient fertilizer use has led the United Nations Economic Commission for Europe (UNECE) to review its so-called "Gothenburg Protocol to Abate Acidification, Eutrophication and Ground-level Ozone". Nitrogen use efficiency (NUE) and N balance will be used as two key indicators in this international convention in order to assess the efficacy of measures to decrease nitrogen (N) losses while maintaining agricultural productivity [46]. Recently, Pires et al. [47] demonstrated that increase in NUE would lead to reduced N fertilization in cereal production as well as improve agronomic, economic and environmental benefits. Considering that increase in global fertilizer consumption is expected to reach 69 million tons in 2030, and the increased use of nitrogen (N) fertilizers is responsible for 67% of this amount. Commercial medicinal plant cultivation will certainly exacerbate this problem in the future. Therefore, it is important for countries to develop efficient policies and guidelines for use of N fertilization going forward.

10. Conclusion and recommendation

Nitrogen fertilization will be an important factor in commercial medicinal crop cultivation. In order to ensure sustainable commercial cultivation of medicinal plant, it is, therefore, necessary to develop efficient N fertilization management programmes as well protocols and policies. It is important that caution is exercised when implementing N fertilization in commercial farming of medicinal plants and the following important aspects should be addressed:

- Determine N requirements of each medicinal plant species and cultivar to ensure optimum quality and yield of medicinal materials while minimizing toxicity to plants, environment and consumers.

- Establish the best types (organic or inorganic) and source (salt, compost and manure) of N fertilizers, which will ensure optimum plant growth with reduced financial and environmental costs.

- Good knowledge of the physical and chemical properties of plant growth media is important because physical and chemical properties of soil vary geographically and this will certainly impact on cation exchange, porosity and organic matter which will in turn affect plant uptake of N. In hydroponics, substrates influence water retention and uptake of nutrients.

- Assess the cost of fertilizer inputs and selling price of medicinal produce in the short- and long-term.

- It is recommended that collaborative partnerships between research, training institutions and commercial farmers be established. Further studies that seek to develop optimized cultivation protocols and policies might be carried out in order to achieve high yield and high quality materials, and sustainable commercial cultivation of medicinal plant.

- Farmers should familiarize themselves with the relevant policies and regulatory frameworks.

Author details

Felix Nchu*, Yonela Matanzima and Charles P. Laubscher

*Address all correspondence to: felixnchu@gmail.com

Department of Horticultural Sciences, Faculty of Applied Sciences, Cape Peninsula University of Technology, Bellville, Cape Town, South Africa

References

[1] Balick, M.J. and Cox, P.A. 1997. Ethnobotanical research and traditional health care in developing countries. In: Bodeker, G., Bhat, K.K.S., Burley, J. and Vantomme, P. Medicinal Plants for Forest Combination and Healthcare. Food and Agriculture Organization of the United Nations. pp. 12-23. http://www.fao.org/3/a-w7261e.pdf

[2] Motaleb, M. A. 2010. Approaches to conservation of medicinal plants and traditional knowledge: A focus on the Chittagong Hill Tracts. IUCN (International Union for Conservation of Nature), Bangladesh Country Office, Dhaka, Bangladesh, pp. 8-30. https://www.iucn.org/sites/dev/files/import/downloads/medicinalplant11book.pdf

[3] Dold, A. P. and Cocks, M.L. 2002. The trade in medicinal plants in the Eastern Cape Province of South Africa. South African Journal of Science **98**, 589-597.

[4] Wiersum, K.F., Dold, A.P., Husselman, M. and Cocks, M. 2006. Cultivation of medicinal plants a tool for biodiversity conservation and poverty alleviation in the Amathole region, South Africa. Medicinal and Aromatic Plants **3**, 43-57.

[5] Mander, M. 1998. Marketing of Indigenous Medicinal Plants in South Africa: A Case Study in Kwazulu-Natal. Food and Agricultural Organization of the United Nations, Rome, Italy. pp. 1-12.

[6] Van Wyk, B.-E., van Oudtshoorn, B. and Gericke, N. 2009. Medicinal Plants of South Africa. Briza Publications, Pretoria, South Africa. pp. 1-250

[7] Bone, K. and Mills, S. 2013. Principles of herbal pharmacology. In: Principles and Practice of Phytotherapy (Second Edition), edited by Kerry Bone and Simon Mills, Churchill Livingstone, Saint Louis. 2013. pp. 17-82, http://dx.doi.org/10.1016/B978-0-443-06992-5.00002-5

[8] Kumar, A., Singh, S.P. and Bhakuni, R.S. 2005. Secondary metabolites of *Chrysnthemum* genus and their biological activities. Current Science, **89**, 1489-1502.

[9] Vu, T.D., Tran, T.L.M., Biteau, F., Mignard, B., Fevre, J.P., Guckert, A. and Bourgaud, F. 2006. Improvement of secondary metabolites production in hydroponic cultures by mechanical and biological processes. Proceedings of International Workshop on Biotechnology in Agriculture, Nong Lam University Ho Chi Minh City October 20, 21 2006. pp. 195-200.

[10] Masclaux-Daubresse, C., Daniel-Vedele, F., Dechorgnat, J., Chardon, F., Gaufichon, L. and Suzuki, A. 2010. Nitrogen uptake, assimilation and remobilization in plants: challenges for sustainable and productive agriculture. Annals of Botany, **105**(7), 1141-1157. doi: 10.1093/aob/mcq028.

[11] Sobiecke, J.-F. 2014. The intersection of culture and science in South African traditional medicine. Indo-Pacific Journal of Phenomenology, **14**(1), 1-13.

[12] Modak, M., Dixit, P., Londhe, J., Ghaskadbi, S., Paul, A. and Devasagayam, T. (2007). Indian herbs and herbal drugs used for the treatment of diabetes. Journal of Clinical Biochemistry and Nutrition, **40**(3), 163-173. http://doi.org/10.3164/jcbn.40.163.

[13] Maroyi, A. 2013. Traditional use of medicinal plants in south-central Zimbabwe: review and perspectives. Journal of Ethnobiology and Ethnomedicine, **9**, 31. doi: 10.1186/1746-4269-9-31.

[14] Shinwari, Z.K., Rehman, M., Watanabe, T. and Yoshikawa, Y. 2006. Medicinal and Aromatic Plants of Pakistan (A Pictorial Guide). Kohat University of Science and Technology, Kohat, Pakistan, 492p.

[15] Street, R.A. and Prinsloo, G. 2013. Commercially important medicinal plants of South Africa: a review. Journal of Chemistry, **2013**, 1-15. http://dx.doi.org/10.1155/2013/205048

[16] Loundou, P.-M. and Watts, S. 2008. Medicinal Plant Trade and Opportunities for Sustainable Management in the Cape Peninsula, South Africa. Unpublished Thesis, University of Stellenbosch, South Africa. pp. 1-103.

[17] Rashid, A.Z.M.M., Tunon, H., Khan, N.A. and Mukul, S.A. 2014. Commercial cultivation of medicinal plants in Northern Bangladesh. European Journal of Environmental Sciences, **4**, 60-68.

[18] Schippmann, U., Cunningham, A.B., Leaman, D.J 2002. Impact of Cultivation and Gathering of Medicinal Plants on Biodiversity: Global Trends and Issues. Case Study No. 7. Biodiversity and the Ecosystem Approach in Agriculture, Forestry and Fisheries, FAO, Rome, pp. 12-13. Visited 20th November 2014. http://www.fao.org/docrep/005/y4586e/y4586e08.htm#TopOfPage

[19] Boczulak, S.A., Hawkins, B.J., and Roy, R. 2014. Temperature effects on nitrogen form uptake by seedling roots of three contrasting conifers. Tree Physiology, **34**, 513-523.

[20] Johanna, B.M. 2007. Variation of Active Constituents in *Euclea nataliensis* Based on Seedling Stages, Seasons and Fertilizers. Unpublished MSc dissertation, University of Pretoria, South Africa. pp. 1-93.

[21] White, A.G., Davies-Coleman, M.T. and Ripley, B.S. 2008. Measuring and optimizing umckalin concentrations in wild harvested and cultivated *Pelargonium sidoides* (Geraniaceae). South African Journal of Botany, **74**, 260-267.

[22] Jehnson, M.H. 1999. Hydroponics worldwide. Acta Horticulture, **48**, 719-730

[23] Hayden, A.L. 2006. Aeroponic and hydroponic systems for medicinal herb, rhizome and root crops. HortiScience, **41**, 536-538.

[24] Koohakan, P., Ikeda, H., Jeanaksorn, T., Tojo, M., Kusakari, S., Okada, K. and Sato, S. 2004. Evaluation of the indigenous micro-organisms in soilless culture: occurrence and quantitative characteristics in different growing systems. Scientia Horticulturae, **101**, 179-188.

[25] Gontier, E., Clement, A., Tran, T.L.M., Gravot, A., Lie'vre, K., Guckert, A. and Bourgaud, A. 2002. Hydroponic combined with natural or forced root permeabilization: a promising technique for plant secondary metabolite production. Plant Science, **163**, 723-732.

[26] Canter, M., Adeline, D. and Teadora, P. 2007. Researches concerning rooting technology of *Pelargonium* genus. Bulletin USAMV-CV, **64**, 1-2.

[27] Argyropoulou, K., Salahas, G, Hela, D. and Papasavvas, A. 2015. Impact of nitrogen deficiency on biomass production, morphological and biochemical characteristics of sweet basil (*Ocimum basilicum* L.) plants, cultivated aeroponically. Agriculture & Food, **3**, 32-42. ISSN: 1314-8591.

[28] Abdolzadeh, A., Hosseinian, F., Aghdasi, M and Sadgipoor, H. 2006. Effects of nitrogen sources and levels on growth and alkaloid content of periwinkle. Asian Journal of Plant Sciences, **5**, 271-276.

[29] Lou, Y. and Baldwin, I. 2004. Nitrogen supply influences herbivore-induced direct and indirect defences and transcriptional responses in *Nicotiana attenuata*. Plant Physiology, **135**, 496-506.

[30] Le Bot, J., Benard, C., Robin, C., Bourgaud, F. and Adamowicz, S. 2009. The "trade-off" between synthesis of primary and secondary compounds in young tomato leaves is altered by nitrate nutrition: experimental evidence & model consistency. Journal of Experimental Botany, **60**, 4301-4314.

[31] Yañez-Mansilla, E. Cartes, P., Reyes-Díaz, M., Ribera-Fonseca, A., Rengel, Z., Lobos, W. and, Alberdi, M. 2015. Leaf nitrogen thresholds ensuring high antioxidant features of *Vaccinium corymbosum* cultivars. Journal of Soil Science and Plant Nutrition, **15**(3), 574-586.

[32] Bottoms, T.G. 2012. Nitrogen requirements and N status determination of lettuce. HortScience, **47**, 1768-1774.

[33] Fontes, P.C.R., Braun, H., Busato, C. and Cecon, P.R. 2010. Economic optimum nitrogen fertilization rates and nitrogen fertilization rate effects on tuber characteristics of potato cultivars. Potato Research, **53**, 167-179. doi: 10.1007/s11540-010-9160-3

[34] Farquharson, R.J., Chen, D., and Li, Y. 2010. What is the impact on farmer nitrogen fertilizer use of incorporating the effects of nitrous oxide emissions? 19th World Congress of Soil Science, Soil Solutions for a Changing World 1 – 6 August 2010, Brisbane, Australia.

[35] Nyborg, M., Malhi, S.S. Mumy, G. Kryzanoski, L. Pennt, D.C. and Laverty, D.H. 2008. Economics of nitrogen fertilization of barley and rapeseed as influenced by nitrate-nitrogen level in soil. Communications in Soil Science and Plant Analysis, **30**, 589-598. http://dx.doi.org/10.1080/00103629909370229

[36] Xego, S. Kambizi, L. and Nchu, F. 2016. Threatened medicinal plants of South Africa: case of the family hyacinthaceae. African Journal of Traditional Complement Alternative Medicine, **13**(3), 169-180.

[37] Bargmann, C.J. 2005. Geology and wine in South Africa. Geoscientist, **15**, 4-8.

[38] The Non-Affiliated Soil Analysis Work Committee. 1990. Handbook of Standard Soil Testing Methods for Advisory Purposes. Soil Science Society of South Africa, Pretoria.

[39] Campbell, C. R. and Plank, C.O. 1998. Preparation of plant tissue for laboratory analysis. In: Kalra, Y.P. (ed.). Handbook of Reference Methods for Plant Analysis. CRC Press, Boca Raton, Fla, USA. pp. 37-49.

[40] Miller, F. C.1992. Composting as a process based on the control of ecologically selective factors. In: Blaine-Metting, F. (ed.). Soil Microbial Ecology: Applications in Agriculture Environment Management, Marcel Dekker Inc., New York. p. 646.

[41] Eloff, J.N. 1998. Which extractant should be used for the screening and isolation of antimicrobial components from plants? Journal of Ethnopharmacology, **60**, 1-8.

[42] Qiu, W., Wang, Z., Huang, C., Chen, B. and Yang, R. 2014. Nitrate accumulation in leafy vegetables and its relationship with water. Journal of Soil Science and Plant Nutrition, **14**(4), 761-768.

[43] Walker, R. 1990. Nitrates, nitrites and N-nitrosocompounds, a review of the occurrence in food and diet and the toxicological implications. Food Additives and Contaminants, **7**, 717-768.

[44] Stanton, L.T. 2001. Nitrate Poisoning. Colorado State University Extension, No.1.610.

[45] Ju, X., Liu, X., Zhang, F. and Roelcke, M. 2004. Nitrogen fertilization, soil nitrate accumulation, and policy recommendations in several agricultural regions of China. AMBIO: A Journal of the Human Environment, **33**(6), 300-305.

[46] Brentrup, F. and Palliere, C. 2010. Nitrogen Use Efficiency as an Agro-environmental Indicator. Accessed 19 June 2015. http://www.oecd.org/tad/sustainable-agriculture/44810433.pdf

[47] Pires, M.V., da Cunha, D.A., de Matos Carlos, S. and Costa, M.H. 2015. Nitrogen-use efficiency, nitrous oxide emissions, and cereal production in Brazil: current trends and forecasts. PLOS ONE, **10**(8), e0135234. doi: 10.1371/journal.pone.0135234.

5

The Effect of N Fertilization on Wheat under Inoculation with *Azospirillum brasilense*

Marcelo Carvalho Minhoto Teixeira Filho,
Fernando Shintate Galindo, Salatiér Buzetti and
Eduardo Henrique Marcandalli Boleta

Abstract

The biological nitrogen fixation (BNF) process in wheat (*Triticum aestivum L.*) occurs by diazotrophic bacteria, particularly *Azospirillum* brasilense. However, researches are lacking on BNF efficiency to define how much mineral nitrogen (N) can be applied to achieve more sustainable high yields, and if urea with the urease enzyme inhibitor is less harmful, benefiting BNF in grasses (cereals). Therefore, the objective was to evaluate the effect of N sources (urea and Super N, urea with urease enzyme inhibitor N-(n-butyl thiophosphoric triamide) (NBPT) and N rates (0, 50, 100, 150, and 200 kg ha^{-1}) applied in topdressing associated with inoculation with *A. brasilense*, regarding the leaf N concentration, leaf chlorophyll index (LCI), accumulation of N in the straw and grains, the nitrogen utilization efficiency (NUE), recovery of the applied nitrogen (RAN), physiological efficiency (FE), agronomic efficiency (AE), and wheat grain yield in the Brazilian Cerrado (tropical savanna) region. The N sources provide similar N accumulations in straw and grains, and wheat grain yield. Inoculation with *A. brasilense* afforded higher N grain concentration (increase in protein content more sustainably) by applying less N fertilizer in topdressing. Inoculation with *A. brasilense* increased the AE, RAN, and NUE.

Keywords: *Triticum aestivum* L., nitrogen sources, biological nitrogen fixation, urease inhibitor, no-tillage system

1. Introduction

The management of nitrogen fertilization is performed in order to ensure adequate productivity, and depending on the N dynamics in the soil, large amounts of N are added to the soil, raising production cost for the farmers. This applies to wheat (*Triticum aestivum* L.) that is an annual cycle plant, considered, among the winter season's cereal, one that has the greater economic importance with large grain yield capacity [1]. This cereal has great relevance in the diet and is cultivated in a wide range of environments and geographic regions. The cereal occupies over 17% of cultivable land in the world and represents approximately 30% of the world's grain production. In the period from 2012 to 2016, the annual average area of wheat cultivated worldwide was approximately 220 million hectares, reaching 734 million tons in the 2015/2016 harvest [2].

The final crop yield is defined according to the cultivar used, the amount of agricultural supplies and management techniques employed. The increasing use of the high yield potential of wheat has implicated in a more frequent use of agricultural supplies, among which nitrogen fertilization is shown to be important in defining the grain yield [3]. Several authors reported a positive response of nitrogen fertilization on the grain yield of wheat [4–6]. Therefore, there is a need to study wheat cultivars verifying their response in the uptake and utilization of nutrients in the soil and their performance and cultural practices in different environments [7].

Nitrogen fertilization includes one of the highest costs of the production process of non-leguminous crops [8]. Wheat, corn, and rice crops utilize approximately 60% of the N fertilizers produced in the world [9]. The use of N fertilizers must be carefully controlled to ensure good yield and manage N in the soil; N fertilizer increases production costs for farmers [10]. Also, both nitrogen fertilizer production and application contribute to the emission of gases (CO_2 and NO_2) that contribute to the increase of the greenhouse effect on the Earth. In a report developed by the International Fertilizer Industry Association and the United Nations Environment Program, 873 m^3 of natural gas was used to produce 1 metric ton of nitrogen fertilizer synthesized by the Haber-Bosch process [11].

It is estimated that there may be a reduction in grain yield due to the volatilization of N-NH_3 at the rate of 10 kg ha^{-1} of grains for each 1% N that is volatilized [12]. In this context, one possibility to increase the nitrogen fertilization efficiency is the use of the N-(n-butyl) thiophosphoric triamide (NPBT) inhibitor, which can delay the hydrolysis of urea and significantly reduce NH_3 losses depending on the weather, that is, heat and rain as well as the chemical characteristics of the soil [13, 14]. Due to the climate in Brazil, urea with urease enzyme inhibitor and conventional urea are equally effective in terms of N nutrition and grain yield of cereals. Studies in countries with milder weather have had different results [15].

Due to the high cost of fertilizers and awareness in support of sustainable agriculture and less pollution, in which the research is growing, another possibility would be to use inoculants containing bacteria that promote growth and increase the productivity of plants. Studies on biological nitrogen fixation (BNF) by *Azospirillum* species in grass have been carried out in Brazil. Until recently, no commercial inoculants with these bacteria are available in the country [16].

Although the plant genotype performs an essential role in the colonization of bacteria, existing cultivars with high and low potential of association [17]. Several studies have been published confirming that *Azospirillum* produces phytohormones that stimulate root growth in many plant species. The components released by *Azospirillum brasilense* responsible for stimulating root growth are indole acetic acid (IAA), gibberellins, and cytokinins [18]. Inoculation with *Azospirillum* can improve the leaf photosynthetic parameters, including chlorophyll content and stomata conductance, greater proline content in shoots and roots, improvement in water potential, an increase in water content in the apoplast, more elasticity of the cell wall, more biomass production, and greater plant size as reported by Barassi et al. [19]. Increases in photosynthetic pigments such as chlorophyll a and b, and auxiliary photoprotective pigments, such as violaxantine, zeaxantine, ateroxantine, lutein, neoxantine, and beta-carotene, which result in greener plants without water-related stress, were verified by Bashan et al. [20].

In addition, the increase in root development caused by inoculation with *Azospirillum* is involved with several other effects. Increases in water and mineral uptake have been reported, as well as greater tolerance to stresses such as salinity and drought, resulting in a more vigorous and productive plant [21, 22]. According to Dobbelaere et al. [23], positive responses to inoculation with *A. brasilense* are obtained even when the crops are grown in soils with high N content, which indicates that the plant responses occur not only due to the fixed N_2 but also depending on the production of phytohormones growth promoters such as cytokinin, gibberellin, and indole acetic acid. Lemos et al. [24], studying five wheat cultivars, found a positive interaction between *A. brasilense* and N fertilization only for one wheat cultivar (CD 150). Increases in N fertilization efficiency associated with inoculation with *A. brasilense* were reported by Galindo et al. [14] but in the corn grain yields in the *Brazilian Cerrado*.

Considering the benefits attributed to several crops by inoculation with *A. brasilense*, with an emphasis on biological nitrogen fixation, greater development of the root system, and, consequently, greater uptake of water and nutrients, the inoculation can improve crop performance allowing greater efficiency of nitrogen fertilization. Thus, more experiments of this type should be carried out in order to evaluate the effect on plant nutrition. In addition, there are still few studies which define how much of N minerals can be applied for BNF to be successful in increasing the yield. It would be interesting to analyze urea with an NBPT urease enzyme inhibitor to verify whether it causes damage to BNF in grass.

The hypothesis of this study was that inoculation with *A. brasilense* can increase the efficiency of N fertilization and N plant nutrition. The objective was to evaluate the effect of N sources and N rates associated, or not, with *A. brasilense* inoculation regarding the leaf N concentration, leaf chlorophyll index (LCI), accumulation of N in the straw and grains, the nitrogen utilization efficiency (NUE), recovery of the applied nitrogen (RAN), physiological efficiency (FE), agronomic efficiency (AE), and wheat grain yield in the *Brazilian Cerrado* (tropical savanna) region.

2. Material and methods

2.1. Location and soil-climatic conditions

The wheat experiment was conducted in 2015, in the field in an experimental area that belongs to the Paulista State University (UNESP), Engineering Faculty, located in Selvíria—MS/Brazil, with the following geographical coordinates 20°22'S, 51°22'W and an altitude of 335 m. Soil in this experimental area was classified as distroferric red oxisol with clay texture (with values of particle sizes as 420, 50, and 530 g kg^{-1} of sand, silt, and clay, respectively), according to Embrapa [25], which has been cultivated with annual cultures over 27 years and with no-tillage system in the past 11 years. The area was under corn cultivation before sowing wheat. The annual average temperature was 23.5°C, annual average pluvial precipitation was 1370 mm, and annual average relative air humidity was between 70 and 80%. Weather data recorded during the experimental period are shown in **Figure 1**.

Glyphosate [1800 g ha^{-1} of active ingredient (a.i.) and 2,4-D (670 g ha^{-1} of a.i.)] herbicides were used for desiccation, applied 2 weeks prior to sowing wheat. Chemical attributes of the soil in the tillable layer were determined before the wheat experiment began. The methods proposed by Raij et al. [26] showed the following results: 13 mg dm^{-3} of P (resin), 6 mg dm^{-3} of S-SO_4, 23 g dm^{-3} of organic matter (OM), pH ($CaCl_2$) of 4.8, 2.6 $mmol_c$ dm^{-3} of K^+, 13.0 $mmol_c$ dm^{-3} of Ca^{2+}, 8.0 $mmol_c$ dm^{-3} of Mg^{2+}, 42.0 $mmol_c$ dm^{-3} of H + Al, 5.9 mg dm^{-3} of Cu, 30.0 mg dm^{-3} of Fe, 93.9 mg dm^{-3} of Mn, 1.0 mg dm^{-3} of Zn (DTPA), 0.24 mg dm^{-3} of B (hot water), and 36% of base saturation. After soil chemical analysis, 2.5 t ha^{-1} of dolomitic limestone (with 88% of relative total neutralizing power) was directly applied as topdressing for 80 days before the wheat was sown in 2015 in order to elevate base saturation to 70%, as recommended by Cantarella et al. [27].

2.2. Treatments and experimental design

A randomized block experimental design in a 2 × 5 × 2 arrangement with four replications was used for both crops. There were two N sources–conventional urea with 45% N and Super N, urea with a urease enzyme inhibitor, NBPT with 45% N; five rates of N (0, 50, 100, 150,

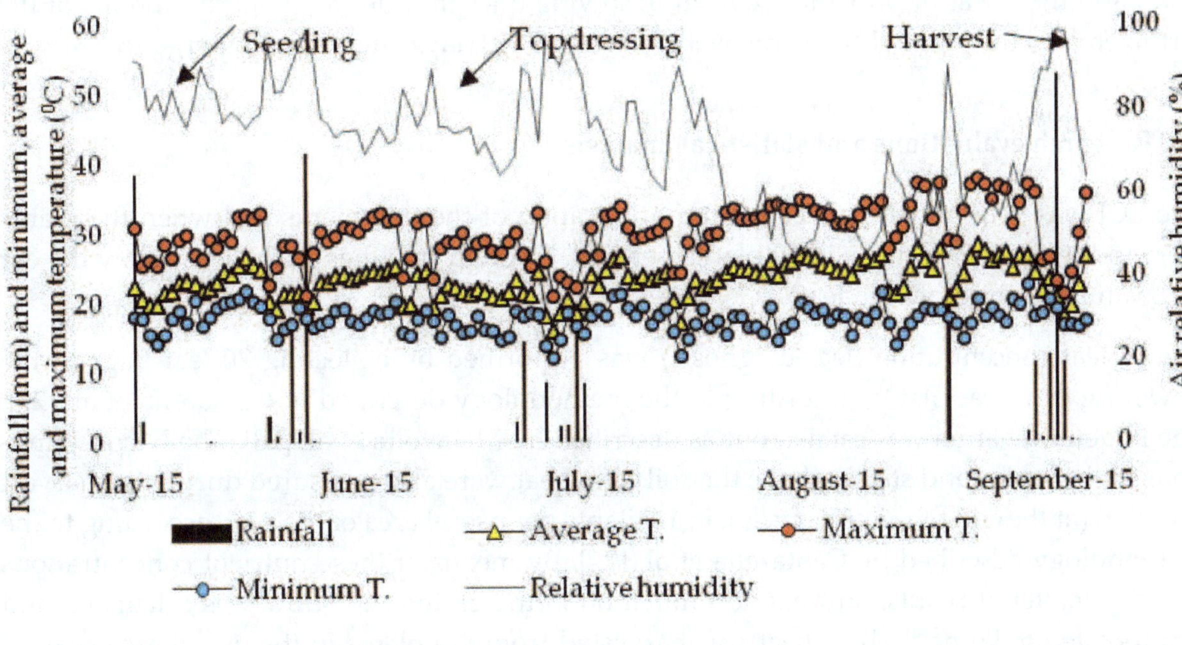

Figure 1. Rainfall, air relative humidity, and maximum and minimum average temperature obtained from the weather station located on the Education and Research Farm of FE/UNESP during wheat cultivation in the period from May to September 2015.

and 200 kg ha^{-1}) applied in topdressing at the growth stage 3.2 on Zadok scale; and two seed inoculation treatments—half of the tests were carried out with seeds inoculated with *A. brasilense*, whereas the other half did not have this inoculation. Each plot consisted of 5 m in length with 12 lines and an inter-row spacing of 0.17 m. The usable area of the plot was 8 center lines, excluding 0.5 m extremities. The plot size was 10.20 m^2.

2.3. Wheat crop management

Wheat seeds were inoculated with 300 mL of inoculant liquid of *A. brasilense* bacteria, AbV5 and AbV6 strains (guaranteed minimum analysis of 2 × 10^8 UFC mL^{-1}) per sack of 40 kg of wheat seeds. The inoculant was mixed with the seeds using a cement mixer, 1 h before planting, and after that, the seed treatments were carried out with carbendazim and thiram fungicides (45 and 105 g a.i. per 100 kg of seeds) and thiodicarb and imidacloprid insecticides (45 and 135 g a.i. per 100 kg of seeds).

Was applied 350 kg ha^{-1} of a 08-28-16 formulation in the forms of urea, triple superphosphate and potassium chloride, respectively, at wheat sowing. The experiments were conducted in a no-tillage system. The area was irrigated by a central pivot sprinkler system. The water coverage was 14 mm over a period of around 72 h. The cultivar used was the CD 116 and sowing was done with an experimental machine on May 15, and 80 seeds per meter were being sown. Metsulfuron methyl (3.0 g a.i. ha^{-1}), a post-emergence herbicide, was applied 20 days after emergence (DAE) to control weeds like *Ipomoea grandifolia, Tridax procumbens,* and *Spermacoce latifolia.*

The seedling emergence was 6 days after sowing. Topdressing with nitrogen fertilization was performed at 35 DAE, manually distributing the fertilizer on the soil surface (no incorporation), and there was approximately 8 cm of sowing lines in order to avoid the contact of the fertilizer with the plants. The plants were harvested 110 days after wheat emergence.

2.4. Research evaluations and statistical analysis

The LCI was determined indirectly after application of the treatments and when the plants were in the flowering stage, in 10 plants per plot, through readings in the leaf below the ear (in the middle third of each leaf).

The N leaf concentration (leaf diagnosis) was performed by collecting 20 leaf flags in the flowering of wheat plants, according to the methodology described in Cantarella et al. [27]. The N determination was carried out as described by Malavolta et al. [28]. The N concentrations in the grains and straw (above the soil) of wheat were also measured during the harvest occasion (at the end of the crop cycle), in 10 plants per useful area of the plot, according to the methodology described by Cantarella et al. [27]. By means of these nutrient concentrations and dry matter of plants, nutrient accumulation in the grains and straw was calculated and extrapolated to kg ha^{-1}. The wheat was harvested from the plants in the useful area of each plot and the grain yield was calculated after mechanical threshing. Data were transformed into kg ha^{-1} and corrected for 13% moisture (on a wet basis).

The accumulation values of N were obtained by the N concentrations in the plant and the dry matter production (DM). With the data of dry mass and accumulation of N, the following indices were calculated:

1. Nitrogen utilization efficiency = total dry matter in kg/accumulation of N in g, in kg of DM/kg of accumulated N [29],
2. Recovery of the applied nitrogen = accumulation of N in kg with fertilization–accumulation of N in kg without fertilization/N rate applied in kg × 100, in percentage [30],
3. Physiological efficiency (FE) = biological yield (straw and grains) without fertilization in kg/N accumulation with fertilizer (straw and grains) in kg–N accumulation without fertilization (straw and grains) in kg, in kg of DM/kg of accumulated N [30], and
4. Agronomic efficiency (AE) = grain yield with fertilization in kg–grain yield without fertilization in kg/N rate in kg, in kg of DM/kg of applied N [30].

The results were subjected to analysis of variance and the Tukey test at 5% probability to compare the averages of N sources and plants that had been inoculated with *A. brasilense* with those that had not been inoculated. Regression equations were fitted for the effect of N rates using the Sisvar program [31].

3. Results and discussions

LCI increased linearly with increasing doses of N (**Table 1**, **Figure 2**). The increase in LCI values, as a consequence of the N rates, resembles those reported by Theago et al. [6], who observed an increase in LCI in wheat up to the dose of 200 kg ha^{-1}, and Teixeira Filho et al. [5] up to 147 kg ha^{-1} of N. This behavior is due to the increase in chlorophyll concentration, promoted by the greater availability of total N in the tissues. This relationship is attributed mainly to the fact that 50–70% of the total leaf N is integral to enzymes that are associated with chloroplasts [32].

Regarding the sources, the Super N provided greater readings of LCI compared to urea (**Table 1**). However, inoculation with *A. brasilense* did not influence LCI, unlike that reported by Galindo et al. [14], working with the N rates (0, 50, 100, 150 and 200 kg ha^{-1} in topdressing) and the sources urea and Super N in the corn crop.

The increase of N rates influenced N leaf concentration (**Table 1**). There was an adjustment of the quadratic function, with maximum concentration point obtained of the N rate at 181 kg ha^{-1} approximately (**Figure 2**). The increase of N concentration in the leaf tissue was expected since by providing more nutrient quantity, the uptake of the crop would be greater, with reflection in the concentration in the leaf and in the aerial part, even in a consolidated system of no tillage in the area of study, over 10 years, which could supply the need of this nutrient due to the decomposition of the straw and as a result of the crop sequence after wheat was the maize crop, the wheat crop response to N rates was more evident.

Similar results were obtained by Theago et al. [6], who observed a linear increase in foliar N concentration with the increment of the doses in coverage (0, 50, 100, 150, and 200 kg ha^{-1}). The increase of N concentration in the wheat leaf, due to the increase of N rates, was also reported by Teixeira Filho et al. [33], with a quadratic response up to the dose of 100 kg ha^{-1}

	LCI	N leaf concentration	N straw	N grains
N rates (kg ha^{-1})		(g kg^{-1})		
0	50.92	38.89	24.12	44.86
50	52.31	44.20	31.39	78.35
100	54.50	47.39	37.08	92.22
150	55.23	48.46	38.59	77.67
200	56.16	49.95	54.89	96.10
N sources				
Urea	52.66 b	46.08 a$^{\Psi}$	38.87 a	78.35 a
Super N	54.99 a	45.47 a	35.56 a	77.33 a
L.S.D. (5%)	1.31	1.33	4.91	9.81
Inoculation				
With *Azospirillum*	54.26 a	45.24 a	34.82	79.20 a
Without *Azospirillum*	53.39 a	46.31 a	39.60	76.48 a
L.S.D. (5%)	1.31	1.33	4.91	9.81
Overall mean	53.82	45.78	37.21	77.84
C.V. (%)	4.64	5.55	25.26	19.04

Selvíria—MS, Brazil, 2015.

$^{\Psi}$Means followed by the same letters in the column do not differ by Tukey at 0.05 probability level.

Table 1. Leaf chlorophyll index (LCI), N leaf concentration, N accumulated in straw and grains of wheat in function of N rates, and sources and inoculation with *Azospirillum brasilense* in wheat crops.

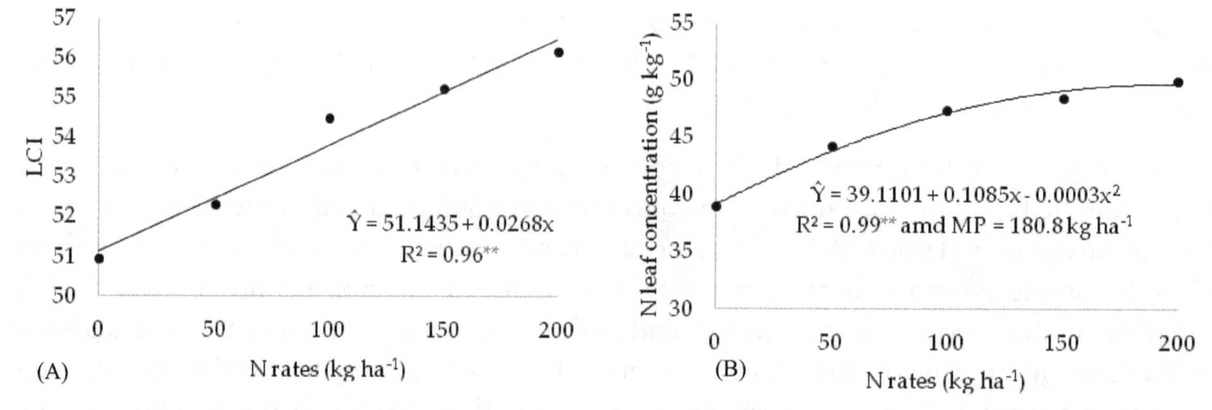

Figure 2. LCI (A) and N leaf concentration (B) in wheat crops in the function of N rates. Selvíria—MS, Brazil, 2015.

of N, and Teixeira Filho et al. [34], with a point of maximum concentration of N being reached with the application estimate of 163 kg ha^{-1} of N. It should be noted that the concentration of N in the leaf tissue, even in the absence of nitrogen fertilization (0 kg ha^{-1}), was above that as recommended by Cantarella et al. [27], ranging from 20–34 g kg^{-1}.

N sources did not differ in N concentration, indicating that Super N was not efficient for N nutrition, even in the area with remaining straw of corn or wheat (**Table 1**). Similar results were verified by Meira et al. [35], using ammonium sulfonitrate, ammonium sulfate, and urea sources for the N concentration of irrigated wheat in the *Cerrado*. However, Teixeira Filho et al. [5] verified higher N concentrations in leaf tissue when the sources of ammonium sulfonitrate and ammonium sulfate compared to urea were used.

On the other hand, Megda et al. [36] (ammonium sulfonitrate, ammonium sulfate, and urea), in the corn crop, verified that the ammonium sulfonitrate source had a higher N concentration in the leaf, compared to the other sources, unlike the results obtained in the present research. It should be noted that Super N acts on the inhibition of the urease enzyme, whereas the ammonium sulfonitrate has in its composition dimethylpyrazolophosphate (DMPP) molecules that act to inhibit nitrification. This makes the fertilizer less susceptible to leaching, since there is a longer residence time of N, as ammonium in the soil, and, under conditions of tropical climate and high temperatures, presents different responses to Super N. Also, there is evidence of active urea transport by high affinity transporters (symport) located in the plasmatic membrane of the root epidermis cells, which would allow uptake of some urea applied before urease has acted and NH_3 has been formed, especially when urea concentration in the soil and soil pH is low [37].

According to Pankievicz et al. [38], there is a greater development and growth of the root system of the *Setaria viridis* grass inoculated with *A. brasilense* as a function of associative fixers, with greater CO_2 fixation and less accumulation of photoassimilated carbon in the leaves, which would favor the plant with greater growth in aerial part, greater accumulation of water, and conditions of less stress caused by the greater accumulation and metabolism of carbon, thus, increasing the concentration of nutrients in the plant. On the other hand, Bashan et al. [22] reported that the production of plant hormones, mainly indole acetic acid by bacteria of the genus *Azospirillum*, plays an essential role in the promotion of plant growth and, according to Hungria et al. [39], can improve the uptake of several macro and micronutrients, increasing the efficiency of the use of the available nutrients; however, this result was not verified in the present work, which raises the question of the difference of efficiency of the inoculation in grass crops.

Analyzing the split between with or without *A. brasilense* inoculation and N rates for N accumulation in the straw (**Table 2**), there was no difference between inoculation and not with *A. brasilense* for any of the N rates. With the increase in N rates, there was a linear increase in the N accumulation in straw (**Figure 3**), regardless of inoculation or not with *A. brasilense*. Dobbelaere et al. [23] also reported positive responses to inoculation with *A. brasilense* even when the crops are grown in soils with high N content available, which indicates that the plant responses occur not only due to the fixed N_2 but mainly depending on the production of phytohormone growth promoters such as cytokinin, gibberellin, and indole acetic acid.

Inoculation	N rates (kg ha⁻¹)				
	0	50	100	150	200
With	22.38 a[Ψ]	30.29 a	35.28 a	34.20 a	51.97 a
Without	25.86 a	30.29 a	38.88 a	42.99 a	57.81 a
L.S.D. (5%)			10.98		

Selvíria—MS, Brazil 2015.

[Ψ]Means followed by the same letters in the column do not differ by Tukey at 0.05 probability level.

Table 2. Inoculation with *A. brasilense* and N rates interaction in the N accumulation in wheat straw.

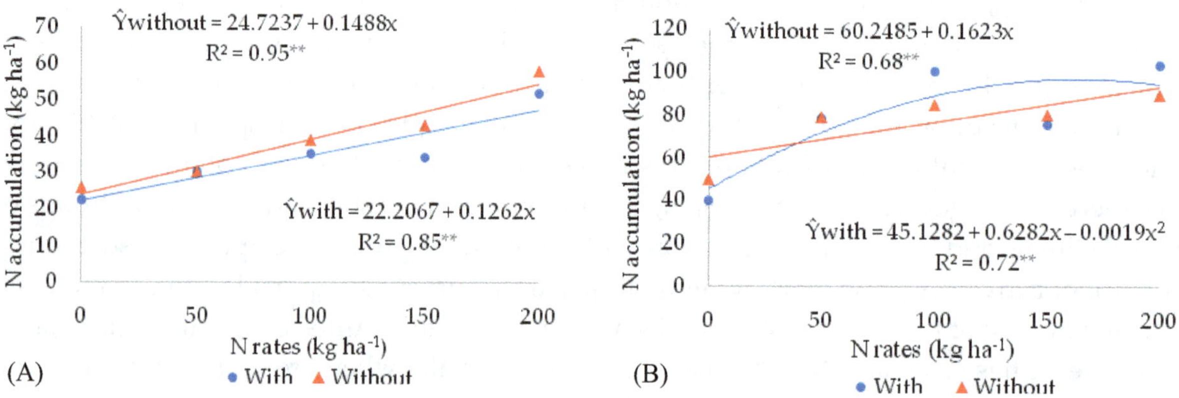

Figure 3. N rates and inoculation with *A. brasilense* interaction in the N accumulation in straw (A) and in grains (B) of wheat. Selvíria—MS, Brazil 2015.

The greater accumulation of nutrients provided by the increase of the N rates associated or not with the inoculation with *A. brasilense* is a very significant result considering that conservation practices benefit the productive systems and environmental conservation, such as no-tillage system, whose straw will provide nutrients to subsequent crops, recycling nutrients, and minimizing losses and environmental pollution, such as nutrient leaching.

N is the nutrient that most interferes in the development and productivity of crops, especially grasses. This mineral nutrient is found in higher concentrations, in vegetative tissues and grains, which characterizes it as being the element most demanded by the wheat plant. Thus, the higher availability of this nutrient in the plants favored the development of the root system, which, by exploiting a larger volume of soil, may have an uptake of a greater amount of nutrients and water, reflecting the accumulation in aerial part, both in the straw and in the grains. Since N is involved in the synthesis of proteins, chlorophyll, coenzymes, phytohormones, nucleic acids, and secondary metabolites (Amines, amides, amino sugars, purines, pyrimidines and alkaloids) [40].

The interaction between N rates and inoculation with *A. brasilense* was significant for N accumulation in grains; however, there was no difference between with or without *A. brasilense* inoculation for any of the N rates (**Table 3**). However, in treatments inoculated with *A. brasilense*,

Inoculation	N rates (kg ha^{-1})				
	0	50	100	150	200
With	40.13 a$^{\Psi}$	78.03 a	99.90 a	75.13 a	102.81 a
Without	49.59 a	78.66 a	84.54 a	80.20 a	89.40 a
L.S.D. (5%)			21.94		

Selvíria—MS, Brazil, 2015.
$^{\Psi}$Means followed by the same letters in the column do not differ by Tukey at 0.05 probability level.

Table 3. Inoculation with *A. brasilense* and N rates interaction in the N accumulation in wheat grains.

there was an adjustment to the quadratic function for the dose of 165.3 kg ha^{-1} of N, whereas in the treatments without inoculation with this bacterium, the adjustment was linearly increasing for N in the grains (**Figure 3**) but reaching lower N concentrations in the highest rate of N.

These bacteria can act on plant growth by producing substances promoting development (auxins, gibberellins, and cytokinins) which provide better root growth [41] and, therefore, help in greater uptake of water and nutrients [42], resulting in a more vigorous and productive plant [16, 22]; to be free-living organisms with endophytic characteristics, it is possible to perform some of the metabolic and vital use of nutrients in the plant, which would then be made available to reflect in increased concentrations in the grains.

Galindo et al. [14] studied N rates (0, 50, 100, 150, and 200 kg ha^{-1}, in topdressing), N sources (Super N and urea), with and without inoculation with *A. brasilense* in maize, and found positive influence of inoculation on nutrient concentration in leaf tissue, which may be indicative of the phytohormonal effect cited in the literature, confirming that *Azospirillum* produces phytohormones that stimulate root growth of several plant species and that this greater development of the roots may imply several other effects such as increases in the water and nutrient uptakes and greater tolerance to stresses such as salinity and drought, resulting in a more vigorous and productive plant [22]. In addition, Barassi et al. [19] reported improvement in leaf photosynthetic parameters, including chlorophyll content and stomatal conductance, higher proline content in shoots and roots, improvement in water potential, increase in water content of apoplast, greater cell wall elasticity, and higher production of plant biomass.

The increase in N rates influenced significantly the number of kernels per spike and spikes per meter (**Table 4**). For the number of kernels per spike, the data adjusted the quadratic function with maximum point in 151 kg ha^{-1} of N (**Figure 4**). On the other hand, the N rates influenced the number of spikes per meter, adjusted to the quadratic function, up to the dose of 110 kg ha^{-1} of N (**Figure 4**). These results explain why grain yield was influenced positively by the increase of N rates, independently of the source of N and inoculation with *A. brasilense* (**Figure 4**).

For the hectoliter weight and 100-kernel weight, there was no influence of N rates (**Table 4**). Similar results were obtained by Teixeira Filho et al. [33] and Theago et al. [6] that did not verify the influence of the N rates on the mentioned parameters in the irrigated wheat crop. Nunes et al. [8] and Souza et al. [43], who found no influence of N rates in topdressing for 100-kernel

	Number of kernels per spike	Number of spikes per meter	Hectoliter weight	100-kernel weight	Grain yield
N rates (kg ha^{-1})			(kg 100 L^{-1})	(g)	(kg ha^{-1})
0	34.40	69.17	87.23	4.20	1906
50	35.75	76.50	86.83	4.10	3027
100	42.48	84.42	87.42	4.19	3363
150	37.88	73.75	86.14	4.13	3167
200	38.91	74.50	85.96	4.12	3263
N sources					
Urea	38.60 a$^{\Psi}$	75.50 a	87.14 a	4.17 a	2930 a
Super N	37.17 a	75.83 a	86.29 a	4.13 a	2960 a
L.S.D. (5%)	2.70	5.90	1.40	0.06	212
Inoculation					
With *Azospirillum*	39.04 a	76.00 a	86.61 a	4.14 a	3007
Without *Azospirillum*	36.73 a	75.33 a	86.82 a	4.15 a	2883
L.S.D. (5%)	2.70	5.90	1.40	0.06	212
Overall mean	37.88	75.67	86.71	4.15	2945
C.V. (%)	13.65	14.92	3.61	3.36	16.08

Selvíria—MS, Brazil, 2015.

$^{\Psi}$Means followed by the same letters in the column do not differ by Tukey at 0.05 probability level.

Table 4. Number of kernels per spike, number of spikes per meter, hectoliter weight, 100-kernel weight, and wheat grain yield in function of N rates and sources and inoculation with *Azospirillum brasilense* in wheat crop.

weight in an area with high availability of N. According to Frank & Bauer (1996), in the period between the emergence phase of the seedlings and the differentiation of the floral primordium, N deficiency reduces the mass of 100-kernel weight. Therefore, the results obtained can be explained by the fact that N deficiency was not observed in the plants in any of the treatments by the fact of the irrigated wheat crop and because the number of kernels per spike increased as a function of increasing N rates, thereby providing increment on the competition for photo assimilates inside the spike [5] but not to the point of reducing the mass of grains. This can be proved by the high hectoliter weight to be obtained (>78 kg 100 L^{-1}) in the experiment, which classified (by individual analysis) the type 1 wheat with the best quality [10].

Conventional urea and urea with NBPT provided similar results for number of kernels per spike, number of spikes per meter, hectoliter weight, 100-kernel weight, and, consequently, for wheat grain yield (**Table 4**) in agreement with Megda et al. [36] and Theago et al. [6]. Prando et al. [44], evaluating the sources of urea, urea + NBPT, and coated urea, in another climatic condition (temperate climate), also did not observe changes in grain yield. It is worth noting that N sources may have presented similar behavior due to irrigation during the entire

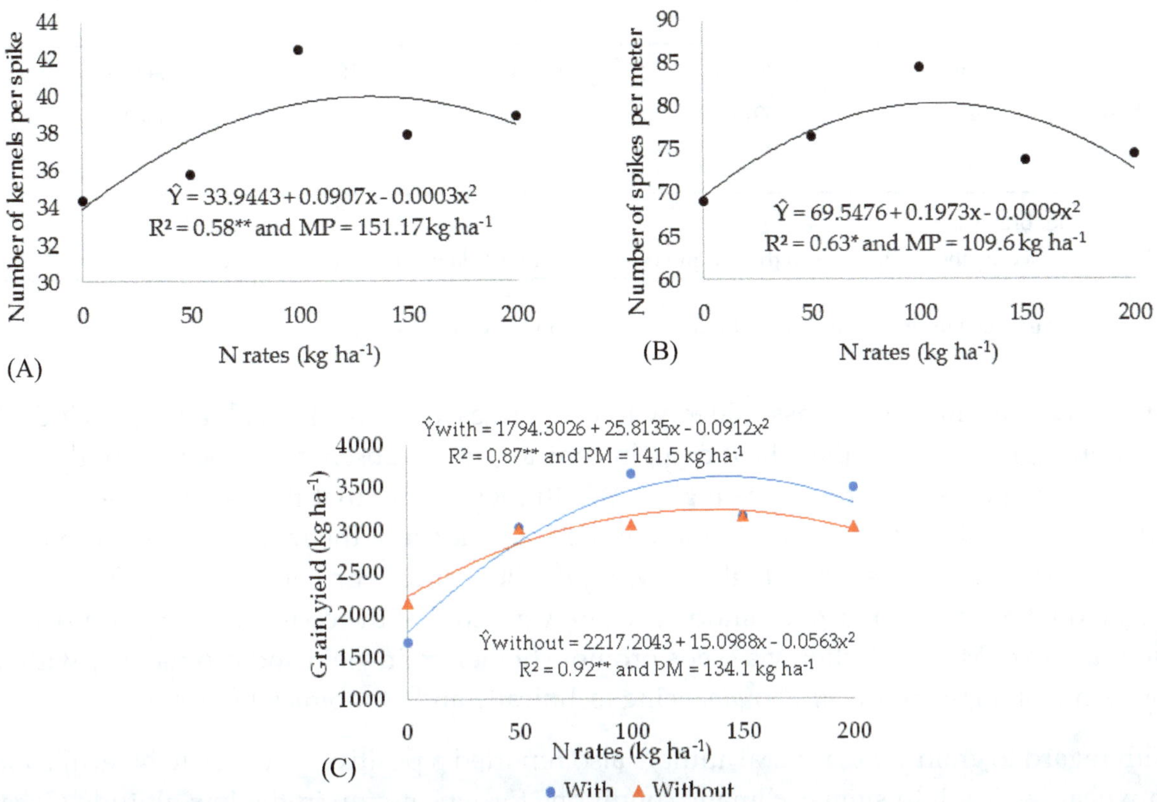

Figure 4. Number of kernels per spike (A) and number of spikes per meter (B) in the function of N rates and N rates and inoculation with *A. brasilense* interaction in the wheat grain yield (C). Selvíria—MS, Brazil 2015.

crop cycle, which would have reduced volatilization losses, mainly in the form of NH_3, which occur at a higher proportion in the first 4 days after urea application.

As regards inoculation with *A. brasilense*, this alone did not significantly interfere in the number of kernels per spike, number of spikes per meter, hectoliter weight, and 100-kernel weight (**Table 4**). Most probably, there was no effect of this inoculation on these production components because the number of kernels per spike is mainly determined by the genetic characteristics of the cultivar, and from the other evaluations mentioned above, it is due to the cultivation of irrigated wheat, fertilization, and adequate plants stand and soil fertility.

There was a significant interaction between N rates and inoculation with *A. brasilense* for wheat grain yield. At the dose of 100 kg ha^{-1} of N, the inoculation provided higher productivity than the uninoculated treatment (**Table 5**). Grain yield adjusted the quadratic function for N rates in the treatments with and without inoculation with *A. brasilense*, with a positive response up to 142 and 134 kg ha^{-1} of N, respectively (**Figure 4**). However, in relation to the control (without N), the optimal N dose of the treatment inoculated with this diazotrophic bacterium provided a higher grain yield of 391 kg ha^{-1} in relation to the best N rate of the treatment without inoculation; that is, this increase was higher by 7%.

The N rates mentioned above for obtaining the maximum grain yield were high since the wheat was cultivated on corn straw (high C/N ratio), that is, part of the applied N was immobilized by the decomposing/mineralizing of microorganisms.

Inoculation	2015				
With	1671 a[Ψ]	3036 a	3663 a	3167 a	3497 a
Without	2141 a	3018 a	3063 b	3166 a	3029 a
L.S.D. (5%)	474				

Selvíria—MS, Brazil, 2015.

[Ψ]Means followed by the same letters in the column do not differ by Tukey at 0.05 probability level.

Table 5. Inoculation with *A. brasilense* and N rates interaction in the wheat grain yield.

The *A. brasilense* inoculation associated with the N rates in the 140 kg ha^{-1} range provided the maximum grain yield of the wheat crop; in contrast, in the absence of inoculation, the magnitude of response to N rates was higher. Worth mentioning that the estimated grain yield of the inoculated treatments was higher, numerically, most of the rates tested in both crops, even using N rates considered high, showing the benefit of inoculation with *A. brasilense* in irrigated wheat crop. In turn, Galindo et al. [45] verified that the co-inoculation with *A. brasilense,* and Co + Mo application via seeds promote higher grain yield and profitability with the soybean crop in the *Brazilian Cerrado,* being technically and economically viable.

With regard to grain yield, several authors also reported a positive response to N fertilization on wheat [4, 7, 46]. In similar climatic conditions for wheat crop in the low-altitude *Cerrado region,* it was reported that the maximum grain yield was 78 [33], 90 [7], and 120 kg ha^{-1} of N [5]. This difference in rates of N that provides maximum wheat grain yield is due to different N requirements of cultivars as well as the variation in soil and climatic conditions of these researches. Anwar et al. [47] found that maximum spikes m^{-2}, grains per spikes, thousand grains weight, and grain yield (4061 kg ha^{-1}) were produced by 125 kg ha^{-1} of N for two wheat cultivars in Pakistan.

Lemos et al. [24] studied five wheat cultivars (CD 104, CD 108, CD 119, CD 120, and CD 150), with and without inoculation and associated with nitrogen rates, and found that response to inoculation with *A. brasilense* in wheat crop occurs satisfactorily when held in conjunction with the nitrogen fertilization, as observed in this study at a dose of 100 kg ha^{-1} N (**Table 4**). However, *A. brasilense* alone is not effective enough to replace entire nitrogen fertilization but is associated with N fertilization, which makes it possible to achieve the highest yields of irrigated wheat grains in *Brazilian Cerrado.*

On the other hand, Ferreira et al. [48], working with foliar application of *A. brasilense* and nitrogen rates in the wheat crop in the *Brazilian Cerrado,* observed that inoculation had no effect on grain yield. Similarly, Nunes et al. [8] studied inoculation with *A. brasilense* in soils with high and low availability of N, and Galindo et al. [49] in research with application times by the leaf of *A. brasilense* with the application of 100 kg ha^{-1} N found no effect of inoculation in the production components and grain yields of wheat in the *Brazilian Cerrado.*

It is noteworthy that bacteria of *Azospirillum* genus are native from soil [50], so it is possible that these were at a high population in the soil under study and, therefore, cancel or minimize the effect of inoculation. Moreover, the affinity of the cultivar with the strains of this bacteria diazotrophic may vary and determine the success or failure of *A. brasilense* inoculation.

The NUE was negatively affected by the increment of N rates with adjustment to the linear function decreasing (**Table 6, Figure 5**). This result can be attributed to the losses of N portrayed clearly in the literature. The increase of N rates culminates in greater losses and less use by the crops, since there is a limit in the nutritional demand of the plant, that is, the plants uptake certain amount of nutrients for a given time; thus, the N that is applied and is not taken can be lost, decreasing the efficiency of fertilization with the higher rates of N, as portrayed in the literature by the law of decreasing increments. The results are similar to those reported by Silva et al. [51], studying N rates (0, 100, 200, and 300 kg ha^{-1}) in the Marandu palisadegrass and Sant'Ana et al. [52], working with common bean crop in the rates of 0, 30, 60, 120, and 240 kg ha^{-1} of N in topdressing.

	NUE	**RAN**	**FE**	**AE**
	(kg kg^{-1})	**(%)**	**(kg D.M. kg^{-1} of N accumulated)**	**(kg grains kg^{-1} N applied)**
N rates (kg ha^{-1})				
0	–	–	–	–
50	38.34	78.82	50.27	22.41
100	30.07	62.08	48.90	14.57
150	13.92	31.08	55.24	8.40
200	15.30	39.52	39.00	6.79
N sources				
Urea	29.65 a	56.87	51.63 a$^{\Psi}$	14.14
Super N	19.16 b	48.87	45.08 a	11.95
L.S.D. (5%)	6.51	12.78	16.97	3.05
Inoculation				
With *Azospirillum*	26.15 a	63.11	42.01 a	16.58
Without *Azospirillum*	22.67 a	42.64	54.70 a	9.50
L.S.D. (5%)	6.51	12.78	16.97	3.05
Overall mean	24.41	52.87	48.35	13.04
C.V. (%)	17.25$^{\#}$	15.33$^{\#}$	19.44$^{\#}$	24.24$^{\#}$

Selvíria—MS, Brazil 2015.

$^{\#}$Corrected data following equation $(x + 0.5)^{0.5}$.

$^{\Psi}$Means followed by the same letters in the column do not differ by Tukey at 0.05 probability level.

Table 6. Nitrogen utilization efficiency (NUE), recovery of the applied nitrogen (RAN), physiological efficiency (FE), and agronomic efficiency (AE) in the function of N rates and sources and inoculation with *Azospirillum brasilense* in the wheat crop.

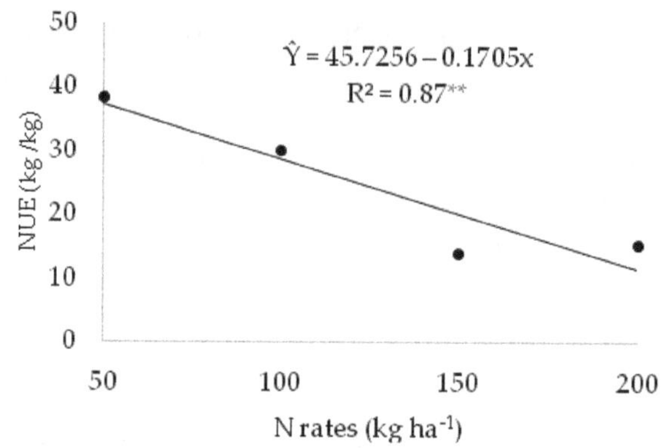

Figure 5. Nitrogen utilization efficiency (NUE) in wheat crop in the function of N rates. Selvíria—MS, Brazil 2015.

In relation to the N sources, urea presented higher NUE compared to Super N, differently from what was expected due to the possibility of mitigation of volatilization of the ammonia provided by Super N fertilizer (urea with urease enzyme inhibitor NBPT) (**Table 6**). However, Dupas et al. [53] evaluated in the *Brazilan Cerrado* the dry-matter yield, RAN, and NUE of palisade grass in response to sources of N (ammonium nitrate, ammonium sulfate, ammonium sulfate-nitrate, urea, urea with urease inhibitor NBPT, polymer-coated urea, and control) in seven harvests (100 kg ha^{-1} N applied after each harvest) and reported for RAN and NUE that the use of N fertilizers that minimizes N loss, such as urea with urease inhibitor NBPT, and polymer-coated urea, was very promising, especially for minimizing the environmental impact of N fertilization. However, they found no difference in DMY due to N sources.

The inoculation with *A. brasilense* did not influence the NUE, although it gave 15.4% greater efficiency compared to the non-inoculated treatments (**Table 6**), which again may be indicative of the phytohormonal effect cited in the literature, confirming that *Azospirillum* produces phytohormones that stimulate root growth of several plant species and that this greater development of the roots may be implied in several other effects, such as increases in the water and nutrient uptakes like N.

The interaction between N rates and inoculation, and N sources and inoculation, was significant for RAN. Analyzing the split between N rates and inoculation, at 50 and 100 kg ha^{-1} of

Inoculation[#]	**N rates (kg ha^{-1})**			
	50	**100**	**150**	**200**
With	96.29 a[Ψ]	79.21 a	33.11 a	43.83 a
Without	61.34 b	44.95 b	29.05 a	35.21 a
L.S.D. (5%)		25.56		

Selvíria—MS, Brazil 2015.

[#]Corrected data following equation $(x + 0.5)^{0.5}$.

[Ψ]Means followed by the same letters in the column do not differ by Tukey at 0.05 probability level.

Table 7. Inoculation with *A. brasilense* and N rates interaction in the recovery of the applied nitrogen (RAN) in wheat crop.

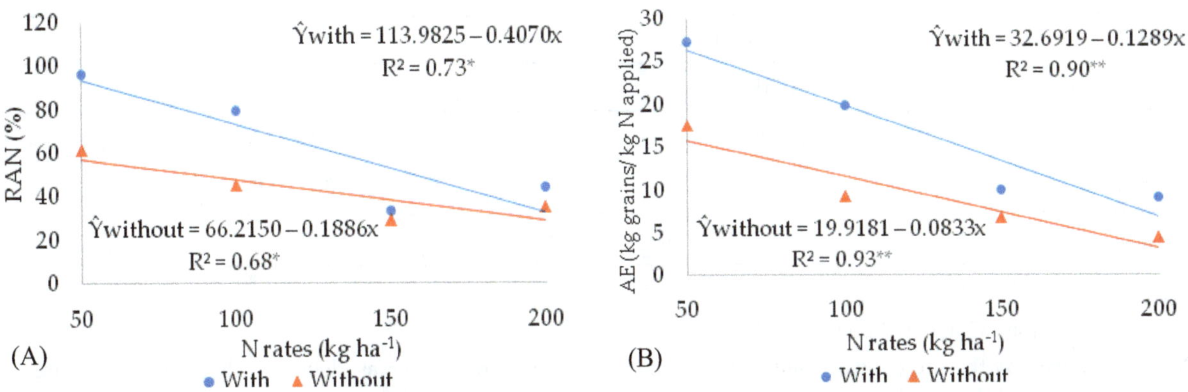

Figure 6. N rates and inoculation with *A. brasilense* interaction in the recovery of the applied nitrogen (RAN) (A) and agronomic efficiency (AE) (B) in wheat crop. Selvíria—MS, Brazil, 2015.

N applied in topdressing, the treatments inoculated provided higher values in the recovery of the applied N (**Table 7**, **Figure 6**). There was adjustment to the linear decreasing function, regardless of inoculation or not with *A. brasilense* (**Figure 6**). In the unfolding between N sources and inoculation, in the absence of inoculation with *A. brasilense*, urea provided higher RAN compared to Super N. When the N source used was Super N, inoculated treatments provided higher RAN compared to those not inoculated (**Table 8**), that is, this compensated the lower efficiency of Super N.

FE was not influenced by N rates, N sources, and inoculation with *A. brasilense* (**Table 6**), which is explained by the adequate growth of the plant (dry-matter accumulation) even when no N was supplied due to adequate N leaf concentration in all treatments (**Table 1**), as previously mentioned.

The interaction between inoculation and rates was also significant in AE, when the rates of 50, 100, and 200 kg ha^{-1} of N were applied; the inoculated treatments with *A. brasilense* had higher AE compared to the non-inoculated treatments, which is a very good result (**Table 9**) because it indicates that there were smaller losses of N with this diazotrophic bacteria inoculation. There was adjustment to the linear decreasing function of AE with and without *A. brasilense* as a function of N rates (**Figure 6**). Pankievicz et al. [38], studying FBN with ammonia release by associative fixers, verified increase and development of the *S. viridis* root system and greater CO_2 fixation through inoculation with *A. brasilense* in such a way that plants cultivated in a nitrate-limited environment similarly developed under N sufficient conditions, elucidating the ability of some mutant strains to increase BNF and positively interfering with the carbon metabolism of C4 plants, which has a close relationship with the assimilatory metabolism of N in the plant. Therefore, considering that in the experimental area, there was a favorable condition to the microbial immobilization of the applied N, and this explains why there was a higher AE of the treatments inoculated as a function of the N rates.

The interaction between N sources and inoculation in AE was significant. In treatments inoculated with *A. brasilense*, urea and Super N provided higher AE compared to uninoculated treatments. In the absence of inoculation, urea provided higher AE compared to Super N (**Table 10**).

Sources#	Inoculation	
	With	Without
Urea	63.36 aAΨ	50.39 aA
Super N	62.86 aA	34.89 bB
L.S.D.	18.08	

Selvíria—MS, Brazil 2015.

#Corrected data following equation $(x + 0.5)^{0.5}$.

ΨMeans followed by the same lowercase letters in the column and same uppercase letters in the line do not differ by Tukey at 0.05 probability level.

Table 8. Inoculation with *A. brasilense* and N sources interaction in the RAN in wheat crop.

Inoculation#	N rates (kg ha^{-1})			
	50	100	150	200
With	27.30 a	19.92 a	9.98 aΨ	9.13 a
Without	17.53 b	9.22 b	6.83 a	4.44 b
L.S.D. (5%)	6.09			

Selvíria—MS, Brazil 2015.

#Corrected data following equation $(x + 0.5)^{0.5}$.

ΨMeans followed by the same letters in the column do not differ by Tukey at 0.05 probability level.

Table 9. Inoculation with *A. brasilense* and N rates interaction in the AE in the wheat crop.

Sources#	Inoculation	
	With	Without
Urea	16.42 aAΨ	11.85 aB
Super N	16.74 aA	7.16 bB
L.S.D.	4.31	

Selvíria—MS, Brazil, 2015.

#Corrected data following equation $(x + 0.5)^{0.5}$.

ΨMeans followed by the same lowercase letters in the column and same uppercase letters in the line do not differ by Tukey at 0.05 probability level.

Table 10. Inoculation with *A. brasilense* and N sources interaction in the AE in wheat crop.

Regarding AE, according to Dobbelaere et al. [23], positive responses to inoculation with *A. brasilense* are obtained even when the crops are cultivated in soils with high levels of available N, which indicates that the plant responses do not only occur due to the fixed N_2 but mainly as a function of the production of growth-promoting phytohormones such as cytokinin, gibberellin, and indole acetic acid. This fact could possibly have affected the root development of wheat, which according to Novakowiski et al. [54] would improve the efficiency of utilization of residual N, water, and other nutrients uptake, directly reflecting a greater agronomic efficiency of the wheat crop with inoculation with *A. brasilense*, as verified in the present work.

N sources did not differ in AE (**Table 6**), which is due in part to the similar concentrations of foliar nutrients obtained with urea and Super N and can be explained by the non-efficacy of NBPT action due to high activity of the urease enzyme as a function of the straw of the predecessor cultures and the high temperatures that are recorded (**Figure 1**). Another possible explanation would be the uptake of a small part of the urea applied before the action of urease and NH_3 formation [37]. Comparisons between several nitrogen fertilizers were made by several authors, and, in general, with satisfactory conditions of soil moisture, no differences have been found in the efficiency of these sources such as grain yield of wheat in the Cerrado for sources of N ammonium sulfonitrate, uran, and urea [36] between urea and ammonium sulfonitrate in the no-tillage system [55] and between urea, urea + NBPT, and coated urea [56].

The efficiency of the use of N sources by annual crops, such as wheat, is low, around 50%, and the causes for this low value are related to the inadequate dose and timing of application associated with volatilization, leaching, as well as degradation, immobilization, and soil erosion [57] and differs with cultivars [58]. Thus, N fertilization strategy should aim to improve the synchronization between the season of application and the season of greater demand for the plant, in order to maximize N uptake and grain yield [59]. The improvement of N use and recovery efficiencies is desirable to increase productivity, reduce production costs, and maintain environmental quality [44].

It is worth noting that the tendency of agriculture is to seek to enrich food from the nutritional point of view, that is, to increase the availability of nutrients in the parts that will be used as food for humans and animals such as wheat grains. This research demonstrated that inoculation with *A. brasilense* associated with nitrogen fertilization in topdressing is beneficial not only to N nutrition and wheat yield but also to increase the nutritional quality of the grains more sustainably, like the protein content of this important cereal. Therefore, as the inoculation is a low-cost technique, easy to apply and use, non-polluting, and which falls within the desired sustainable context in actuality, the trend is that this technology can be increasingly used in wheat crops.

4. Conclusions

Urea provides higher N utilization efficiency, while the Super N obtains greater LCI and recovery of the applied nitrogen, being the last one only when inoculated. However, the N sources provide similar N accumulations in straw and grains yield of wheat; thus, it is recommended to use urea at the best cost-benefit ratio.

N leaf concentration, LCI, and N straw accumulation increase with the nitrogen fertilization increment, regardless of the N source or *A. brasilense* inoculation.

The increment in N rates in association with *A. brasilense* inoculation increases the N grain concentration up to 165 kg ha^{-1} N, whereas without this inoculation occurred a linear increase with lower maximum N grain concentration. That is, the inoculation afforded higher N grain concentration applying less nitrogen fertilizers in topdressing. Therefore, it can increase more sustainably the protein content in the wheat grain.

With *A. brasilense* inoculation, the increment in N rates increases the wheat yield up to 142 kg ha^{-1} N, whereas without this inoculation increases occurred up to 134 kg ha^{-1}. However, even at the highest doses, the inoculation afforded higher grain yield.

Inoculation with *A. brasilense* increased the agronomic efficiency, apparent N recovery, and N utilization efficiency. This research demonstrated that inoculation with *A. brasilense* associated with nitrogen fertilization in topdressing is beneficial to N nutrition and wheat yield, increasing nitrogen fertilization efficiency.

For further increasing the efficiency of nitrogen fertilization, new researches of complementary inoculation with *A. brasilense* during the vegetative phase of the plant would be interesting.

Author details

Marcelo Carvalho Minhoto Teixeira Filho*, Fernando Shintate Galindo, Salatiér Buzetti and Eduardo Henrique Marcandalli Boleta

*Address all correspondence to: mcmteixeirafilho@agr.feis.unesp.br

Universidade Estadual Paulista (UNESP), Campus de Ilha Solteira, Ilha Solteira, São Paulo State, Brazil

References

[1] Marini N, Tunes LM, Silva JI, Moraes DM, Cantos FAA. Carboxim Tiram fungicide effect in wheat seeds physiological quality (*Triticum aestivum* L.). Revista Brasileira de Ciência Agrárias. 2011;**6**:17-22. DOI: 10.5039/agraria.v6i1a737 (in Portuguese with abstract in English)

[2] USDA. United States Department of Agriculture. Databases: Production, Supply and Distribution Online. 2016. Available from: http://apps.fas.usda.gov/psdonline/ [Accessed: December 23, 2016]

[3] Zagonel J, Venancio WS, Kunz RP, Tanamati H. Nitrogen doses and plant densities with and without a growth regulator affecting wheat, cultivar OR-1. Ciência Rural. 2002;**32**:25-29. DOI: dx.doi.org/10.1590/S0103-84782002000100005 (in Portuguese with abstract in English)

[4] Povh FP, Molin JP, Gimenez LM, Pauletti V, Molin R, Salvi JV. Behavior of NDVI obtained from an active optical sensor in cereals. Pesquisa Agropecuária Brasileira. 2008;**43**:1075-

1083. DOI: dx.doi.org/10.1590/S0100-204X2008000800018 (in Portuguese with abstract in English)

[5] Teixeira Filho MCM, Buzetti S, Andreotti M, Arf O, Benett CGS. Doses, sources and time of nitrogen application on irrigated wheat under no-tillage. Pesquisa Agropecuária Brasileira. 2010;**45**:797-804. DOI: dx.doi.org/10.1590/S0100-204X2010000800004 (in Portuguese with abstract in English)

[6] Theago EQ, Buzetti S, Teixeira Filho MCM, Andreotti M, Megda MM, Benett CGS. Nitrogen application rates, sources, and times affecting chlorophyll content and wheat yield. Revista Brasileira de Ciência do Solo. 2014;**38**:1826-1835. DOI: dx.doi.org/10.1590/S0100-06832014000600017

[7] Teixeira Filho MCM, Buzetti S, Alvarez RCF, Freitas JG, Arf O, Sá ME. Response of wheat cultivars irrigated by sprinkler to side dressing nitrogen under savannah soil. Acta Scientiarum: Agronomy. 2007;**29**:421-425. DOI: 10.4025/actasciagron.v29i3.471 (in Portuguese with abstract in English)

[8] Nunes PHMP, Aquino LA, dos Santos LPD, Xavier FO, Dezordi LR, Assunção NS. Yield of the irrigated wheat crop subjected to nitrogen application and to inoculation with *Azospirillum brasilense*. Revista Brasileira de Ciência do Solo. 2015;**39**:174-182. DOI: 10.1590/01000683rbcs20150354 (in Portuguese with abstract in English)

[9] Espindula MC, Rocha VS, Souza MA, Campanharo M, Pimentel AJB. Urease inhibitor (NBPT) and efficiency of single or split application of urea in wheat crop. Revista Ceres. 2014;**61**:273-279. DOI: dx.doi.org/10.1590/S0034-737X2014000200016

[10] Teixeira Filho MCM, Buzetti S, Andreotti M, Benett CGS, Arf O, Sá ME. Wheat nitrogen fertilization under no till on the low altitude *Brazilian Cerrado*. Journal of Plant Nutrition. 2014;**37**:1732-1748. DOI: dx.doi.org/10.1080/01904167.2014.889150

[11] Xu G, Fan X, Miller AJ. Plant nitrogen assimilation and use efficiency. Annual Review of Plant Biology. 2012;**63**:153-182. DOI: 10.1146/annurev-arplant-042811-105532

[12] Lara Cabezas WAR, Trivelin PCO, Korndorfer GH, Pereira S. Nitrogen balance of solid and liquid nitrogen sources on side-dressed no-till corn system at Triângulo Mineiro (MG). Revista Brasileira de Ciência do Solo. 2000;**24**:363-376. DOI: http://dx.doi.org/10.1590/S0100-06832000000200014 (in Portuguese with abstract in English)

[13] Cantarella H, Trivelin PCO, Contin TLM, Dias FLF, Rossetto R, Marcelino R, Coimbra RB, Quaggio JA. Ammonia volatilization from urease inhibitor-treated urea applied to sugarcane trash blankets. Scientia Agricola. 2008;**65**:397-401. DOI: http://dx.doi.org/10.1590/S0103-90162008000400011

[14] Galindo FS, Teixeira Filho MCM, Buzetti S, Santini JMK, Alves CJ, Nogueira LM, Ludkiewicz MGZ, Andreotti M, Bellotte JLM. Corn yield and foliar diagnosis affected by nitrogen fertilization and inoculation with *Azospirillum brasilense*. Revista Brasileira de Ciência do Solo. 2016;**40**:e0150364. DOI: 10.1590/18069657rbcs20150364

[15] Watson CJ. Urease Activity and Inhibition: Principles and Practice. No. 454. London: The International Fertilizer Society Meeting; 2000. p. 39

[16] Hungria M. Inoculation with *Azospirillum brasilense*: Innovation in performance at low cost [documents, 325]. Londrina: Embrapa Soja; 2011. p. 37 (in Portuguese)

[17] Quadros PD. Inoculation of *Azospirillum* spp. in seeds of maize genotypes grown in Rio Grande do Sul [dissertation]. Porto Alegre: Federal University of Rio Grande do Sul, Brazil; 2009. p. 74 (in Portuguese)

[18] Tien TM, Gaskins MH, Hubbell DH. Plant growth substances produced by *Azospirillum brasilense* and their effect on the growth of pearl millet (*Pennisetum americanum* L.). Applied and Environmental Microbiology. 1979;**37**:1016-1029

[19] Barassi CA, Sueldo RJ, Creus CM, Carrozzi L, Casanovas EM, Pereyra MA. *Azospirillum* potential to optimize plant growth under adverse conditions. In: Cassán FD, Garcia de Salomone I, editors. *Azospirillum* ssp.: Cell Physiology, Plant Interactions and Agronomic Research in Argentina. Argentina: Asociación Argentina de Microbiologia; 2008. pp. 49-59 (in Spanish)

[20] Bashan Y, Bustillos JJ, Leyva LA, Hernandez JP, Bacilio M. Increase in auxiliary photoprotective photosynthetic pigments in wheat seedlings induced by *Azospirillum brasilense*. Biology and Fertility of Soils. 2006;**42**:279-285. DOI: 10.1007/s00374-005-0025-x

[21] Dobbelaere S, Croonenborghs A, Thys A, Ptacek D, Vanderleyden J, Dutto P, Labandera-Gonzalez C, Caballero-Mellado J, Aguirre JF, Kapulnik Y, Brener S, Burdman S, Kadouri D, Sarig S, Okon Y. Response of agronomically important crops to inoculation with *Azospirillum*. Australian Journal of Plant Physiology. 2001;**28**:871-879. DOI: 10.1071/PP01074

[22] Bashan Y, Holguin G, De-Bashan LE. *Azospirillum*-plant relations physiological, molecular, agricultural, and environmental advances (1997-2003). Canadian Journal of Microbiology. 2004;**50**:521-577. DOI: 10.1139/w04-035

[23] Dobbelaere S, Vanderleyden J, Okon Y. Plant growth-promoting effects of diazotrophs in the rhizosphere. Critical Reviews in Plant Sciences. 2003;**22**:107-149. DOI: org/10.1080/713610853

[24] Lemos JM, Guimarães VF, Vendruscolo ECG, Santos MF, Offemann LC. Response of wheat cultivars to inoculation of seeds with *Azospirillum brasilense* and to nitrogenous fertilizer side dressed to the plants. Científica. 2013;**41**:189-198. DOI: dx.doi.org/10.15361/1984-5529.2013v41n2p189-198 (in Portuguese with abstract in English)

[25] Embrapa—Empresa Brasileira de Pesquisa Agropecuária. National Center for Soil Research. Brazilian System of Soil Classification. 3rd ed. Brasília: Embrapa; 2013. p. 353. (in Portuguese)

[26] van Raij B, Andrade JC, Cantarella H, Quaggio JA. Chemical Analysis to Evaluate the Fertility of Tropical Soils. 2nd ed. Campinas: IAC; 2001. p. 285 (in Portuguese)

[27] Cantarella H, van Raij B, Camargo CEO. Cereals. In: van Raij B, Cantarella H, Quaggio JA, Furlani AMC, editors. Recommendations Liming and Fertilization for the State of São Paulo. 2nd ed. Campinas: IAC; 1997. p. 285 (in Portuguese)

[28] Malavolta E, Vitti GC, Oliveira SA. Evaluation of the Nutritional Status of Plants: Principles and Applications. 2nd ed. Piracicaba: Brazilian Association for Research of Potash and Phosphate; 1997. p. 319 (in Portuguese)

[29] Moll RH, Kamprath EL, Jackson A. Analysis and interpretation of factors which contribute to efficiency of nitrogen utilization. Agronomy Journal. 1982;**74**:562-564

[30] Fageria NK. Optimizing nutrient use efficiency in crop production. Revista Brasileira de Engenharia Agrícola e Ambiental. 1998;**2**:6-16. DOI: http://dx.doi.org/10.1590/1807-1929/agriambi.v02n01p6-16 (in Portuguese with abstract in English)

[31] Ferreira DF. Sisvar: A computer statistical analysis system. Ciência e Agrotecnologia, 2011;**35**:1039-1042. DOI: dx.doi.org/10.1590/S1413-70542011000600001

[32] Chapman SC, Barreto HJ. Using a chlorophyll meter to estimate specific leaf nitrogen of tropical maize during vegetative growth. Agronomy Journal. 1997;**89**:557-562

[33] Teixeira Filho MCM, Buzetti S, Alvarez RCF, Freitas JG, Arf O, Sá ME. Response of wheat cultivars to plant population and nitrogen fertilization in a Cerrado region. Científica. 2008;**36**:97-106. DOI: dx.doi.org/10.15361/1984-5529.2008v36n2p97+–+106 (in Portuguese with abstract in English)

[34] Teixeira Filho MCM, Buzetti S, Arf O, Alvarez RCF, Maeda AS, Sá ME. Response of wheat cultivars to nitrogen with and without fungal treatment. Revista Brasileira de Ciências Agrárias. 2012;**7**:626-634. DOI: http://www.redalyc.org/articulo.oa?id=119024993015 (in Portuguese with abstract in English)

[35] Meira FA, Buzetti S, Andreotti M, Arf O, Sá ME, Andrade JAC. Sources and times of nitrogen application on irrigated corn crop. Semina: Ciências Agrárias. 2009;**30**:275-284 (in Portuguese with abstract in English)

[36] Megda MM, Buzetti S, Andreotti M, Teixeira Filho MCM, Vieira MX. Response of wheat cultivars to nitrogen related to sources and application time under irrigation in no-tillage system. Ciência e Agrotecnologia. 2009;**33**:1055-1060. DOI: http://dx.doi.org/10.1590/S1413-70542009000400016 (in Portuguese with abstract in English)

[37] Liu L, Ludewig U, Frommer WB, von Wirén N. AtDUR3 encodes a new type of high-affinity urea/H^+ symporter in Arabidopsis. The Plant and Cell. 2003;**15**:790-800. DOI: https://www.ncbi.nlm.nih.gov/pubmed/12615950

[38] Pankievicz VCS, Amaral FP, Santos KFDN, Agtuca B, Xu Y, Schueller MJ, Arisi ACM, Steffens MBR, Souza EM, Pedrosa FO, Stacey G, Ferrieri RA. Robust biological nitrogen fixation in a model grass-bacterial association. The Plant Journal. 2015;**81**:907-919. DOI: 10.1111/tpj.12777

[39] Hungria M, Campo RJ, Souza SEM, Pedrosa FO. Inoculation with selected strains of *Azospirillum brasilense* and *A. lipoferum* improves yields of maize and wheat in Brazil. Plant and Soil. 2010;**331**:413-425. DOI: 10.1007/s11104-009-0262-0

[40] Marschner P. Marschner's Mineral Nutrition of Higher Plants. 3rd ed. New York: Academic Press; 2012. p. 651

[41] Okon Y, Vanderleyden J. Root-associated *Azospirillum* species can stimulate plants. Applied and Environment Microbiology. 1997;**6**:366-370. DOI: A1997XJ86700015

[42] Correa OS, Romero AM, Soria MA, Estrada M. *Azospirillum brasilense*-plant genotype interactions modify tomato response to bacterial diseases, and root and foliar microbial communities. In: Cassán FD, Garcia SI, editors. *Azospirillum* ssp.: Cell Physiology, Plant Interactions and Agronomic Research in Argentina. Argentina: Asociación Argentina de Microbiologia; 2008. pp. 87-95

[43] Souza JA, Buzetti S, Teixeira Filho MCM, Moreira A. Sources, rates and time of nitrogen application on maize crops under no-tillage system. Communications in Soil Science and Plant Analysis. 2016;**47**:2200-2207. DOI: http://dx.doi.org/10.1080/00103624.2016.1228948

[44] Prando AM, Zucarelli C, Fronza V, Bassoi MC, de Oliveira FA. Forms of urea and nitrogen levels in top dressing in the agronomic performance of wheat genotypes. Semina: Ciências Agrárias. 2012;**33**:621-632. DOI: 10.5433/1679-0359.2012v33n2p621 (in Portuguese with abstract in English)

[45] Galindo FS, Teixeira Filho MCM, Buzetti S, Santini1 JMK, Ludkiewicz MGZ, Baggio G. Modes of application of cobalt, molybdenum and *Azospirillum brasilense* on soybean yield and profitability. Revista Brasileira de Engenharia Agrícola e Ambiental. 2017;**21**:180-185. DOI: http://dx.doi.org/10.1590/1807-1929/agriambi.v21n3p180-185

[46] Cazetta DA, Fornasieri Filho D, Arf O. Response of cultivars of wheat and triticale to nitrogen in a no-till system. Científica. 2007;**35**:155-165. DOI: dx.doi.org/10.15361/1984-5529.2007v35n2p155+—+165 (in Portuguese with abstract in English)

[47] Anwar S, Iqbal B, Khan S, Faraz M, Ali N, Hussain S, Anjum MM. Nitrogen and phosphorus fertilization of improved varieties for enhancing yield and yield components of wheat. Pure and Applied Biology. 2016;**5**:183-192. DOI: http://dx.doi.org/10.19045/bspab.2016.50091

[48] Ferreira JP, Andreotti M, Arf O, Kaneko FH, Nascimento V, Sabundjian MT. Inoculation with *Azospirillum brasilense* and nitrogen in top dressing in wheat in Cerrado region. Tecnologia e Ciência Agropecuária. 2014;**8**:27-32 (in Portuguese with abstract in English)

[49] Galindo FS, Ludkiewicz MGZ, Bellote JLM, Santini JMK, Teixeira Filho MCM, Buzetti S. Leaf application times of *Azospirillum brasilense* in yield of wheat irrigated. Tecnologia e Ciência Agropecuária. 2015;**9**:43-48 (in Portuguese with abstract in English)

[50] Sala VMR, Cardoso EJBN, Freitas JG, Silveira APD. Wheat genotypes response to inoculation of diazotrophic bacteria in field conditions. Pesquisa Agropecuária Brasileira. 2007;**42**:833-842. DOI: http://dx.doi.org/10.1590/S0100-204X2007000600010 (in Portuguese with abstract in English)

[51] Silva DRG, Costa KAP, Faquin V, de Oliveira IP, de Souza MRF, Souza MAS. Nutritional efficiency and nitrogen uptake by capim-marandu at pasture in moderate stage of degradation under doses and sources of nitrogen. Ciência e Agrotecnologia. 2011;**35**:242-249. DOI: http://dx.doi.org/10.1590/S1413-70542011000200003 (in Portuguese with abstract in English)

[52] Sant'Ana EVP, dos Santos AB, da Silveira PM. The efficiency of use of nitrogen applied in top dressing in irrigated bean. Revista Brasileira de Engenharia Agrícola e Ambiental. 2011;**15**;458-462. DOI: http://www.scielo.br/pdf/rbeaa/v15n5/v15n5a04.pdf (in Portuguese with abstract in English)

[53] Dupas E, Buzetti S, Rabêlo FHS, Sarto AL, Cheng NC, Teixeira Filho MCM, Galindo FS, Dinalli RP, Gazola RN. Nitrogen recovery, use efficiency, dry matter yield, and chemical composition of palisade grass fertilized with nitrogen sources in the Cerrado biome. Australian Journal of Crop Science. 2016;**10**:1330-1338. DOI: 10.21475/ajcs.2016.10.09. p7854

[54] Novakowiski JH, Sandini IE, Falbo MK, Moraes A, Novakowiski JH, Cheng NC. Residual effect of nitrogen fertilization and *Azospirillum brasilense* inoculation in the maize culture. Semina: Ciências Agrárias. 2011;**32**:1687-1698. DOI: dx.doi.org/10.5433/1679-0359.2011v32n4Sup1p1687 (in Portuguese with abstract in English)

[55] Silva SA, Arf O, Buzetti S, Silva MG. Nitrogen sources and application times in no-till wheat on Cerrado soil. Revista Brasileira de Ciência do Solo. 2008;**32**:2717-2722. DOI: http://dx.doi.org/10.1590/S0100-06832008000700015 (in Portuguese with abstract in English)

[56] Prando AM, Zucarelli C, Fronza V, Oliveira FA, Oliveira Júnior A. Productive characteristics of wheat according to nitrogen sources and levels. Pesquisa Agropecuária Tropical. 2013;**43**:34-41. DOI: http://dx.doi.org/10.1590/S1983-40632013000100009 (in Portuguese with abstract in English)

[57] Fageria NK, Baligar VC. Enhancing nitrogen use efficiency in crop plants. Advances in Agronomy. 2005;**88**:97-185. DOI: http://dx.doi.org/10.1016/S0065-2113(05)88004-6

[58] Furtini IV, Ramalho MAP, Abreu AFB, Furtini Neto AEF. Differential response of breeding lines of common bean to nitrogen. Ciência Rural. 2006;**36**:1696-1700. DOI: http://dx.doi.org/10.1590/S0103-84782006000600005 (in Portuguese with abstract in English)

[59] dos Santos AB, Fageria NK. Physiological characteristics of common bean in tropical varzea soils as affected by rate and nitrogen management. Ciência e Agrotecnologia. 2008;**32**:23-31. DOI: http://www.scielo.br/pdf/cagro/v32n1/a03v32n1.pdf (in Portuguese with abstract in English)

6

Nitrogen-Fixation by Endophytic Bacteria in Agricultural Crops: Recent Advances

Akshit Puri, Kiran Preet Padda and Chris P. Chanway

Abstract

Endophytic bacteria represents a unique class of bacteria that can colonize interior tissues of plant and provide a range of benefits to the plant similar to those provided by the rhizospheric bacteria. Certain endophytic bacteria can provide nitrogen to the plants through biological nitrogen fixation, which is an important source of nitrogen input in agriculture and represents a promising substitute for chemical fertilizers, and are known as endophytic diazotrophic bacteria. Besides fixing nitrogen, endophytic bacteria can produce plant growth hormones like auxin and gibberellin, help in nutrient uptake, and increase the plant's tolerance to biotic and abiotic stresses. Various direct and indirect methods have been used to quantify the amount of nitrogen fixed by these bacteria, including the acetylene reduction assay, which is a quick but indirect method, and the ^{15}N isotopic dilution assay, which is a robust and accurate method. Research on endophytic diazotrophic bacteria has come a long way, and in this chapter, we have briefly discussed the mechanisms of biological nitrogen fixation and methods to quantify the fixed nitrogen along with reviewing recent studies focused on evaluating the role of endophytic diazotrophic bacteria in promoting plant growth in both native and nonnative crop hosts.

Keywords: endophytic bacteria, diazotroph, biological nitrogen fixation, plant growth promotion, agricultural crops

1. Introduction

Nitrogen (N) is an essential component of all proteins and enzymes, nucleic acids that make up DNA, and chlorophyll that enables the process of photosynthesis in plants [1]. It is a very common element in nature that is present in abundant amounts in atmosphere, lithosphere, and hydrosphere of the earth [2]. However, much of this N is in the form of dinitrogen (N_2), which is inert and cannot be used by plants. In order for plants to use this dinitrogen, it has to be reduced/fixed into forms like nitrate (NO_3^-) and ammonium (NH_4^+). N fixation, the process by which dinitrogen is reduced to plant-available forms, is, therefore, a vital process for the sustenance of life on earth. A major industrial process by which dinitrogen is converted into ammonia is known as the Haber-Bosch process. This artificial N-fixation process was established in 1913 and uses a catalyst (iron with a small amount of aluminum added) at high pressure (as much as 5.06×10^7 Pa) and high temperature (600–800 K) consuming large amounts of fossil fuel. Ammonia produced through this highly expensive process is combined with other elements to produce nitrogenous fertilizers like urea and ammonium nitrate. Although the use of these fertilizers is inevitable in meeting rising food demand to sustain the growing global population, their indiscriminate use has set off very negative effects on the environment [3]. Naturally, N is commonly fixed by two processes. The first is atmospheric N fixation by lightning, in which the enormous amount of energy contained in lightning breaks dinitrogen molecules and enables their atoms to combine with oxygen in the air forming N oxides that dissolve in rain. These oxides of N then form nitrates that are carried to the earth in rainfall [4]. The second is biological N fixation (BNF), in which certain prokaryotic microorganisms, known as diazotrophs, fix N by breaking down the triple bond of dinitrogen using a highly specialized enzyme complex called nitrogenase enzyme and convert it to ammonia [4]. This chapter mainly focuses on diazotrophic bacteria that can fix N while living in the internal tissues of plants. In this chapter, only recent developments (from last 5 years) related to this subject have been discussed.

2. Biological nitrogen fixation (BNF)

Farmers since ancient Chinese and Roman civilizations practiced crop rotation with legumes to increase soil fertility and agricultural productivity. However, the science behind such practice was first revealed by Boussingault in 1838, who established that legumes can fix N. But it was not until 1886 when Hellriegel and Wilfarth provided a firm evidence that microbes are responsible for N fixation occurring in leguminous plants [5].

2.1. Chemistry and genetics of BNF

The overall chemical reaction of BNF catalyzed by the nitrogenase enzyme is represented below:

$$N_2 + 8H^+ + 8e^- + 16MgATP \rightarrow 2NH_3 + H_2 + 16MgADP + 16Pi \qquad (1)$$

Nitrogenase is a complex enzyme comprised of two metalloproteins: the Mo-Fe protein, also called dinitrogenase protein, and the Fe protein, also called dinitrogenase reductase protein. The dinitrogenase protein is a heterotetramer composed of two α- and two β-subunits with an overall molecular weight of 240kDa. This protein contains two types of metal centers, the FeMo-cofactor and the P-cluster pair, of which the FeMo cofactor is the active site where dinitrogen binds, whereas the P-cluster mediates electron transfer between the Fe protein and the FeMo cofactor. The dinitrogenase reductase protein is a homodimer of two identical subunits, with an overall molecular mass of ~60 kDa. It contains two ATP/ADP molecules and one Fe_4-S_4 cluster [6, 7].

The overall functioning of nitrogenase can be summarized as a key biochemical cycle that involves five steps [6, 7]: (i) the reduction of Fe protein by electron carriers such as flavodoxin or ferredoxin; (ii) association of the reduced Fe protein (including two MgATP complexes) with the Mo-Fe protein in preparation for electron transfer; (iii) hydrolysis of MgATP, which enables transfer of one electron to the Mo-Fe protein (via Fe_4S_4 and the P-cluster); (iv) electron transfer to dinitrogen and thus its reduction, while it is bound to the active site within the Mo-Fe protein; and (v) dissociation of the two protein molecules, exchange of ATP back into the Fe protein, and rereduction of the Fe protein.

The structure and function of nitrogenase enzyme are encoded by ~20 genes, known as N-fixation genes (*nif* genes), organized in 7 operons (*nif* cluster) spanning over 24 kb. These genes fall into three categories, structural, regulatory, and supplementary, and can be housed either in genomic DNA or on plasmids. The Fe protein is encoded by the *nifH* gene and the Mo-Fe protein is encoded by *nifD* and *nifK* genes [8, 9]. The *nifD, nifH,* and *nifK* genes are recognized as structural *nif* genes since they are responsible for encoding the aforementioned structural subunits [10]. The *nif* cluster of the free-living bacterium *Klebsiella pneumoniae* is the most studied of *nif* genes and serves as a model for understanding the regulation, synthesis, and assembly of nitrogenase enzyme [11].

2.2. Quantification of biologically fixed N

BNF can be measured using various methods, the most common being: N balance method, xylem solute analysis, acetylene reduction assay, and stable isotope (^{15}N) method [12]. In the N balance method, the amount of N fixed is estimated by calculating the difference between total N content of plants inoculated by diazotrophs and those that are not inoculated. In this method, it is assumed that both inoculated and noninoculated plants absorb equal amounts of N from the soil, which is hard to justify as there are differences in root morphology and physiological attributes [12]. In the xylem solute analysis, the composition of N compounds flowing through the xylem sap to the shoot of the plant is determined. The N absorbed by plants from the soil is predominantly nitrate, whereas the fixed N is primarily in the form of amides and ureides [13]. This difference in composition of N compounds is used to make quantitative measurements of N fixation [14]. However, its major disadvantage is that only a very small proportion of N-fixing plants export fixed N in the form of ureides [15]. The acetylene reduction assay is a popular technique used to indirectly measure BNF by estimating the nitrogenase enzyme activity. It is based on the ability of nitrogenase to reduce acetylene (H—C≡C—H) to ethylene by breaking the triple bond between carbon atoms. Samples are incubated in a gas-tight chamber and a portion of the head space is injected with acetylene. After incubation, gas samples are collected from the chamber and analyzed for ethylene production using gas chromatography [16]. It is a simple, low cost, and sensitive assay that can measure BNF in bacterial cultures, detached nodules, plant parts, or even whole plants.

The major disadvantage is the short-term nature of the assay and the autoinhibition of acetylene conversion to ethylene [17]. The stable isotope method using ^{15}N is a widely used and accepted method. This method is based on the principle that soil has a noticeably different ^{15}N to ^{14}N ratio as compared to the atmosphere, which has a constant ratio (0.3663%). Therefore, plants absorbing fixed N from the atmosphere will have a different ^{15}N to ^{14}N ratio as compared to the ones absorbing N only from the soil. When plants inoculated with diazotrophs are grown in air labeled with ^{15}N, they are expected to have an enhanced ratio as compared to the noninoculated ones (^{15}N incorporation method). When available soil N is labeled with ^{15}N, a reduction in the ratio is expected since the inoculated plants tend to incorporate fixed N from the air as compared to the noninoculated plants, which take up labeled N from the soil (^{15}N isotope dilution method) [17].

2.3. N-fixing organisms

The ability to fix N, in other words, the presence of nitrogenase enzyme, is only limited to certain bacteria and archaea [18]. Within these groups, it is quite widely distributed revealing considerable phylogenetic diversity among diazotrophs. A comprehensive list of N-fixing bacteria and archaea, under 12 broad phylogenetic groups based on 16S rDNA phylogeny was prepared by Young [19]. Diazotrophs are also widely distributed ecologically. They can be found living in soils and water freely, in the rhizosphere and phyllosphere and inside the plant tissues, in symbiotic association with legumes and actinorhizal association with woody plants, and in cyanobacterial symbiosis with phytoplankton, fungi, and terrestrial plants [19]. Free-living diazotrophs are those that do not associate with plants and are found in soils that are free from the direct influence of plant roots. These microorganisms are ubiquitous in terrestrial and aquatic environments and are physiologically very diverse [20]. Many diazotrophs can be found dwelling in the rhizosphere of a plant. Due to their ability to fix N, diazotrophs can have a competitive advantage over other microbes in the rhizosphere. They prevail in the rhizosphere particularly when soil N is limited [21]. The phyllosphere (leaf surface) is another microsite known to be colonized by diazotrophs [22]. The symbiotic association between legume and *Rhizobium* is a well-known mutualistic relationship involving *Leguminosae* plants and *Rhizobiaceae* bacteria [23]. This symbiosis has been studied widely from ecological, agronomic, and molecular biological perspectives not only to enhance the N-fixing efficacy of existing symbioses but also to determine if similar associations might be developed with nonleguminous plants [24, 25]. The actinorhizal association is functionally analogous to the legume and *Rhizobium* association but is restricted between a small group of woody plant species known as Actinorhizal plants and diazotrophs belonging to a genus, *Frankia* [26]. Many diazotrophic cyanobacteria also form symbiotic association with eukaryotes and are known to contribute a significant portion of N required for growth of both organisms through BNF in N-limited aquatic and terrestrial environments [27, 28].

The presence of diazotrophs in nonleguminous plants was first detected by Brazilian researchers in the rhizosphere and rhizoplane of sugarcane (*Saccharum officinarum*) [29, 30]. In subsequent studies, various diazotrophs like *Azospirillum lipoferum, Azospirillum amazonense, Bacillus azotofixans, Enterobacter cloacae, Erwinia herbicola,* and *Bacillus polymyxa* [31–34] were isolated from the rhizosphere of sugarcane. Initially, it was postulated that nitrogenase activity only occurs in the rhizosphere soil but not in roots [35, 36]. However, later it was determined that rhizospheric N fixation does not occur at sufficient rates to facilitate high sugarcane yields. Cavalcante and Döbereiner [37] were the first to report the isolation of a diazotroph (*Gluconacetobacter diazotrophicus*) from internal tissues of a nonleguminous plant (stem and root tissues of sugarcane) and postulated that this bacterium might be involved in fixing high amounts of N biologically. This bacterium was able to multiply considerably and fix N at high sucrose concentrations [38] and in low pH conditions typically found in internal tissues of sugarcane [38, 39]. This led to the postulation that it can satisfy almost all of the sugarcane N requirements while living inside their tissues. Such bacteria that were able to multiply inside the tissues of a live plant and promote its growth through one or more mechanisms had already been discovered many years ago and are known as 'endophytic bacteria.'

3. Endophytic bacteria

The term 'endophyte' was first coined more than 150 years ago by de Bary [40] for pathogenic fungi entering the internal tissues of leaves. Since then, many authors have redefined this term, but each has its own restrictions. Taken literally, the word endophyte means 'in the plant' (endon = within; phyton = plant) [41]. Since our main focus in this chapter is on 'endophytic bacteria,' we would like to reiterate the definition notated by Chanway et al. [42]: "bacteria that can be detected at a particular moment within the tissue of apparently healthy plant hosts without inducing disease or organogenesis are known as endophytic bacteria." The occurrence of endophytic bacteria in internal tissues was first reported inside a healthy potato plant [43]. Since then, many scientific studies have been focused on isolating the endophytic bacteria from a variety of plant species and evaluating their benefits for agricultural plants [44–47]. In contrast to free-living, rhizosphere or phyllosphere microorganisms, endophytic bacteria are better protected from abiotic stresses such as extreme variations in temperature, pH, nutrient, and water availability as well as biotic stresses such as competition [48–50]. In addition, endophytic bacteria colonize niches that are more conducive to forming mutualistic relationships with plants [51], for example, providing fixed N to the plant and getting photosynthate in return [52–54]. Following the rhizospheric colonization, endophytic bacteria can colonize various plant organs such as roots, stem, leaves, flowers, fruits, and seeds [55–61], indicating different capacities of endophytic bacteria to colonize various plant compartments. They can even colonize legume nodules [62] and tubercles of mycorrhizal fungi [63]. The endophytic bacterial population is extremely variable in different plant organs and tissues and have been shown to vary from as low as hundreds to as high as 10^9 cfu per gram plant tissue [64–67].

Localization of endophytic bacteria within plant tissues requires techniques that facilitate observation on a tiny spatial scale. Various methods have been used to locate bacteria *in planta* and visualize them at their sites of colonization, but each one has its own limitations. Most methods require either chemical or physical treatment of plant tissues for *in situ* detection and visualization of endophytic bacteria [68]. However, the use of autofluorescent proteins in conjunction with confocal laser scanning microscopy (CLSM) eliminates the need for any chemical treatment of plant tissues and requires minimal physical preparation of plant tissue samples before microscopic visualization. The green fluorescent protein (GFP) gene found in the jellyfish *Aequorea aequorea* is the most popular autofluorescent protein used for localization of endophytic bacteria. GFP is a useful biomarker because it does not require any substrate or cofactor in order to fluoresce. GFP cassettes can be integrated into the bacterial

Endophytic diazotrophic bacteria	Isolated from	Colonized into	Method used to confirm N-fixing ability	References
Pseudomonas aeruginosa PM389	Pearl millet (*Pennisetum glaucum*)	Wheat (*Triticum aestivum* L.)	Amplification of *nifH* genes; acetylene reduction assay	[104]
Azospirillum amazonense AR3122; *Burkholderia vietnamiensis* AR 1122;	Rice (*Oryza sativa* L.)	Rice (*Oryza sativa* L.)	Acetylene reduction assay	[97]
Paenibacillus kribbensis HS-R01, HS-R14; *Bacillus aryabhattai* HS-S05; *Bacillus megaterium* KW7-R08; *Klebsiella pneumoniae* KW7-S06, KW7-S22, KW7-S27, KW7-S33; *Bacillus subtilis* CB-R05; *Microbacterium binotii* CB-S18; *Microbacterium trichotecenolyticum* SW521-L21, SW521-L37;	Rice (*Oryza sativa* var. *Japonica*)	Rice (*Oryza sativa* var. *Japonica*)	Amplification of *nifH* genes	[106]
Bacillus subtilis EB-04; *Bacillus pumilus* EB-64, EB-169; *Paenibacillus* sp. EB-144	Banana tree cultivar 'Prata Anã' (*Musa acuminata* × *balbisiana*)	—	Amplification of *nifH* genes; acetylene reduction assay	[103]
Bacillus sp. CNPSo 2476, CNPSo 2477, CNPSo 2478; *Enterobacter* sp. CNPSo 2480	Corn (*Zea mays* L.)	Corn (*Zea mays* L.)	Amplification of *nifH* genes; acetylene reduction assay	[102]
Gluconacetobacter diazotrophicus Pal5T-BR11281; *Amazon Azospirillum* Cbamc-BR11145; *Herbaspirillum seropedicae* HRC54-BR11335; *Herbaspirillum rubrisubalbicans* HCC103-BR11504; *Burkholderia tropica* PPe8T-BR11366	Sugarcane (*Saccharum officinarum*)	Sugarcane (*Saccharum officinarum*)	Kjeldahl method; natural abundance of ^{15}N in leaf samples; isotopic ^{15}N dilution	[98]
Burkholderia spp.; *Klebsiella* spp.; *Novosphingobium* spp.; *Sphingomonas* spp.	Rice (*Oryza sativa*)	Rice (*Oryza sativa*)	Acetylene reduction assay	[105]
Paenibacillus polymyxa P2b-2R	Lodgepole pine (*Pinus contorta var. latifolia*)	Corn (*Zea mays*), canola (*Brassica napus* L.), tomato (*Solanum lycopersicum*)	Amplification of *nifH* genes; acetylene reduction assay; isotopic ^{15}N dilution	[109, 113, 117, 119]

Table 1. List of endophytic diazotrophic bacteria recently isolated and associated with agricultural crops.

chromosome and expressed through an inducible or constitutive promoter of indigenous or exogenous origin [69–72]. Alternatively, a plasmid-borne GFP gene can be introduced into bacterial cells of interest [73–75]. Bacterial cells expressing GFP can be visualized by epifluorescence microscopy or CLSM [76, 77]. This technique has been used with various agricultural crops including wheat (*Triticum* spp.) [78], rice (*Oryza sativa*) [78–80], corn (*Zea mays*) [78, 81], tomato (*Solanum lycopersicum*) [82], ryegrass (*Lolium multiflorum*) [83], creeping bentgrass (*Agrostis stolonifera*) [84], and grapevine (*Vitis vinifera*) [72].

3.1. Endophytic diazotrophic bacteria

A few years after the discovery of diazotrophs by Cavalcante and Döbereiner [37] in the stem and root tissues of sugarcane plant, Döbereiner [85] coined the term "endophytic diazotrophic bacteria" to designate all diazotrophs able to colonize primarily the root interior of graminaceous plants, survive very poorly in soil and fix N in association with these plants [86]. Since the discovery of endophytic diazotrophic bacteria in sugarcane, other agronomically important crop species like rice [87–89], corn [90–93], wheat [94], canola (*Brassica napus* L.) [95], and Kallar grass (*Leptochloa fusca* L.) [96] have been postulated to receive significant amounts of fixed N in this way. In the following section, recent studies (from last 5 years) about endophytic diazotrophic bacteria and their role in promoting the growth of agricultural crops primarily by providing N nutrition as a result of BNF and secondarily through other plant growth–promotion (PGP) mechanisms have been discussed in detail (listed in **Table 1** as well).

4. Recent studies highlighting the role of endophytic diazotrophic bacteria in agricultural crops

Rice is a major staple crop in many countries around the world. It is a highly N-demanding crop; thus, it becomes extremely important to find alternatives to reduce the use of chemical N fertilizers applied to rice without decreasing the productivity. Endophytic diazotrophic strains were isolated from root, culm, and leaf tissues of traditional rice varieties (Zebu Branco and Manteiga) cultivated traditionally by the local farmers in the Maranhão state, Brazil [97]. Ten strains showing consistent acetylene reduction activity and capable of producing indole-3-acetic acid (IAA) were identified as belonging to the genera *Azospirillum*, *Sphingomonas*, and *Burkholderia*. These endophytic diazotrophic strains were inoculated into 10 different traditional varieties of rice to select the best strain/rice variety interaction by growing them in gnotobiotic, greenhouse, and field conditions. Although a strain belonging to the genus *Azospirillum* showed highest biomass enhancement (48%) under gnotobiotic conditions, *Burkholderia vietnamiensis* strain AR1122 inoculated into a traditional variety Arroz 70 showed best results as compared to other strain/variety combinations when grown under greenhouse and field conditions. The grain yield of Arroz 70 variety was also significantly enhanced when inoculated with the strain AR1122 in comparison to a control treatment that was provided with sufficient amounts of N fertilizer. These results clearly indicate that *Burkholderia vietnamiensis* strain AR1122 is a candidate biofertilizer for traditional rice varieties in Brazil and should be investigated with other genotypes of rice for a sustainable rice crop production. In Brazil, sugarcane has been one of the fastest growing crops, reaching new frontiers and decisively influencing the economic, social, and cultural development. However, similar to rice, it is also one of the most N-demanding crops that makes it crucial to invest in research on alternatives other than chemical N fertilizers like biofertilizers with diazotrophs, so as to ensure a competitive and sustainable development of sugarcane industry. A study conducted in 2014 reported the effects of inoculating the sugarcane plants with a consortium of five different endophytic diazotrophic bacteria of *Gluconacetobacter diazotrophicus*, *Herbaspirillum*, and *Burkholderia* [98]. In this study, the consortium was evaluated with regard to the agronomic performance and N nutrition of sugarcane in field against chemical N fertilizer and it was found that the consortium of inoculant increased the stalk yield of sugarcane similar to the chemical fertilization. However, authors did not find any evidence of BNF in sugarcane by the consortium of diazotrophic strains, which indicates that the diazotrophic strains used in this study may possess other PGP characteristics that could have resulted in increased yields of sugarcane. In another study, *Gluconacetobacter diazotrophicus* strain PAL 5, which has been studied extensively for its N-fixing and PGP abilities [99], and a strain belonging to the genus *Herbaspirillum* were inoculated into sugarcane plants to evaluate their drought stress recovery [100]. After being subjected to 21 days of drought stress, bacteria-inoculated plants had significantly higher shoot and root dry weight (50 and 70%, respectively) and total N content in leaves (77%). Authors also reported that these diazotrophic strains induce preservation of leaf water potential and relative water content by closing stomata efficiently resulting in plant water preservation during the drought, which highlights the ability of these endophytic diazotrophic bacteria to protect the plant from abiotic stresses. Another type of abiotic stress, that is, salinity, has been recently reported to stimulate the population and diversity of endophytic diazotrophic bacteria in forage cactus (*Opuntia stricta*) [101]. In this study, the population density of endophytic diazotrophic strains in root tissues was evaluated by using the most probable number method (MPN) and strains were characterized phenotypically to evaluate the diversity. Authors reported that the forage cactus plants that received the highest amount of saline water had the highest population density of putative endophytic diazotrophic bacteria with high phenotypic diversity. These findings indicate that endophytic diazotrophic bacteria thrive when conditions are adverse by assisting the host plant through direct or indirect mechanisms to flourish in poor conditions.

Corn is an agriculturally important crop that is extensively grown and consumed by a large population around the world. Szilagyi-Zecchin et al. [102] isolated and identified six endophytic strains from roots of corn growing in the southern Brazilian region of Campo Largo, PR. Out of these six endophytic isolates, four were able to grow on N-free media, consistently reducing acetylene, and were found positive for the presence of *nifH* gene. Apart from showing positive results for N-fixing activity, two out of these four strains (identified as *Bacillus* sp.) also showed other PGP characteristics, like production of IAA, siderophores, and lytic enzymes and antagonism against the common pathogenic fungi. When all endophytic isolates were reinoculated into corn to check for *in vivo* plant growth promotion, another endo-

germination by 47% and root volume by 44% [102]. In yet another study conducted in Brazil, 40 endophytic strains were isolated from roots of banana (*Musa* L.) tree cultivar 'Prata Anã' [103]. Banana is a very common edible fruit (botanically a berry), produced primarily in the tropics but consumed all around the world. Banana trees grow rapidly and require substantial amount of nutrients in the soil for their development and fruit production. Out of the 40 strains isolated in that study, 20 strains were able to grow on N-free media, but only four isolates showed positive results for N-fixing activity when analyzed using acetylene reduction assay and Kjeldahl method. All four isolates were identified as belonging to the genus *Bacillus* and were also tested positive for *in vitro* phosphate solubilization and IAA production, thus, indicating their potential to be used as growth-promoting microbial inoculants for banana trees pending *in vivo* greenhouse or field experiments.

Pearl millet (*Pennisetum glaucum* (L.) R. Br.) is a staple cereal crop of the hottest and driest areas of tropics and subtropics. Pearl millet is commonly grown in Rajasthan, India, which has an arid climate and uncertain and erratic rainfall season. In a study reported in 2013, endophytic diazotrophic strains were isolated from pearl millet plants growing in a field with a nutrient-deficient sandy clay loam soil located in Rajasthan [104]. *Pseudomonas aeruginosa* strain PM389 was the most dominant diazotrophic strain in pearl millet plants harvested from this field, whose upward migration and establishment in the stem tissues were later tracked by using enterobacterial repetitive intergenic consensus sequences-PCR (ERIC-PCR) as a biomarker. Efficient reduction of acetylene during the acetylene reduction assay and presence of *nifH* gene indicated the N-fixing potential of the strain PM389. As reported in the study, this strain possesses other PGP characteristics as well, like mineral phosphate solubilization, siderophore production, and antagonistic activity against many pathogenic bacterial and fungal species. In addition, when inoculated into a nonnative plant species (wheat), strain PM389 significantly increased seed germination rate, root and shoot length, and vigor index, which highlights its ability to infect other crop hosts and promote their growth [104]. Local cultivars that have been grown traditionally for many years could serve as a source for potential endophytic diazotrophic bacteria that could be applied to modern commercial varieties as biofertilizers. This theory was proved by scientists from Thailand, who isolated 396 potential endophytic diazotrophic strains from 6 different landraces of rice growing in Chiang Mai, Thailand [105]. Based on the results of acetylene reduction assay, authors chose 21 isolates that were further screened to 10 on the basis of tests conducted for other PGP characteristics. These strains belonged to genera *Burkholderia, Klebsiella, Novosphingobium,* and *Sphingomonas* and were able to recolonize the tissues of a commercial rice cultivar Khao Dawk Mali 105 along with increasing the N content in the seedlings and promoting seedling length and dry weight. Korean rice cultivars have also been evaluated for the presence of endophytic diazotrophic bacteria [106]. Twelve potential endophytic diazotrophic strains were isolated and identified as belonging to the genera *Paenibacillus* [107], *Bacillus, Microbacterium,* and *Klebsiella* and were tested positive for the presence of *nifH* gene. When reinoculated into rice plants, these strains improved plant growth, increased height and dry weight, and showed antagonistic effects against fungal pathogens, thus, establishing their potential role as biofertilizer and biocontrol agents for Korean rice cultivars.

Our lab group has been working with endophytic diazotrophic bacteria from many years and has published several reports regarding the role of these bacteria in fixing N and promoting plant growth in both agricultural and forest ecosystems [108]. In 2012, our lab discovered an endophytic diazotrophic bacterium, *P. polymyxa* P2b-2R, from stem tissues of lodgepole pine (*Pinus contorta var. latifolia*) trees naturally regenerating at a site located near Williams Lake, BC, Canada [109]. Strain P2b-2R was able to grow on N-free media and consistently reduced significant amounts of acetylene in the acetylene reduction assay [109]. This bacterial strain was able to fix significant amounts of atmospheric N (up to 79%) when reinoculated into lodgepole pine and evaluated using foliar ^{15}N isotope dilution method [110–112]. It was also observed that strain P2b-2R possesses *nif* genes required to encode the nitrogenase enzyme, thus confirming the N-fixing ability of this strain [113].

Host plant	Harvest (days)	%Ndfa[a]		% growth promotion						References
				Foliar nitrogen concentration[b]		Seedling length[c]		Seedling biomass[d]		
		P2b-2R	P2b-2R*gfp*	P2b-2R	P2b-2R*gfp*	P2b-2R	P2b-2R*gfp*	P2b-2R	P2b-2R*gfp*	
Corn	10	6.65	—	5.42	—	10.0	—	20.9	—	[117]
	20	10.8	10.9	13.6	25.0	13.8	41.3	26.1	34.0	[117, 121]
	30	19.6	14.1	14.2	22.6	35.3	36.3	30.9	55.5	[117, 121]
	40	15.7	18.0	17.1	27.6	24.7	27.6	28.4	48.9	[121]
	90	30.2	32.2	27.3	31.8	51.9	68.4	52.7	66.9	[122]
Canola	20	8.08	13.0	28.7	37.8	17.8	37.4	57.0	91.6	[119]
	30	12.9	15.1	18.0	36.1	20.5	48.7	53.7	93.5	[119]
	40	16.2	22.1	23.4	40.8	28.4	69.4	37.1	108	[119]
	60	21.8	—	40.3	—	24.9	—	30.1	—	[118]
	90	27.1	35.1	11.7	25.0	70.7	102.5	100.8	159.1	[120]
Tomato	20	10.0	8.32	33.3	25.5	40.6	48.4	56.1	44.1	[119]
	30	12.3	11.2	30.6	23.2	36.5	37.5	69.0	61.4	[119]
	40	18.1	16.7	30.0	22.5	24.9	28.3	93.0	82.9	[119]

[a]Percent nitrogen derived from the atmosphere (%Ndfa).
[b]Percent increase in foliar nitrogen concentration by inoculation with *P. polymyxa* strains P2b-2R and P2b-2R*gfp*.
[c]Percent seedling length promoted by inoculation with *P. polymyxa* strains P2b-2R and P2b-2R*gfp*.
[d]Percent seedling biomass promoted by inoculation with *P. polymyxa* strains P2b-2R and P2b-2R*gfp*.
These parameters were calculated using the formulas described in Puri et al. [122].

Table 2. Plant growth promotion and biological nitrogen fixation by *Paenibacillus polymyxa* strain P2b-2R and its GFP-tagged derivative, P2b-2R*gfp*, when inoculated into agricultural crops, namely, corn, canola, and tomato.

Endophytic colonization of lodgepole pine by P2b-2R strain was confirmed by constructing a GFP-tagged derivative of P2b-2R and visualizing the sites of colonization using CLSM [75]. It was found that this strain can colonize both intercellular and intracellular spaces of lodgepole pine interior tissues possibly by degrading major cell wall components [75, 114]. Strain P2b-2R was able to colonize internal tissues of another gymnosperm tree species, western red cedar (*Thuja plicata*), and fix considerable amounts of N from the atmosphere along with enhancing seedling length and biomass of cedar [115, 116]. Subsequently, Puri et al. [117] hypothesized that strain P2b-2R could provide similar benefits to angiosperms, specifically agricultural crop species, by colonizing them endophytically. They tested this hypothesis by inoculating strain P2b-2R into agriculturally important crops, namely corn, canola, and tomato, and found that P2b-2R was able to colonize internal tissues of these crop species, fix substantial amounts of atmospheric N, and increase seedling length and biomass (see **Table 2**) [117–119]. These reports indicate the ability of strain P2b-2R to symbiotically associate with a broad range of hosts and promote their growth primarily by fixing atmospheric N. An interesting observation with the GFP-tagged P2b-2R strain (P2b-2R*gfp*) was reported recently where P2b-2R*gfp* inoculation significantly enhanced corn and canola seedling length and biomass as compared to the wild-type P2b-2R inoculation [119–122]. In addition, strain P2b-2R*gfp* fixed significantly higher amounts of N as compared to the wild-type strain. Subsequently, similar results were reported when both strains were inoculated into their original host, that is, lodgepole pine [123]. To the best of our knowledge, these were the very first *in planta* studies in literature reporting that GFP tagging of a bacterial strain could significantly enhance its ability to promote plant growth. Enhancement of these abilities *in vitro* after GFP-tagging were reported previously in *Azospirillum brasilense* [124]. A plausible reason for increased N fixing and plant growth–promoting efficacy of P2b-2R after GFP tagging could be the overexpression of structural *nif* genes (*nifH*, *nifD*, and *nifK*), which play an important role in the N-fixation process [121]. However, it is still unclear how GFP tagging affects the expression of structural *nif* genes of strain P2b-2R. Also, other plausible reasons behind the increased plant growth–promoting efficacy after GFP tagging need to be investigated.

5. Conclusions

Since their discovery in sugarcane tissues decades ago, endophytic diazotrophic bacteria have been characterized for their role in performing BNF. Studies have suggested that these bacteria can act as N biofertilizer for highly N-demanding crops like sugarcane, corn, and rice. Most recent studies have also focused their attention on testing the PGP characteristics of isolated endophytic diazotrophic strains other than N fixation, which indicates the growing concern of agricultural scientists to develop bacterial inoculants that can enhance plant growth through a variety of mechanisms, so as to decrease the dependence on chemical fertilizers. Endophytic diazotrophic strains like *P. polymyxa* P2b-2R that are able to colonize nonnative host and fix atmospheric N and promote their growth have great potential as biofertilizers for sustainable crop production.

Acknowledgements

Authors would like to dedicate this work to Late Mr. Darshan K. Puri (1956–2014). You were, are and always will be an inspirational figure for us.

Author details

Akshit Puri[1], Kiran Preet Padda[1] and Chris P. Chanway[1,2]*

*Address all correspondence to: chris.chanway@ubc.ca

1 Department of Soil Science, Faculty of Land and Food Systems, The University of British Columbia, Vancouver, Canada

2 Department of Forest and Conservation Sciences, Faculty of Forestry, The University of British Columbia, Vancouver, Canada

References

[1] Leghari SJ, Wahocho NA, Laghari GM, Laghari AH, Bhabhan GM, Talpur KA, Bhutto TA, Wahocho SA, Lashari AA. Role of nitrogen for plant growth and development: A review. Advances in Environmental Biology. 2016;**10**(9):209-219

[2] Greenwood NN, Earnshaw A. Chemistry of the Elements. 2nd ed. Oxford: Butterworth-Heinemann; 1997

[3] Galloway JN, Cowling EB. Reactive nitrogen and the world: 200 years of change. AMBIO: A Journal of the Human Environment. 2002;**31**(2):64-71. DOI: 10.1579/0044-7447-31.2.64

[4] Myrold DD, Bottomley PJ, Biological N. Inputs. In: Paul EA, editor. Soil Microbiology, Ecology and Biochemistry. Burlington: Elsevier Academic Press; 2007. pp. 365-388

[5] McCosh FWJ. The plant and nitrogen. Boussingault. Dordrecht: Springer Netherlands. 1984:123-138. DOI: 10.1007/978-94-009-6297-2_10

[6] Dance I. Elucidating the coordination chemistry and mechanism of biological nitrogen fixation. Chemistry - An Asian Journal. 2007;**2**(8):936-946. DOI: 10.1002/asia.200700131

[7] Kim J, Rees DC. Nitrogenase and biological nitrogen fixation. Biochemistry. 1994;**33**(2): 389-397. DOI: 10.1021/bi00168a001

[8] Rubio LM, Ludden PW. Biosynthesis of the iron—Molybdenum cofactor of nitrogenase. Annual Review of Microbiology. 2008;**62**:93-111. DOI: 10.1146/annurev.micro.62.081307. 162737

[9] Seefeldt LC, Hoffman BM, Dean DR. Mechanism of Mo-dependent nitrogenase. Annual Review of Biochemistry. 2009;**78**:701-722. DOI: 10.1146/annurev.biochem.78.070907.103812

[10] Brill WJ. Biochemical genetics of nitrogen fixation. In: Ciferri O, Dure L, editors. Structure and Function of Plant Genomes. New York: Springer; 1983. pp. 231-237. DOI: 10.1007/978-1-4684-4538-1_23

[11] Dixon R, Eady RR, Espin G, Hill S, Laccarino M, Kahn D, Merrick M. Analysis of regulation of *Klebsiella pneumoniae* nitrogen fixation (*nif*) gene cluster with gene fusions. Nature. 1980;**286**(5769):128-132. DOI: 10.1038/286128a0

[12] Danso SKA. Assessment of biological nitrogen fixation. Nutrient Cycling in Agroecosystems. 1995;**42**(1-3):33-41. DOI: 10.1007/BF00750498

[13] McClure PR, Israel DW. Transport of nitrogen in the xylem of soybean plants. Plant Physiology. 1979;**64**(3):411-416. DOI: 10.1104/pp.64.3.411

[14] Herridge DF. Effects of nitrate and plant development on the abundance of nitrogenous solutes in root-bleeding and vacuum-extracted exudates of soybean. Crop Science. 1984;**24**(1):173-179. DOI: 10.2135/cropsci1984.0011183X002400010041x

[15] Kessel CV, Roskoski JP, Keane K. Ureide production by N_2-fixing and non-N_2-fixing leguminous trees. Soil Biology and Biochemistry. 1988;**20**(6):891-897. DOI: 10.1016/0038-0717(88)90100-9

[16] Hardy RWF, Holsten RD, Jackson EK, Burns RC. The acetylene-ethylene assay for N_2 fixation: Laboratory and field evaluation. Plant Physiology. 1968;**43**(8):1185-1207. DOI: 10.1104/pp.43.8.1185

[17] Hardarson G, Danso SKA. Methods for measuring biological nitrogen fixation in grain legumes. Plant and Soil. 1993;**152**(1):19-23. DOI: 10.1007/BF00016330

[18] Galloway JN, Townsend AR, Erisman JW, Bekunda M, Cai Z, Freney JR, Martinelli LA, Seitzinger SP, Sutton MA. Transformation of the nitrogen cycle: Recent trends, questions, and potential solutions. Science. 2008;**320**(5878):889-892. DOI: 10.1126/science.1136674

[19] Young JPW. Phylogenetic classification of nitrogen-fixing organisms. In: Stacey G, Burris RH, Evans HJ, editors. Biological Nitrogen Fixation. New York: Chapman and Hall; 1992. pp. 43-86

[20] Reed SC, Cleveland CC, Townsend AR. Functional ecology of free-living nitrogen fixation: A contemporary perspective. Annual Review of Ecology, Evolution, and Systematics. 2011;**42**:489-512. DOI: 10.1146/annurev-ecolsys-102710-145034

[21] Döbereiner J, Pedrosa FO. Nitrogen-Fixing Bacteria in Non-Leguminous Crop Plants. Madison: Science Tech; 2011

[22] Lindow SE, Brandl MT. Microbiology of the phyllosphere. Applied and Environmental Microbiology. 2003;**69**(4):1875-1883. DOI: 10.1128/AEM.69.4.1875-1883.2003

[23] Postgate JR. Nitrogen Fixation. 3rd ed. Cambridge: Cambridge University Press; 1998

[24] Oldroyd GED, Murray JD, Poole PS, Downie JA. The rules of engagement in the legume-rhizobial symbiosis. Annual Review of Genetics. 2011;**45**:119-144. DOI: 10.1146/annurev-genet-110410-132549

[25] Udvarte M, Poole PS. Transport and metabolism in legume-rhizobia symbioses. Annual Review of Plant Biology. 2013;**64**:781-805. DOI: 10.1146/annurev-arplant-050312-120235

[26] Huss-Danell K. Actinorhizal symbioses and their N_2 fixation. New Phytologist. 1997; **136**(3):375-405

[27] Schell DM, Alexander V. Nitrogen fixation in Arctic coastal tundra in relation to vegetation and micro-relief. Arctic. 1973;**26**:130-137

[28] Hobara S, McCalley C, Koba K, Giblin AE, Weiss MS, Gettel GM, Shaver GR. Nitrogen fixation in surface soils and vegetation in an Arctic tundra watershed: A key source of atmospheric nitrogen. Arctic, Antarctic, and Alpine Research. 2006;**38**(3):363-372. DOI: 10.1657/1523-0430(2006)38[363:NFISSA]2.0.CO;2

[29] Döbereiner J, Alvahydo R. Sóbre a influénciada canade-acucar na occoréncia de "*Beijerinckia*" no solo II. Influéncia das diversas partes do vegetal. Revista Brasileira de Biologia. 1959;**19**: 401-412

[30] Döbereiner J. Nitrogen fixing bacteria of the genus *Beijerinckia* Drex. in the rhizosphere of sugarcane. Plant and Soil. 1961;**15**(3):211-216. DOI: 10.1007/BF01400455

[31] Rennie RJ, de Freitas JR, Ruschel AP, Vose PB. Isolation and identification of nitrogen fixing bacteria associated with sugarcane (*Saccharum* sp.). Canadian Journal of Microbiology. 1982;**28**(5):462-467. DOI: 10.1139/m82-070

[32] Magalhaes FMM, Baldani JI, Souto SM, Kuykendal JR, Döbereiner JA. New acid tolerant *Azospirillum* species. Anais da Academia Brasileira de Ciências. 1983;**55**:417-430

[33] Seldin L, van Elsas JD, Penido EGC. *Bacillus azotofixans* sp. nov., a nitrogen-fixing species from Brazilian soils and grass roots. International Journal of Systematic Bacteriology. 1984;**34**:451-456. DOI: 10.1099/00207713-34-4-451

[34] Baldani JI, Baldani VLD, Seldin L, Döbereiner J. Characterization of *Herbaspirillum seropedicae* gen. Nov., sp. nov., a root associated nitrogen fixing bacterium. International Journal of Systematic Bacteriology. 1986;**36**:86-93. DOI: 10.1099/00207713-36-1-86

[35] Döbereiner J, Day JM, Dart PJ. Nitrogenase activity in the rhizosphere of sugarcane and some other tropical grasses. Plant and Soil. 1972;**37**(1):191-196. DOI: 10.1007/BF01578494

[36] Ruschel AP. Associative N_2-fixation by sugar cane. In: Vose PB, Ruschel AP, editors. Associative N_2-Fixation. Vol. 2. Boca Raton: CRC; 1981. pp. 81-90

[37] Cavalcante VA, Döbereiner J. A new acid tolerant nitrogen fixing bacterium associated with sugarcane. Plant and Soil. 1988;**108**(1):23-31. DOI: 10.1007/BF02370096

[38] Boddey RM, Urquiaga S, Reis V, Döbereiner J. Biological nitrogen fixation associated with sugar cane. Plant and Soil. 1991;**137**(1):111-117. DOI: 10.1007/BF02187441

[39] Stephan MP, Oliveira M, Teixeira KRS, Martinez-Drets G, Döbereiner J. Physiology and dinitrogen fixation of *Acetobacter diazotrophicus*. FEMS Microbiology Letters. 1991;**77**(1): 67-72. DOI: 10.1111/j.1574-6968.1991.tb04323.x

[40] de Bary A. Morphologie und Physiologie Pilze, Flechten, und myxomyceten. In: Hofmeister W, editor. Handbuch der Physiologischen Botanik. Zweiter Band. Leipzig: Wilhelm Engelmann; 1866. Available from: http://babel.hathitrust.org/cgi/pt?id=hvd.32044053007316 Accessed: 2017-10-02

[41] Chanway CP. Endophytes: They're not just fungi. Canadian Journal of Botany. 1996; **74**(3):321-322. DOI: 10.1139/b96-040

[42] Chanway CP, Anand R, Yang H. Nitrogen fixation outside and inside plant tissues. In: Ohyama T, editor. Advances in Biology and Ecology of Nitrogen Fixation. Croatia: InTech; 2014. pp. 3-23. DOI: 10.5772/57532

[43] Trevet IW, Hollis JP. Bacteria in storage organs of healthy plants. Phytopathology. 1948; **38**:960-967

[44] Hallmann J, Quadt-Hallmann A, Mahaffee WF, Kloepper JW. Bacterial endophytes in agricultural crops. Canadian Journal of Microbiology. 1997;**43**(10):895-914. DOI: 10.1139/m97-131

[45] Kobayashi D, Palumbo J. Bacterial endophytes and their effects on plants and uses in agriculture. In: Bacon CW, White JF, editors. Microbial Endophytes. New York: Marcel Dekker; 2000. pp. 199-233

[46] Sturz AV, Christie BR, Nowak J. Bacterial endophytes: Potential role in developing sustainable systems of crop production. Critical Reviews in Plant Sciences. 2000;**19**(1):1-30

[47] Suman A, Yadav AN, Verma P. Endophytic microbes in crops: Diversity and beneficial impact for sustainable agriculture. In: Singh DP, Abhilash PC, Prabha R, editors. Microbial Inoculants in Sustainable Agricultural Productivity. New Delhi: Springer; 2016. pp. 117-143. DOI: 10.1007/978-81-322-2647-5_7

[48] Loper JE, Haack C, Schroth MN. Population dynamics of soil pseudomonads in the rhizosphere of potato (*Solanum tuberosum* L.). Applied and Environmental Microbiology. 1985;**49**(2):416-422

[49] Cocking E. Endophytic colonization of plant roots by N-fixing bacteria. Plant and Soil. 2003;**252**(1):169-175. DOI: 10.1023/A:1024106605806

[50] Rosenblueth M, Martínez-Romero E. Bacterial endophytes and their interaction with hosts. Molecular Plant-Microbe Interactions. 2006;**19**(8):827-837. DOI: 10.1094/MPMI-19-0827

[51] Richardson A, Barea J-M, McNeill A, Prigent-Combaret C. Acquisition of phosphorus and nitrogen in the rhizosphere and plant growth promotion by microorganisms. Plant and Soil. 2009;**321**(1-2):305-339. DOI: 10.1007/s11104-009-9895-2

[52] Reinhold-Hurek B, Hurek T. Interactions of gramineous plants with *Azoarcus* spp. and other diazotrophs: Identification, localization, and perspectives to study their function. Critical Reviews in Plant Sciences. 1998;**17**(1):29-54

[53] Reinhold-Hurek B, Hurek T. Life in grasses: Diazotrophic endophytes. Trends in Microbiology. 1998b;**6**(4):139-144. DOI: 10.1016/S0966-842X(98)01229-3

[54] Santi C, Bogusz D, Franche C. Biological nitrogen fixation in non-legume plants. Annals of Botany. 2013;**111**(5):743-767. DOI: 10.1093/aob/mct048

[55] James EK, Gyaneshwar P, Mathan N, Barraquio WL, Reddy PM, Iannetta PPM, Olivares FL, Ladha JK. Infection and colonization of rice seedlings by the plant growth-promoting bacterium *Herbaspirillum seropedicae* Z67. Molecular Plant-Microbe Interactions. 2002; **15**(9):894-906. DOI: 10.1094/MPMI.2002.15.9.894

[56] Sessitsch A, Reiter B, Pfeifer U, Wilhelm E. Cultivation-independent population analysis of bacterial endophytes in three potato varieties based on eubacterial and Actinomycetes-specific PCR of 16S rRNA genes. FEMS Microbiology Ecology. 2002;**39**(1):23-32. DOI: 10.1111/j.1574-6941.2002.tb00903.x

[57] Berg G, Krechel A, Ditz M, Sikora RA, Ulrich A, Hallmann J. Endophytic and ectophytic potato-associated bacterial communities differ in structure and antagonistic function against plant pathogenic fungi. FEMS Microbiology Ecology. 2005;**51**(2):215-229. DOI: 10.1016/j.femsec.2004.08.006

[58] Okunishi S, Sako K, Mano H, Imamura A, Morisaki H. Bacterial flora of endophytes in the maturing seed of cultivated rice (*Oryza sativa*). Microbes and Environments. 2005;**20**(3):168-177. DOI: 10.1264/jsme2.20.168

[59] Compant S, Mitter B, Colli-Mull JG, Gangl H, Sessitsch A. Endophytes of grapevine flowers, berries, and seeds: Identification of cultivable bacteria, comparison with other plant parts, and visualization of niches of colonization. Microbial Ecology. 2011;**62**(1):188-197. DOI: 10.1007/s00248-011-9883-y

[60] de Melo Pereira GV, Magalhaes KT, Lorenzetii ER, Souza TP, Schwan RF. A multiphasic approach for the identification of endophytic bacterial in strawberry fruit and their potential for plant growth promotion. Microbial Ecology. 2012;**63**(2):405-417. DOI: 10.1007/s00248-011-9919-3

[61] Trognitz F, Piller K, Nagel M, Borner A, Bacher C-F, Rechlik M, Mayrhofer H, Sessitsch A. Isolation and characterization of endophytes isolated from seeds of different plants and the application to increase juvenile development. In: Tagung Zukünftiges Saatgut—Produktion, Vermarktung, Nutzung und Konzervierung. Future Seed—Production, Marketing, Use and Conservation; 24-26 November 2014; Austria. Irdning: Höhere Bundeslehr- und Forschungsanstalt für Landwirtschaft Raumberg-Gumpenstein; 2014. pp. 25-28

[62] Benhizia Y, Benhizia H, Benguedouar A, Muresu R, Giacomini A, Squartini A. Gamma proteobacteria can nodulate legumes of the genus *Hedysarum*. Systematic and Applied Microbiology. 2004;**27**(4):462-468. DOI: 10.1078/0723202041438527

[63] Paul LR, Chapman WK, Chanway CP. Diazotrophic bacteria reside inside *Suillus tomentosus/Pinus contorta* tuberculate ectomycorrhizae. Botany. 2013;**91**(1):48-52. DOI: 10.1139/cjb-2012-0191

[64] Jacobs MJ, Bugbee WM, Gabrielson DA. Enumeration, location, and characterization of endophytic bacteria within sugar beet roots. Canadian Journal of Botany. 1985;**63**(7): 1262-1265. DOI: 10.1139/b85-174

[65] Misaghi IJ, Donndelinger CR. Endophytic bacteria in symptom-free cotton plants. Phytopathology. 1990;**80**(9):8080-8811

[66] Sturz AV, Christie BR, Matheson BG, Nowak J. Biodiversity of endophytic bacteria which colonize red clover nodules, roots, stems and foliage and their influence on host growth. Biology and Fertility of Soils. 1997;**25**(1):13-19. DOI: 10.1007/s003740050273

[67] Chi F, Shen S, Cheng H, Jing Y, Yanni YG, Dazzo FB. Ascending migration of endophytic rhizobia, from roots to leaves, inside rice plants and assessment of benefits to rice growth physiology. Applied and Environmental Microbiology. 2005;**71**(11):7271-7278. DOI: 10.1128/AEM.71.11.7271-7278.2005

[68] Anand R, Paul L, Chanway C. Research on endophytic bacteria: Recent advances with forest trees. In: Schulz B, Boyle C, Sieber TN, editors. Microbial Root Endophytes. Berlin: Springer Heidelberg; 2006. pp. 89-106. DOI: 10.1007/3-540-33526-9_6

[69] Tombolini R, Unge A, Davey ME, Bruijn FJ, Jansson JK. Flow cytometric and microscopic analysis of GFP-tagged *Pseudomonas fluorescens* bacteria. FEMS Microbiology Ecology. 1997;**22**(1):17-28. DOI: 10.1111/j.1574-6941.1997.tb00352.x

[70] Tombolini R, Jansson JK. Monitoring of GFP tagged bacterial cells. In: La Rossa R, editor. Bioluminescence Methods and Protocols. Totowa: Humana Press; 1998. pp. 285-298. DOI: 10.1385/0-89603-520-4:285

[71] Xi C, Lambrecht M, Vanderleyden J, Michiels J. Bi-functional *gfp*-and *gus*A-containing mini-*Tn*5 transposon derivatives for combined gene expression and bacterial localization studies. Journal of Microbiological Methods. 1999;**35**(1):85-92. DOI: 10.1016/S0167-7012(98)00103-1

[72] Compant S, Reiter B, Sessitsch A, Nowak J, Clement C, Ait Barka E. Endophytic colonization of *Vitis vinifera* L. by plant growth-promoting bacterium *Burkholderia* sp. strain PsJN. Applied and Environmental Microbiology. 2005;**71**(4):1685-1693. DOI: 10.1128/AEM.71.4.1685-1693.2005

[73] Timmusk S, Grantcharova N, Wagner EGH. *Paenibacillus polymyxa* invades plant roots and forms biofilms. Applied and Environmental Microbiology. 2005;**71**(11):7292-7300. DOI: 10.1128/AEM.71.11.7292-7300.2005

[74] Chelius MK, Triplett EW. *Dyadobacter fermentans* gen. Nov., sp. nov., a novel gram-negative bacterium isolated from surface-sterilized *Zea mays* stems. International Journal of Systematic and Evolutionary Microbiology. 2000;**50**:751-758. DOI: 10.1099/00207713-50-2-751

[75] Anand R, Chanway CP. Detection of GFP-labeled *Paenibacillus polymyxa* in auto fluorescing pine seedling tissues. Biology and Fertility of Soils. 2013;**49**(1):111-118. DOI: 10.1007/s00374-012-0727-9

[76] Villacieros M, Power B, Sanchez-Contreras M, Lloret J, Oruezabal RI, Martin M, Fernandez-Pinas F, Bonilla I. Colonization behaviour of *Pseudomonas fluorescens* and *Sinorhizobium meliloti* in the alfalfa (*Medicago sativa*) rhizosphere. Plant and Soil. 2003;**251**(1): 47-54. DOI: 10.1023/A:1022943708794

[77] Germaine K, Keogh E, Garcia-Cabellos G, Borremans B, Lelie D, Barac T, Oeyen L, Vangronsveld J, Moore FP, Moore ERB, Campbell CD, Ryan D, Dowling DN. Colonisation of poplar trees by gfp expressing bacterial endophytes. FEMS Microbiology Ecology. 2004;**48**(1):109-118. DOI: 10.1016/j.femsec.2003.12.009

[78] Sevilla M, Kennedy C. Colonization of rice and other cereals by *Acetobacter diazotrophicus*, an endophyte of sugarcane. In: Ladha JK, Reddy PM, editors. The Quest for Nitrogen Fixation in Rice. Makati: International Rice Research Institute; 2000. pp. 151-165

[79] Rouws LF, Meneses CH, Guedes HV, Vidal MS, Baldani JI, Schwab S. Monitoring the colonization of sugarcane and rice plants by the endophytic diazotrophic bacterium *Gluconacetobacter diazotrophicus* marked with *gfp* and *gusA* reporter genes. Letters in Applied Microbiology. 2010;**51**(3):325-330. DOI: 10.1111/j.1472-765X.2010.02899.x

[80] Alquéres S, Meneses C, Rouws L, Rothballer M, Baldani I, Schmid M, Hartmann A. The bacterial superoxide dismutase and glutathione reductase are crucial for endophytic colonization of rice roots by *Gluconacetobacter diazotrophicus* PAL5. Molecular Plant-Microbe Interactions. 2013;**26**(8):937-945. DOI: 10.1094/MPMI-12-12-0286-R

[81] Mousa WK, Shearer CR, Limay-Rios V, Zhou T, Raizada MN. Bacterial endophytes from wild maize suppress *Fusarium graminearum* in modern maize and inhibit mycotoxin accumulation. Frontiers in Plant Science. 2015;**6**:805. DOI: 10.3389/fpls.2015.00805

[82] Kumar A, Munder A, Aravind R, Eapen SJ, Tümmler B, Raaijmakers JM. Friend or foe: Genetic and functional characterization of plant endophytic *Pseudomonas aeruginosa*. Environmental Microbiology. 2013;**15**(3):764-779. DOI: 10.1111/1462-2920.12031

[83] Sun K, Liu J, Gao Y, Jin L, Gu Y, Wang W. Isolation, plant colonization potential, and phenanthrene degradation performance of the endophytic bacterium *Pseudomonas* sp. Ph6-*gfp*. Scientific Reports. 2014;**4**:5462. DOI: 10.1038/srep05462

[84] Shehata HR, Lyons EM, Jordan KS, Raizada MN. Bacterial endophytes from wild and ancient maize are able to suppress the fungal pathogen *Sclerotinia homoeocarpa*. Journal of Applied Microbiology. 2016;**120**(3):756-769. DOI: 10.1111/jam.13050

[85] Döbereiner J. Recent changes in concepts of plant bacteria interactions: Endophytic N_2 fixing bacteria. Ciência e Cultura. 1992;**44**(5):310-313

[86] Baldani JI, Olivares FL, Hemerly AS, Reis FB Jr, Oliveira ALM, Baldani VLD, Goi SR, Reis VM, Döbereiner J. Nitrogen-fixing endophytes: Recent advances in the association with graminaceous plants grown in the tropics. In: Elmerich EC, editor. Biological Nitrogen Fixation for the 21st Century. Dordrecht: Springer Netherlands; 1998. pp. 203-206. DOI: 10.1007/978-94-011-5159-7_90

[87] Baldani VLD, Baldani JI, Döbereiner J. Inoculation of rice plants with the endophytic diazotrophs *Herbaspirillum seropedicae* and *Burkholderia* spp. Biology and Fertility of Soils. 2000;**30**(5-6):485-491. DOI: 10.1007/s003740050027

[88] Gyaneshwar P, James EK, Mathan N, Reddy PM, Reinhold-Hurek B, Ladha JK. Endophytic colonization of rice by a diazotrophic strain of *Serratia marcescens*. Journal of Bacteriology. 2001;**183**(8):2634-2645. DOI: 10.1128/JB.183.8.2634-2645.2001

[89] Hurek T, Handley LL, Reinhold-Hurek B, Piche Y. *Azoarcus* grass endophytes contribute fixed nitrogen to the plant in an unculturable state. Molecular Plant-Microbe Interactions. 2002;**15**(3):233-242. DOI: 10.1094/MPMI.2002.15.3.233

[90] Olivares FL, Baldani VLD, Reis VM, Baldani JI, Döbereiner J. Occurrence of the endophytic diazotrophs *Herbaspirillum* spp. in roots, stems, and leaves, predominantly of Gramineae. Biology Fertility of Soils. 1996;**21**(3):197-200. DOI: 10.1007/BF00335935

[91] Riggs PJ, Chelius MK, Iniguez AL, Kaeppler SM, Triplett EW. Enhanced maize productivity by inoculation with diazotrophic bacteria. Australian Journal of Plant Physiology. 2001;**28**(9):829-836. DOI: 10.1071/PP01045

[92] Roesch LFW, Camargo FAO, Bento FM, Triplett EW. Biodiversity of diazotrophs within the soil, root and stem of field grown maize. Plant and Soil. 2008;**302**(1-2):91-104. DOI: 10.1007/s11104-007-9458-3

[93] Montañez A, Abreu C, Gill PR, Hardarson G, Sicardi M. Biological nitrogen fixation in maize (*Zea mays* L.) by ^{15}N isotope-dilution and identification of associated culturable diazotrophs. Biology and Fertility of Soils. 2009;**45**(3):253-263. DOI: 10.1007/s00374-008-0322-2

[94] Sabry RS, Saleh SA, Batchelor CA, Jones J, Jotham J, Webster G, Kothari SL, Davey MR, Cocking EC. Endophytic establishment of *Azorhizobium caulinodans* in wheat. Proceedings of the Royal Society of London B: Biological Sciences. 1997;**264**(1380):341-346. DOI: 10.1098/rspb.1997.0049

[95] Germida J, de Freitas J. Nitrogen fixing rhizobacteria as biofertilizers for canola. In: Saskatchewan Canola Development Commission (Project code: CARP 9513). 1998. Available from: http://www.saskcanola.com/quadrant/System/research/reports/report-Germida-nitrogenfixing-long.pdf Accessed: 2017-10-02

[96] Malik KA, Bilal R, Mehnaz S, Rasul G, Mirza MS, Ali S. Association of nitrogen-fixing, plant-growth-promoting rhizobacteria (PGPR) with kallar grass and rice. Plant and Soil. 1997;**194**(1-2):37-44. DOI: 10.1023/A:1004295714181

[97] Araújo AES, Baldani VLD, Galisa PS, Pereira JA, Baldani JI. Response of traditional upland rice varieties to inoculation with selected diazotrophic bacteria isolated from rice cropped at the northeast region of Brazil. Applied Soil Ecology. 2013;**64**:49-55. DOI: 10.1016/j.apsoil.2012.10.004

[98] Schultz N, da Silva JA, Sousa JS, Monteiro RC, Oliveira RP, Chaves VA, Pereira W, da Silva MF, Baldani JI, Boddey RM, Reis VM, Urquiaga S. Inoculation of sugarcane with diazotrophic bacteria. Revista Brasileira de Ciência do Solo. 2014;**38**(2):407-414. DOI: 10.1590/S0100-06832014000200005

[99] Puri A, Padda KP, Chanway CP. Plant growth promotion by endophytic bacteria in nonnative crop hosts. In: Maheshwari DK, Annapurna K, editors. Endophytes: Crop Productivity and Protection. Switzerland: Springer International Publishing; 2017. pp. 11-45. DOI: 10.1007/978-3-319-66544-3_2

[100] Aguiar NO, Medici LO, Olivares FL, Dobbss LB, Torres-Netto A, Silva SF, Novotny EH, Canellas LP. Metabolic profile and antioxidant responses during drought stress recovery in sugarcane treated with humic acids and endophytic diazotrophic bacteria. Annals of Applied Biology. 2016;**168**(2):203-213. DOI: 10.1111/aab.12256

[101] dos Reis AG, da Silva TR, Carvalho BR, Neiva JNM, de Araújo GGL, Júnior PIF. Quantification and characterization of putative diazotrophic bacteria from forage palm under saline water irrigation. Revista Geama. 2017;**3**(4):261-268

[102] Szilagyi-Zecchin VJ, Ikeda AC, Hungria M, Adamoski D, Kava-Cordeiro V, Glienke C, Galli-Terasawa LV. Identification and characterization of endophytic bacteria from corn (*Zea mays* L) roots with biotechnological potential in agriculture. AMB Express. 2014; **4**:26. DOI: 10.1186/s13568-014-0026-y

[103] Andrade LF, De Souza GL, Nietsche S, Xavier AA, Costa MR, Cardoso AM, Pereira MC, Pereira DF. Analysis of the abilities of endophytic bacteria associated with banana tree roots to promote plant growth. The Journal of Microbiology. 2013;**52**(1):27-34. DOI: 10.1007/s12275-014-3019-2

[104] Gupta G, Panwar J, Jha PN. Natural occurrence of *Pseudomonas aeruginosa*, a dominant cultivable diazotrophic endophytic bacterium colonizing *Pennisetum glaucum* (L.) R. Br. Applied Soil Ecology. 2013;**64**:252-261. DOI: 10.1016/j.apsoil.2012.12.016

[105] Rangjaroen C, Rerkasem B, Teaumroong N, Noisangiam R, Lumyong S. Promoting plant growth in a commercial rice cultivar by endophytic diazotrophic bacteria isolated from rice landraces. Annals of Microbiology. 2015;**65**(1):253-266. DOI: 10.1007/s13213-014-0857-4

[106] Ji SH, Gururani MA, Chun SC. Isolation and characterization of plant growth promoting endophytic diazotrophic bacteria from Korean rice cultivars. Microbiological Research. 2014;**169**(1):83-98. DOI: 10.1016/j.micres.2013.06.003

[107] Padda KP, Puri A, Chanway CP. *Paenibacillus polymyxa*—A prominent biofertilizer and biocontrol agent for sustainable agriculture. In: Meena VS, Mishra P, Thakuria D, Bisht J, Pattanayak A, editors. Agriculturally Important Microbes for Sustainable Agriculture. Singapore: Springer; 2017. pp. 165-191. DOI: 10.1007/978-981-10-5343-6_6

[108] Puri A, Padda KP, Chanway CP. Beneficial effects of bacterial endophytes on forest tree species. In: Maheshwari DK, Annapurna K, editors. Endophytes: Crop Productivity

and Protection. Switzerland: Springer International Publishing; 2017. pp. 111-132. DOI: 10.1007/978-3-319-66544-3_6

[109] Bal A, Anand R, Berge O, Chanway C. Isolation and identification of diazotrophic bacteria from internal tissues of *Pinus contorta* and *Thuja plicata*. Canadian Journal of Forest Research. 2012;**42**(4):807-813. DOI: 10.1139/x2012-023

[110] Bal A, Chanway CP. Evidence of nitrogen fixation in lodgepole pine inoculated with diazotrophic *Paenibacillus polymyxa*. Botany. 2012;**90**(9):891-896. DOI: 10.1139/b2012-044

[111] Anand R, Grayston S, Chanway CP. N_2-fixation and seedling growth promotion of lodgepole pine by endophytic *Paenibacillus polymyxa*. Microbial Ecology. 2013;**66**(2):369-374. DOI: 10.1007/s00248-013-0196-1

[112] Yang H, Puri A, Padda KP, Chanway CP. Effects of *Paenibacillus polymyxa* inoculation and different soil nitrogen treatments on lodgepole pine seedling growth. Canadian Journal of Forest Research. 2016;**46**(6):816-821. DOI: 10.1139/cjfr-2015-0456

[113] Anand R, Chanway CP. *nif* gene sequence and arrangement in the endophytic diazotroph *Paenibacillus polymyxa* strain P2b-2R. Biology and Fertility of Soils. 2013;**49**(7):965-970. DOI: 10.1007/s00374-013-0793-7

[114] Yang H, Puri A, Padda KP, Chanway CP. Substrate utilization by endophytic *Paenibacillus polymyxa* that may facilitate bacterial entrance and survival inside various host plants. FACETS. 2017;**2**:120-130. DOI: 10.1139/facets-2016-0031

[115] Bal A, Chanway CP. ^{15}N foliar dilution of western red cedar in response to seed inoculation with diazotrophic *Paenibacillus polymyxa*. Biology and Fertility of Soils. 2012;**48**(8):967-971. DOI: 10.1007/s00374-012-0699-9

[116] Anand R, Chanway C. N_2-fixation and growth promotion in cedar colonized by an endophytic strain of *Paenibacillus polymyxa*. Biology and Fertility of Soils. 2013;**49**(2):235-239. DOI: 10.1007/s00374-012-0735-9

[117] Puri A, Padda KP, Chanway CP. Can a diazotrophic endophyte originally isolated from lodgepole pine colonize an agricultural crop (corn) and promote its growth? Soil Biology and Biochemistry. 2015;**89**:210-216. DOI: 10.1016/j.soilbio.2015.07.012

[118] Puri A, Padda KP, Chanway CP. Evidence of nitrogen fixation and growth promotion in canola (*Brassica napus* L.) by an endophytic diazotroph *Paenibacillus polymyxa* P2b-2R. Biology and Fertility of Soils. 2016;**52**(1):119-125. DOI: 10.1007/s00374-015-1051-y

[119] Padda KP, Puri A, Chanway CP. Effect of GFP tagging of *Paenibacillus polymyxa* P2b-2R on its ability to promote growth of canola and tomato seedlings. Biology and Fertility of Soils. 2016;**52**(3):377-387. DOI: 10.1007/s00374-015-1083-3

[120] Padda KP, Puri A, Chanway CP. Plant growth promotion and nitrogen fixation in canola by an endophytic strain of *Paenibacillus polymyxa* and its GFP-tagged derivative in a long-term study. Botany. 2016;**94**(12):1209-1217. DOI: 10.1139/cjb-2016-0075

[121] Padda KP, Puri A, Zeng Q, Chanway CP, Wu X. Effect of GFP-tagging on nitrogen fixation and plant growth promotion of an endophytic diazotrophic strain of *Paenibacillus polymyxa*. Botany. 2017;**95**(9):933-942. DOI: 10.1139/cjb-2017-0056

[122] Puri A, Padda KP, Chanway CP. Seedling growth promotion and nitrogen fixation by a bacterial endophyte *Paenibacillus polymyxa* P2b-2R and its GFP derivative in corn in a long-term trial. Symbiosis. 2016;**69**(2):123-129. DOI: 10.1007/s13199-016-0385-z

[123] Tang Q, Puri A, Padda KP, Chanway CP. Biological nitrogen fixation and plant growth promotion of lodgepole pine by an endophytic diazotroph *Paenibacillus polymyxa* and its GFP-tagged derivative. Botany. 2017;**95**(6):611-619. DOI: 10.1139/cjb-2016-0300

[124] Rodriguez H, Mendoza A, Cruz MA, Holguin G, Glick BR, Bashan Y. Pleiotropic physiological effects in the plant growth-promoting bacterium *Azospirillum brasilense* following chromosomal labeling in the clpX gene. FEMS Microbiology Ecology. 2006;**57**(2): 217-225. DOI: 10.1111/j.1574-6941.2006.00111.x

7

Inter- and Intra-Annual Variability of Nitrogen Concentrations in the Headwaters of the Mero River

M. Luz Rodríguez-Blanco,
M. Mercedes Taboada-Castro, Ricardo Arias and
M. Teresa Taboada-Castro

Abstract

This study examines the inter- and intra-annual variability of different forms of N [total nitrogen (TN), nitrate-nitrogen (N-NO_3) and total Kjeldahl nitrogen (TKN)] in stream waters of a rural headwater catchment in Galicia (NW Spain) during a 5-year period, covering 2004–2009 water years (October–September). Daily time series were used to verify the temporal variability and to characterize the nitrogen pollution. The TN concentrations were low, although the values constantly exceeded the critical range (0.5–1.0 mg L^{-1}) over which potential risk of eutrophication of water systems exists. Nitrate was the predominant form of nitrogen in the river throughout the study period, accounting for 82–85% of the TN. Significant differences were found for different forms of N between water years and seasons, indicative of wide inter- and intra-annual variability of nitrogen concentrations, mainly related to rainfall and flow oscillations. The seasonal pattern in the concentrations of TN, N-NO_3 and TKN in stream water was similar to many humid and temperate catchments, with higher concentrations in winter, when variability was also the highest in the period, and lower values in summer.

Keywords: nitrogen, variability, rural headwater catchment, temperate humid climate, NW Spain

1. Introduction

A large number of biological, chemical and physical processes are involved in the production, transformation and translocation of N compound in terrestrial and aquatic systems [1]. The most important natural factors controlling nitrogen losses from soils to surface waters are climatology, topography and hydrology, and soil characteristics. For example, the temperature acts on N transformation processes, whereas rainfall distribution and intensity plays a decisive role in the leachate. The distribution of water flow over and through the soil has a great influence on nitrate discharges into the waters [2, 3]. These factors are dynamic and highly variable in space and time, and their action can result in heavy discharges of nitrogen into the watercourses. The impact of natural factors on nitrogen losses can be modified by anthropogenic factors, including soil use and crop rotation, the application of slurry and other fertilizers, as well as the amount and method of application [4–6]. The diffuse pollution of water resources by nitrogen is a problem that affects the whole world, with agriculture being the main primary source of diffuse pollution of water resources [6, 7]. Ultimately, the interaction between the factors involved in the N loss from the soils and those related to N transport determine the final losses of nitrogen to surface waters, as well as the N concentrations and N forms in the rivers; hence, these interactions must be understood in order to manage diffuse losses of agricultural nutrients efficiently.

The dynamics and distribution of N forms in the aquatic system is not easy to understand since N concentrations vary markedly from one catchment to another, depending not only on land use, agricultural management, the type and quantity of fertilizer applied but also on the climatological and hydrological characteristics of the basins [8]. Several authors relate variability in N concentrations to catchment size and, in general, attribute small spatial and temporal variability to small river basins [9, 10]. In addition, the N concentrations display wide variability within the same catchment. For example, for a given rainfall event, the N concentrations may differ greatly at different times of the year [11, 12]. Climate and land use change is expected to alter the transfer of nutrients from land to water, although the results are controversial [13–16]. Some studies show that nitrogen loads will increase under climate change and may be affected equally by climate and land use change [15]. However, other studies indicate that nitrogen loads will decrease under climate change mainly due to decreases in runoff, which may be runoff, may be more affected by climate change than changes in land use [13, 16].

A good knowledge of the temporal variation of nitrogen concentrations is a key for understanding the N transport at the catchment scale, for modelling their behaviour and, ultimately for designing agricultural practices management to reduce potential N losses in a more effective manner under current and changing climate and land uses conditions. Therefore, it is essential to have good data on N for long-term periods. Most of studies on temporal changes of N concentration were mainly focused on catchments with a marked anthropogenic influence and only on one form of N, generally nitrate [9, 10]. Less attention has been paid to understanding the N dynamics at headwater streams with minimal human impact, even though they serve as a reference point against which data from more disturbed catchments can be compared. All this makes clear the need for more detailed long-term studies in small, minimally disturbed, headwater catchments to clarify the behaviour of different forms of N and understand the factors governing the temporal variability of N concentrations. Therefore, there is a clear need to monitor N concentrations in detail over a long-term period to investigate the N dynamics.

In this context, the aim of this study was to investigate the inter-annual and seasonal variations of different forms of nitrogen (TN, N-NO_3 and TKN) concentrations in the headwaters of Mero River basin, a typical groundwater-dominated rural catchment in Galicia (NW Spain) located upstream from the Cecebre reservoir, which has unique characteristics, since it is important both strategically (the only drinking water source of the city of A Coruña and its metropolitan area, approximately 450,000 inhabitants) and ecologically (site of community importance included in the Nature 2000 network). Additionally, the data collected in the catchment study are of great use in evaluating of several models.

2. Material and methods

2.1. Study site description

The study was conducted on a small headwater catchment of the Mero River basin (Galicia, NW Spain; **Figure 1**). The Corbeira has a catchment area of 16 km^2 and a total length of approximately 10 km. The geology of the catchment is homogeneous and dominated by basic metamorphic schist of the Órdenes Complex [17]. The main soil types are umbrisols and cambisols [18], which are relatively deep because of heavy weathering. The surface soil layer has a silt and silt-loam texture, high organic content (2.5–11.5%) and acid pH (4.5–5.6).

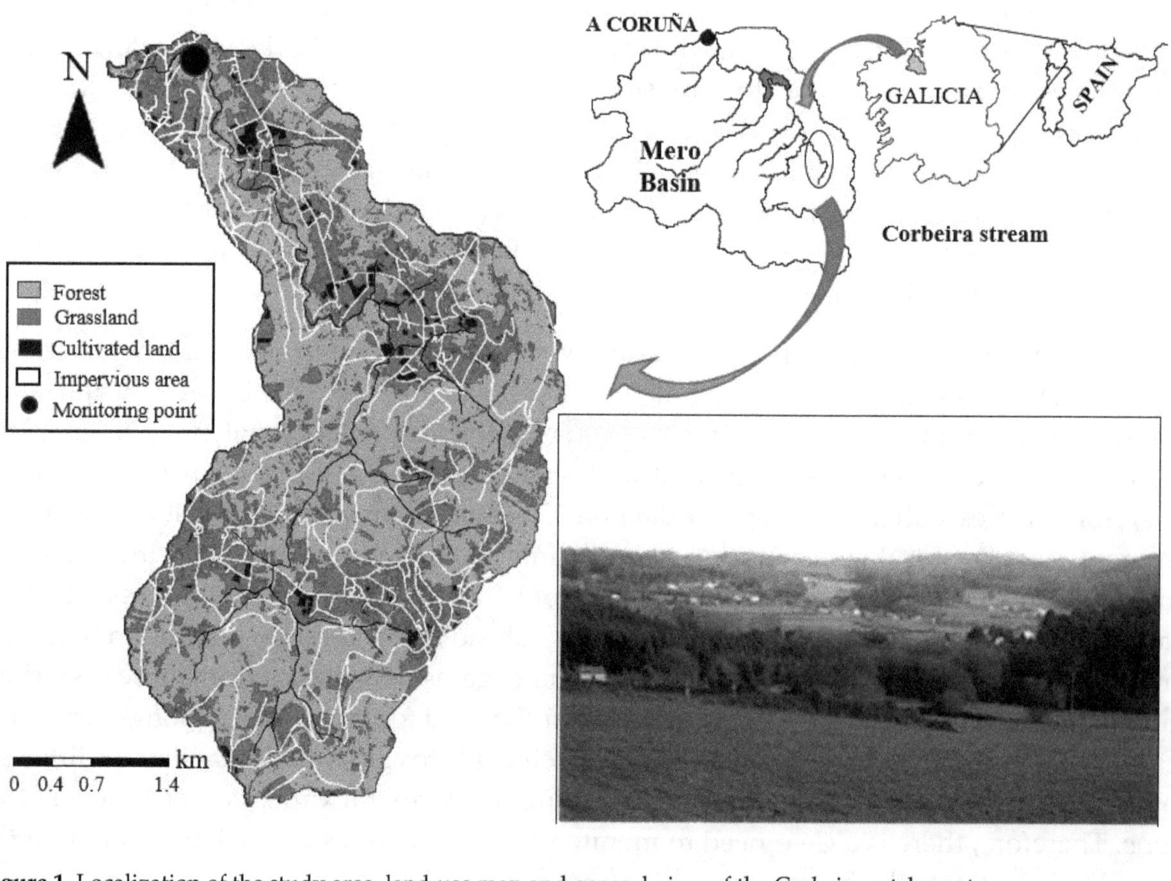

Figure 1. Localization of the study area, land-use map and general view of the Corbeira catchment.

The C/N of the upper layer is almost 10 (9.2–12). The Corbeira catchment is minimally disturbed by human activity; it is predominantly covered by forest (65%) and agricultural land (30%). The agricultural area is primarily used for meadows and natural pastures (83% of agricultural land), with the latter being located in more or less humid areas, mainly in flood areas, where they produce fodder for consumption in green, dry or as silage. Croplands comprise 3.8% of the catchment area and are mainly used for fodder crops, such as maize, and for winter cereals. Other crops—potatoes, turnips, vegetables or fruit trees—are consigned to small family farms oriented to self-consumption. Finally, a small percentage (about 5%) of the catchment is made up of areas occupied by buildings and roads. Agriculture and forestry are interspersed throughout the catchment (**Figure 1**). Forest areas (dominated mainly by commercial eucalyptus and pines) occupy areas with steep slopes. Two distinct agricultural areas can be distinguished in the catchment (**Figure 1**): one is located in the upper part of the catchment, where the slopes are moderately steep, with some cropland near the drainage network and the largest cattle load in the study area, which presents a great risk of erosion, with sediments and nutrients delivery to the stream. The other zone, located on the right margin of the river, in the middle-lower part of the catchment, is where most agricultural activity in the catchment takes place and is where the widest area of cultivated land is found, some with good connectivity to the drainage network. It also highlights the mosaic vegetation in agricultural areas, with most of the croplands bordered by meadows and even, in some cases, by small stone walls, which can obstruct connectivity of croplands with the drainage network, resulting in a decrease in the potential pollutant effect in surface waters.

The Corbeira catchment is a rural area characterized by the presence of small population centres (population density 35 inhabitants km^2). The vast majority of households lack sewerage, so domestic sewage, mainly faecal water, is stored in individual septic tanks. Livestock density was estimated at 0.29 livestock units ha^{-1}, with a predominance of cattle heads (75%). Organic wastes (slurries and manures) generated by agricultural and livestock activity are the most frequently used fertilizers in agriculture in the catchment, with mineral fertilizers supporting these organic ones. Slurries (mainly cattle) are primarily used in meadows and grasslands several times a year, even in the rainy season (October–December), whereas manures are used on croplands, especially in small orchards and in croplands given to maize production, generally prior to planting. Forest areas do not receive fertilizers.

The climate of the study area is Atlantic (temperate oceanic). The mean annual rainfall for the 1983–2009 period reaches a value of 1050 mm. Rainfall is distributed quite evenly throughout the year, although it is concentrated in the autumn and winter months, with October, November and December being the rainiest months and July and August the driest. In general, long-term rainfall causes high rainfall volume but rarely achieves high rainfall intensity, although this can vary substantially, depending on the type of front passing over. The temperature of the study area is characterized by its evenness, with a mean annual of 13°C and a thermal amplitude of 10°C for the period 1983–2009. Concerning the monthly temperature evolution throughout the year, the minimum values occurred in January and the maximum in July. The mean annual discharge of the Corbeira stream amounts to 0.20 m^3 s^{-1}. The hydrological regime is pluvial, with a seasonal pattern in hydrological response characterized by a dry summer period lasting up to first rainfall events in autumn, when the soil reserves recover, followed by a wet period that extends through the autumn and winter seasons [19]. Consequently, the maximum monthly mean discharge is observed in March (0.31 m^3 s^{-1}), whereas the minimum monthly mean discharge is registered in September (0.06 m^3 s^{-1}). The baseflow index, a measure of the proportion of stream flow delivered from groundwater, is about 0.80, indicative of a high contribution of groundwater to stream flow. A more detailed description of the study catchment characteristics can be found in the study of Rodríguez-Blanco et al. [19–21].

3. Hydrological monitoring, sampling and water analysis

Stream discharge and N concentrations have been measured at the catchment outlet from October 2004 to September 2009. Discharge was calculated from measuring water levels (ISCO 720) using the rating curve. Water level data were measured continuously (every min) and recorded at 10-min intervals. Water samples were collected during both baseflow and runoff events. Under baseflow conditions, water samples were manually collected every 10–15 days, whereas under runoff events, they were collected using an automatic sampler (ISCO 6712-FS) at short-time intervals, depending on the characteristics of the runoff events (magnitude, duration).

Stream water samples were analysed for total Kjeldahl nitrogen (TKN), nitrate (NO_3), nitrite (NO_2) and ammonium (NH_4). The TKN concentrations, which represent the sum of ammoniacal and organic nitrogen, were determined by Kjeldahl digestion of unfiltered samples, following the American Public Health Association method [22]. After filtration (0.45 μm), NO_3 and NO_2 concentrations were analysed by capillary electrophoresis, whereas the NH_4+ concentrations were measured using an ammonia-selective electrode. The NO_2 and NH_4 concentrations were below the detection limit in all cases.

Daily flow-weighted mean N concentrations were used to analyse the temporal variability and to characterize the N pollution. Data were organized into four seasons: autumn (October, November and December), winter (January, February and March), spring (April, May and June) and summer (July, August and September). For annual data, the hydrological year was used instead of the calendar year. The flow-weighted mean concentrations were calculated for each year or season by dividing the daily load (N-NO_3 and TKN) by the daily stream flow.

An analysis of variance (ANOVA) was performed to examine the effect of inter-annual variability (five levels: 2004/05, 2005/06, 2006/07, 2007/08 and 2008/09) and intra-annual variability (four levels: autumn, winter, spring and summer) on N (N-NO_3 and TKN) concentrations in the stream. The Tukey test was used as a *post-hoc* test whenever the ANOVA indicated a significance. For all analyses, a significance value of 0.05 was used. All statistical analyses were carried out using the PASW Statistics 18 for Windows program package (SPSS Inc.).

4. Results and discussion

4.1. Weather and discharge

Rainfall and stream discharge were highly variable during the study period (**Table 1**). Annual rainfall for 2004/2005–2008/2009 averaged 1147 mm, 97 mm above the mean of the 1983–2009 (1050 mm), and ranged from 895 mm in 2004/2005 to 1397 mm in 2006/2007. Rainfall distribution throughout the year was also variable. For example, in the hydrological year 2005/2006, more than 80% (963 mm) of annual rainfall occurred in the first half of the year, whereas in 2006/2007 about 55% (750.4 mm) was registered during autumn (October, November and December), reaching record autumn rainfall in the study area. Dry conditions occurred in the second half of the year 2007 and in the autumn 2007/2008, giving rise to the most severe drought episode registered in north-western of Spain over the last half century. Rainfall evolution throughout 2007/2008 was also complex. A dry autumn was followed by a very wet period, mainly January, March, April and May (62% of annual rainfall), in which rainfall events of great magnitude and intensity were registered.

Temperatures during the study period were also variable, although less than for rainfall. The 5-year study period mean temperature was 13.2°C, i.e. 0.2°C above the mean of the 1983–2009 period. Note that 2006/2007 was, together with 2000/2001, the hottest hydrological year in the historical record, whereas 2008/2009 was the coldest in the past 20 years. The annual thermal amplitude, i.e. the difference between the mean temperature of the warmest and the coldest month, varied substantially between years. Thus, the amplitude of the year 2005/2006 (14.1°C) was more than 50% of that registered in the year 2006/2007 (9.2°C). There is a clear seasonality, reaching the lowest mean temperatures in winter and the highest temperatures during the summer (**Figure 2**).

Annual flow, as with rainfall, has a high inter-annual variability, oscillating between 234 mm in 2007/2008 and 449 mm in 2005/2006, with a mean flow during the study period of 364 mm. Direct runoff makes up only a small proportion of streamflow, since it only represents between 20 and 38% of the annual flow. Although the annual flow was highest in 2005/2006, direct runoff was higher in 2006/2007 and 2008/2009 than in 2005/2006, i.e. in the wettest hydrological years, reflecting the role of soil water content in the runoff generation, as has been shown in a previous study in this catchment [19]. Streamflow also showed a clear seasonality, with maximum flows occurring in January–February and minimum flows occurring in September–October (**Figure 2**). This behaviour is typical of baseflow-dominated catchments, such as the Corbeira, resulting from the seasonality in rainfall distribution, although there was a lack of linearity between rainfall and streamflow attributed to differences in soil water balance [19].

	2004/2005	2005/2006	2006/2007	2007/2008	2008/2009
Rainfall (mm)	895	1192	1397	1054	1196
Temperature (°C)	13.2	13.5	13.7	13.0	12.6
Streamflow (mm)	298	449	412	240	421

Table 1. Annual variability of rainfall, temperature and streamflow during the five hydrological years of the study period.

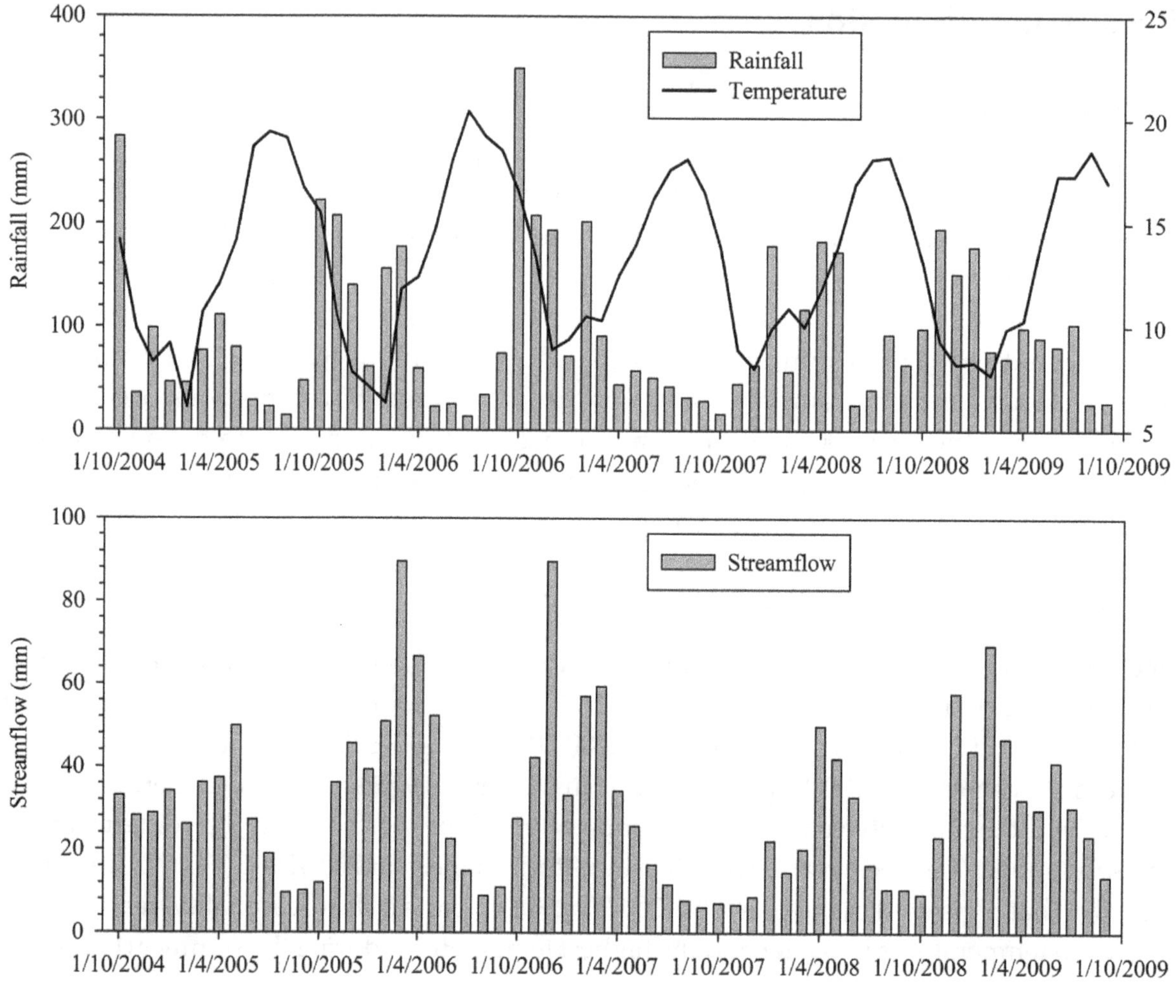

Figure 2. Monthly variation of rainfall, temperature and streamflow during the five hydrological years of the study period.

4.2. Inter-annual variation of N concentrations

Over the study period, the mean annual flow-weighted concentrations were 1.40, 1.15 and 0.26 mg L^{-1} for TN, N-NO_3 and TKN, respectively (**Table 2**). These low N concentrations are typical of streams draining minimally perturbed catchments, i.e. with low N deposition, high percentage of forest area and low agriculture activity, as is the case of Corbeira catchment, so the concentrations there were lower than typical concentrations for catchments with more intense land use, such as agriculture and urban. They were also lower than those reported by Serrano et al. [23] for the Mero basin headwater (mean concentrations of 2.23, 1.88 and 0.35 mg L^{-1} for TN, N-NO_3 and TKN, respectively), which shows similar climatic characteristic, landform and soil management of the study area, but with a higher percentage of agricultural land and livestock density. However, they were considerably higher than the levels in minimally disturbed, natural water draining areas in the north-western region of Spain [24] as well as the reference values proposed by Meybeck et al. [25] for pristine river systems, suggesting some N enrichment in the stream water, probably as a result of the agricultural practices in the catchment. In the Corbeira catchment, in addition to its presence as a natural component, N derives from agriculture, mainly because agricultural lands receive

	2004/2005	2005/2006	2006/2007	2007/2008	2008/2009
TN (mg L $^{-1}$)	1.30 (0.36)	1.50 (0.57)	1.37 (0.40)	1.39 (0.55)	1.45 (0.34)
N-NO$_3$ (mg L^{-1})	1.09 (0.17)	1.21 (0.29)	1.16 (0.22)	1.13 (0.27)	1.15 (0.22)
TKN (mg L^{-1})	0.21 (0.29)	0.29 (0.38)	0.21 (0.24)	0.28 (0.38)	0.30 (0.22)

Table 2. Flow-weighted TN, N-NO$_3$, TKN concentrations and standard deviation during the five hydrological years of study period.

a contribution of nitrogen fertilizers considerably higher than that of the forest areas and the N entry of N coming from the population nuclei is low.

The TN, N-NO$_3$ and TKN concentrations differed among the observed years (**Table 2**). The concentrations of TN ranged from 1.30 to 1.49 mg L^{-1}, whereas N-NO$_3$and NTKN contents varied from 1.09 to 1.21 mg L^{-1} and from 0.21 and 0.29 mg L^{-1}, respectively. For all N forms, the minimum values were registered in 2004/2005, an exceptionally dry water year with rainfall 17% lower than the mean rainfall for the period 1983–2009. Maximum concentrations were recorded in 2006/2007 for N-NO$_3$ and TN, i.e. in the rainiest water year of the study period, and in 2008/2009 for TKN (highest SS yield, data not shown). The observed changes in stream N concentrations most likely reflected the effects of change in hydro-climatic factors, such as hydrology of the catchment and temperature, because potential drivers of N, such as land use and soil management have not varied substantially in the area over the last few decades and, in consequence, fertilization rates have been stable during the monitoring period. However, the trend in mean annual stream N (both N-NO$_3$ and TKN) concentrations were not synchronous with changes in stream flow (**Table 1**), suggesting that changes in the timing and magnitude of rainfall episodes could lead to changes in the relationship between N transport per unit of water volume, in agreement with other studies [9, 26], as they affect N availability and transformation processes.

All of the water samples had higher N-NO$_3$ concentrations than TKN, so the N-NO$_3$ was the predominant form of N in the study catchment, accounting for, on average, 82% of the TN (80–85%). These rates are consistent with those commonly recorded in agricultural and rural groundwater-dominated catchments worldwide, with organic nitrogen comprising 10–20% of TN [23, 27]. However, the fractionation in this study contrasts with results obtained in a widely variety of catchments minimally disturbed catchments with TN concentrations lower than 2 mg L^{-1}, where frequently more than 80–90% of TN is commonly delivered in the form of organic N [28, 29], constituting a substantial component of TN in many ecosystems receiving low N enrichment. Mean annual N-NO$_3$ concentrations as a percentage of TN decreased from 85% in 2006/2007 to 80% in 2008/2009, the water year of highest particulate material which could favour higher N transport in organic form.

Daily flow-weighted TN concentrations were consistently below the threshold level of 2 mg L^{-1}, threshold identified in the European Nitrogen Assessment [30] as an appropriate target for the delivery of good ecological status in European water. However, daily flow-weighted TN concentrations exceeded the critical range of 0.5–1.0 mg N L^{-1} above which there is a potential risk of eutrophication of water systems [31], highlighting a possible negative effect on agricultural practices in the stream waters.

4.3. Intra-annual variation of N concentrations

The daily TN, N-NO_3 and TKN concentrations were characterized by a marked intra-annual variability (**Figure 3**), with the highest concentrations (and standard deviation) in winter, when the supply of N on both landscape and streamflow is usually abundant (**Figure 4**). Fertilization in the catchment takes place in autumn and spring. The lowest concentrations were obtained in summer, in combination with low streamflow, the maximum baseflow contribution to streamflow and higher biological activity within the catchment and stream, reducing the flushing of leaching soil NO_3 and thus reducing the N concentrations in the stream. Significant differences were detected between seasons, except for N-NO_3 between autumn and winter ($p > 0.05$) and for TKN between autumn and spring ($p > 0.05$) and between spring and summer (0.05). This pattern was also observed in many other rivers draining forest and agricultural catchments, irrespective of the importance of groundwater to river flow volume [32], and it is frequently related to water moving through the soil, less plant nitrogen uptake and microbial immobilisation during winter in addition to more intensive erosion, which all contribute to higher N levels during the winter. However, this general pattern varied among years depending on the weather and discharge patterns. For examples, in the 2008/2009 hydrological year, TN, N-NO_3 and TKN concentrations were significantly lower in spring than summer, when it reached the highest daily flow-weighted mean TKN concentration, which was notably influenced by two large rainfall-runoff events in July of 2009. The summer of the 2008/2009 water year was characterized by elevated rainfall in July (170% higher than mean rainfall for July in the period 1983–2009) concentrated in two intense rainfall events, which caused high surface runoff and particulate material delivery to the stream [11] and consequently high TKN concentrations were observed. However, the very dry autumn 2007/2008 led to very low TKN concentrations, very similar to those observed in summer. These results suggest that temperature-dependent biological uptake is not the more important factor controlling the seasonal nitrogen dynamics as frequently it occurs in more natural systems [28], but that delivery of nitrogen to the stream is driving the dynamics, with maximum concentrations when N availability and flow, and thus the transport capacity, is at their highest.

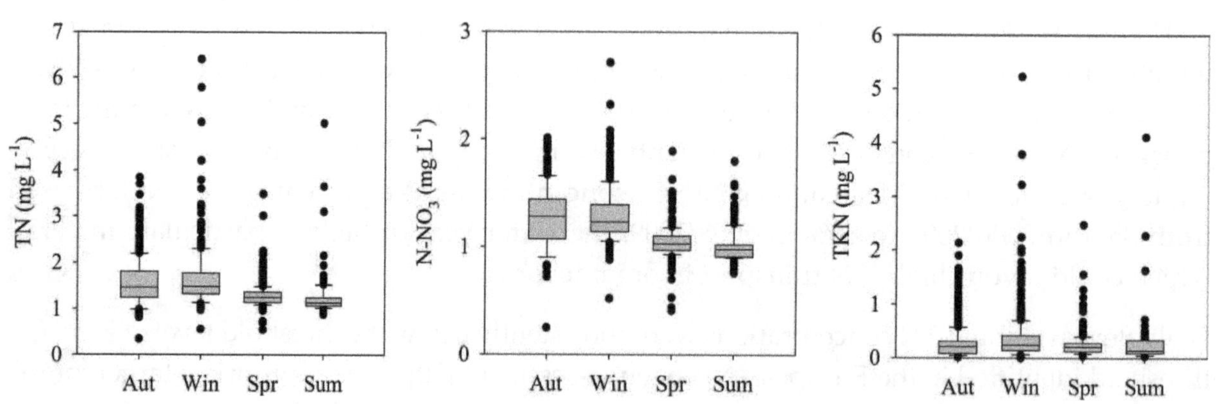

Figure 3. Box plot showing nitrogen seasonal variations. Aut: autumn, Win: winter, Spr: spring, Sum: summer.

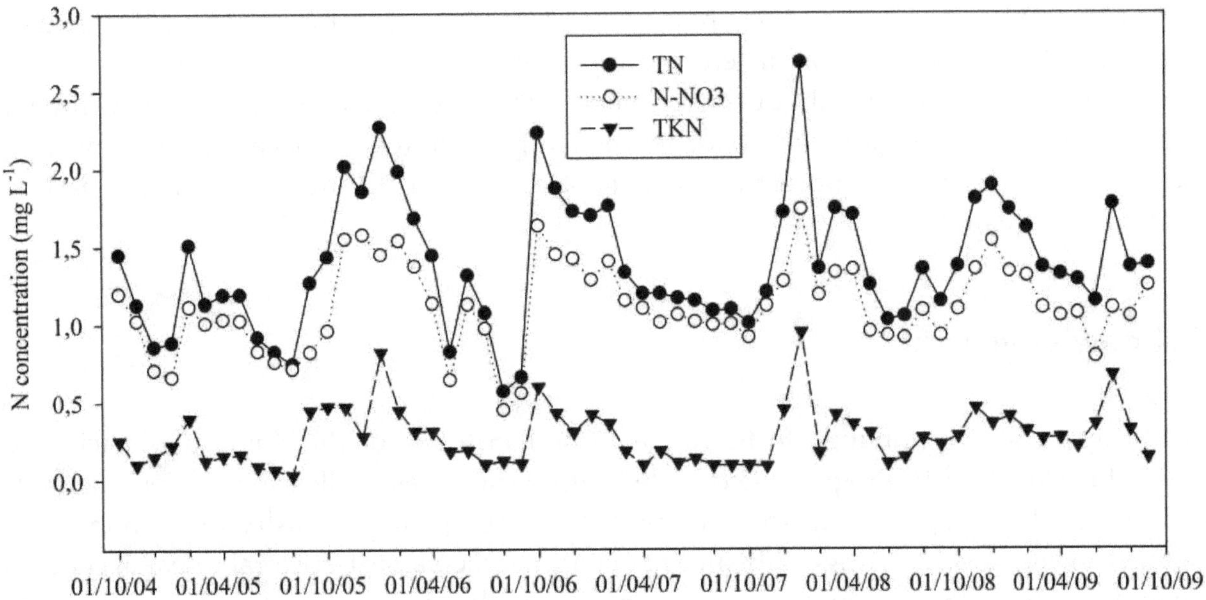

Figure 4. Variation of monthly flow-weighted nitrogen (N-NO_3 and TKN) concentrations during the five hydrological years of study period.

The contribution of the individual fractions made to the TN also showed an intra-annual variability. Thus, N-NO_3 contributed to more than 83% of TN during autumn and spring, representing about 86% in summer, but this contribution decreases during winter (77%), i.e. in the season with higher streamflow. This reflects an increase in the organic nitrogen during winter, probably due to more intense erosion during high flow periods.

In all seasons, a positive relationship exists between daily N concentrations and flow, suggesting a predominantly diffuse source of N. However, the relationship was highly variable between seasons. Thus, in autumn and winter, N-NO_3 was more strongly related to stream flow, whereas in spring and summer, TKN was more strongly related to flow, as expected in view of the role of shear stress in moving particles. These results suggest that although N-NO_3 and TKN may display a similar seasonal pattern, controls of N-NO_3 differ substantially from controls on TKN. Nitrate is originated from subsurface paths whereas TKN is delivered from surface flow paths.

5. Conclusions

Total nitrogen concentrations were relatively low, consistent with the lack of significant anthropogenic pressure in the catchment. However, the N concentrations were always over the critical range over which there is a potential risk of eutrophication of water systems. Most of the N flushed from the Corbeira catchment is in the form of nitrate (82–85%), so management practices could target nitrate, particularly that from fertilizer. In the study area, rainfall and stream flow change annually and seasonally, so the total nitrogen, nitrate and Kjeldahl nitrogen concentrations also showed an important inter- and intra-annual variability linked to rainfall and flow oscillations; showing that N concentrations are mainly controlled by hydrology processes. These findings have particular relevance in the current global context of climate change, which is predicted to bring more frequent heat waves and an increase extreme rainfall events (mainly during winter) in the study area. These changes will have impacts on the hydrological cycle and, consequently, in nitrogen concentrations, and may increase the transfer of nitrogen from soils to stream water during winter and impact on stream water quality.

Acknowledgements

This chapter is a contribution to the projects 10MDS103031 of the Xunta de Galicia and CGL2014-56907-R of the Programa Estatal de Investigación, Desarrollo e InnovaciónOrientada a los Retos de la Sociedad, which was funded by the Spanish Ministry of Economy and Competitiveness. M.L. Rodríguez-Blanco has been awarded a post-doctoral research contract (Juan de la Cierva Programme), which was funded by the Spanish Ministry of Economy and Competitiveness.

Author details

M. Luz Rodríguez-Blanco*, M. Mercedes Taboada-Castro, Ricardo Arias and
M. Teresa Taboada-Castro

*Address all correspondence to: mrodriguezbl@udc.es

Faculty of Sciences, Centre for Advanced Scientific Research (CICA), University of A Coruna, A Coruña, Spain

References

[1] Qualls RG, Richarson CJ. Factors controlling concentration, export, and decomposition of dissolved organic nutrients in the Everglades of Florida. Biogeochemistry. 2003;**62**(2):197-229

[2] Jordan TE, Correl DL, Weller DE. Relating nutrient discharges from watersheds to land use and streamflow variability. Water Resources Research. 1997;**33**(11):2579-2590

[3] Outram FN, Cooper RJ, Sünnenberg G, Hiscock KM, Lovett AA. Antecedent conditions, hydrological connectivity and anthropogenic inputs: Factors affecting nitrate and phosphorus transfers to agricultural headwater streams. Science of the Total Environment. 2016;**545-546**:184-199

[4] Myrbeck A, Stenberg M, Arvidson J, Rydberg T. Effects of autumn tillage of clay soil on mineral N content, spring cereal yield and soil structure over time. European Journal of Agronomy. 2012;**37**:96-104

[5] Taboada-Castro MM, Rodríguez-Blanco ML, Taboada-Castro MT. Assessing the influence of catchment land use patterns on the nutrients and major ions chemistry of drainage water. Communications in Soil Science and Plant Analysis. 2013;**44**(1-4):422-428

[6] Jones C, Kim SW, Schilling K. Use of continuous monitoring to assess stream nitrate flux and transformation patterns. Environment Monitoring Assessment. 2017;**189**(35). DOI: 10.1007/s10661-016-5749-6

[7] Lam QD, Schmalz B, Fohrez N. Assessing the spatial and temporal variations of water quality in lowland areas, Northern Germany. Journal of Hydrology. 2012;**428-439**:137-147

[8] Álvarez-Cobelas M, Angeler DG, Sánchez-Carillo S. Export of nitrogen from catchments: A worldwide analysis. Environmental Pollution. 2008;**156:**261-269

[9] Rusjan S, Vidmar A. The role of seasonal and hydrological conditions in regulating dissolved inorganic nitrogen budgets in a forest catchment in SW Slovenia. Science of the Total Environment. 2017;**575**:1109-1118

[10] Duncan JM, Band LE, Groffman PM, Bernhardt ES. Mechanisms driving the seasonality of catchment scale nitrate export: Evidence for riparian ecohydrologic controls. Water Resources Research. 2015;**51**(6):3982-3997

[11] Rodríguez-Blanco ML, Taboada-Castro MM, Taboada-Castro MT, Oropeza-Mota J. Relating nitrogen export patterns from a mixed land use catchment in NW Spain. Hydrological Processes. 2015;**29**:2720-2730

[12] Dupas R, Jomaa S, Musolff A, Borchardt D, Rode M. Disentangling the influence of hydroclimatic patterns and agricultural management on river nitrate dynamics from sub-hourly to decadal time scales. Science of the Total Environment. 2016;**571**:791-800

[13] Whitehead PG, Wilby RL, Battarbee RW, Kernan M, Wade L. A review of the potential impacts of climate change on surface water quality. Hydrological Sciences. 2009;**54**:101-103

[14] Gabriel M, Knightes C, Cooter E, Dennnis R. Evaluating relative sensitivity of SWAT-simulated nitrogen discharge to projected climate and land cover changes for two watersheds in North Carolina, USA. Hydrological Processes. 2016;**30**:1403-1418

[15] Trang NTT, Shrestha S, Shrestha M, Datta A, Kawasaki A. Evaluating the impact of climate and land use change on the hydrology and nutrient yield in a transboundary river basin: A case study in the 3S river basin. Science of the Total Environment. 2017;**576**:586-598

[16] Serpa D, Nunes JP, Keizer JJ, Abrantes N. Impacts of climate change and land use changes on the water quality of a small Mediterranean catchment with intensive viticulture. Environmental Pollution. 2017;**224**:454-465 DOI: /10.1016/j.envpol.2017.02.026

[17] IGME (Instituto Tecnológico Geominero de España). Mapa Geológico de España, 1:50000. Hoja 45. Betanzos. Madrid: Servicio de Publicaciones del Ministerio de Industria y Energía; 1981

[18] IUSS Working Group WRB. World Reference Base for Soil Resources 2014, update 2015. International soil classification system for naming soils and creating legends for soil maps. Roma: World Soil Resources Reports N 106. FAO; 2015. p. 203

[19] Rodríguez-Blanco ML, Taboada-Castro MM, Taboada-Castro MT. Rainfall-runoff response and event-based runoff coefficients in a humid area (northwest Spain). Hydrological Sciences Journal. 2012;**57**:444-459

[20] Rodríguez-Blanco ML, Taboada-Castro MM, Palleiro L, Taboada-Castro MT. Temporal changes in suspended sediment transport in and Atlantic catchment, NW Spain. Geomorphology. 2010;**123**:181-188

[21] Rodríguez-Blanco ML, Taboada-Castro MM, Taboada-Castro MT. Linking the field to the stream: Soil erosion and sediment yield in a rural catchment, NW Spain. Catena. 2013;**102**:74-81

[22] APHA. Standard Methods for the Examination of Water and Wastewater. 20th ed. Washington: American Public Health Association; 1998. p. 1161

[23] Serrano L, Palleiro L, Rodríguez-Blanco ML, Taboada-Castro MM, Taboada-Castro MT. A preliminary assessment of the riverine nitrogen concentration and load in a agroforestry catchment. Communications in Soil and Plant Analysis. 2015;**46**:326-331

[24] Antelo JM, Arce F. Características fisicoquimicas das aguas superficiais. In: Díaz-Fierros, editor. As Augas de Galicia. Spain: Santiago de Compostela; 1996. pp. 351-446

[25] Meybeck MG, Friedrich R, Thomas R, Chapman D. Rivers. In: Chapman D, Hall, editor. Water Quality Assessments: A Guide to Use of Biota, Sediments and Water in Environmental Monitoring. Cambridge; 1992

[26] Bhat S, Hatfield K, Jacobs JM, Lowrance R, Williams R. Surface runoff contribution of nitrogen during storm events in a forest catchment. Biogeochemistry. 2007;**85**:253-262

[27] Halliday SJ, Skeffington RA, Wade AJ, Neal C, Reynolds B, Norris D, Kirchner JW. Upland streamwater nitrate dynamics across decadal to sub-daily timescales: A case study of Plynlimon, Wales. Biogeosciences. 2013;**10**:8013-8038

[28] Kaushal SS, Lewis WM. Patterns in the chemical fractionation of organic nitrogen in Rocky Mountain streams. Ecosystems. 2003;**6**:483-492

[29] Sarkkola S, Nieminen M, Koivusalo H, Laurén A, Kortelainen P, Mattsson T, Palviainen M, Piirainen S, Starr M, Finér L. Trends in concentrations and export of nitrogen in boreal forest streams. Boreal Environment Research. 2012;**17**:85-101

[30] Durand P, Breuer L, Johnes P, Billen G, Butturini A, Pinay G, Grinsven H,Garnier J, Rivett M, Reasy D, Curtis C, Siemens J, Marbely S, Kaste Ø, Humborg C, Loeb R, Klein J, Hejzlar J, Skoulikidis N, Kortelainen P, Lepistö A, Wright R. Nitrogen processes in aquatic ecosystems. In: Sutton MA, Howard CM, Erisman JW, editors. The European Nitrogen Assessment. Cambridge: Cambridge University Press; 2011. pp. 126-146

[31] Camargo JA, Alonso A. Ecological and toxicological effects of inorganic nitrogen pollution in aquatic ecosystems: A global assessment. Environment International. 2006;**32**:831-849. DOI: 10.1016/j.envint.2006.05.002

[32] Harrington ST, Harrington JR. Dissolved and particulate nutrient transport dynamics of a small Irish catchment: The River Owenabue. Hydrology and Earth System Sciences. 2014;**18**:2191-2200

8

Nitrogen use Efficiency in Rice

Shuangjie Huang, Chunfang Zhao, Yali Zhang and
Cailin Wang

Abstract

Food security is a major global issue because of the growing population and decreasing land area. Rice (*Oryza sativa* L.) is the most important staple cereal crop in the world. Application of nitrogen (N) fertilizer has improved crop yield in the world during the past five decades but with considerable negative impacts on the environment. New solutions are therefore urgently needed to simultaneously increase yields while maintaining or preferably decreasing applied N to maximize the nitrogen use efficiency (NUE) of crops. Plant NUE is inherently complex with each step (including N uptake, translocation, assimilation, and remobilization) governed by multiple interacting genetic and environmental factors. Based on the current knowledge, we propose some possible approaches enhancing NUE, by molecular manipulation selecting candidate genes and agricultural integrated management practices for NUE improvement. Developing an integrated research program combining approaches, mainly based on whole-plant physiology, quantitative genetics, forward and reverse genetics, and agronomy approaches to improve NUE, is a major objective in the future.

Keywords: rice, nitrogen use efficiency, nitrate, ammonium, N uptake, N assimilation, N remobilization

1. Introduction

The global population is predicted to reach 9 billion, and food supplies are projected to increase by 70–100% by 2050 [1, 2]. Given the limited capacity for arable land expansion, it requires sustaining yield improvement in existing land to meet the increasing food demand [3]. Rice is one of the staple food crops for approximately half of the global population. Therefore, rice production must be increased significantly to satisfy the requirements of the growing world population. However, we are facing challenges in increasing rice production under the pressures of decreased arable land area, global climate change, intensified natural disasters, and frequent occurrence of diseases and pests [4]. Nitrogen (N) is one of the essential macroelements required for plant growth and development. Soil N availability usually limits plant yields in most agricultural cropping systems [5]. Thus, application of N fertilizer has become an important, cost-effective strategy to increase crop yields in intensive agricultural systems worldwide [6]. However, traditionally adding N fertilizer to improve crop yields may have reached a plateau. Excessive application of nitrogen fertilizer may not result in yield improvements but will lead to serious environmental problems [7, 8]. From 1960 to 2012, the global N fertilizer consumption increased by 800%, and the annual N consumption in China increased from 8 to 35% of the world's N consumption [4]. Although the rate of cereal grain yield increased by 65% between 1980 and 2010, the consumption of chemical fertilizers increased by 512% [9]. High N fertilizer input leads to low nitrogen use efficiency (NUE) due to the rapid N losses from ammonia volatilization, denitrification, surface runoff, and leaching in the soil-flood water system. As a consequence, significant environmental problems (i.e., soil acidification, air pollution, water eutrophication) occurred [10–12]. To achieve further high crop productivity and high NUE under well-fertilized conditions, new solutions are urgently needed to increase yields while maintaining or preferably decreasing applied N [13].

In this chapter, we outlined the definition of NUE, the genes related to NUE, as well as the effect of the factors on the expression of those genes, with an emphasis on rice research. Based on the current knowledge, we proposed some possible strategies enhancing NUE, by breeding, molecular manipulation selecting candidate genes, and developing a range of optimized crop management practices for NUE improvement.

2. Defining nitrogen use efficiency

NUE is inherently complex determined by the interaction of multiple genes with the environment factors. A number of different definitions and calculations of NUE include N utilization, N content, and N availability as NUE equation components (**Table 1**) [13, 14]. In general, plant NUE comprises two key components: N uptake efficiency (NUpE), which is the efficiency of absorption/uptake of supplied N, and N utilization efficiency (NUtE), which is the efficiency of assimilation and remobilization of plant N to ultimately produce grain [13, 14]. The simplest definition of plant NUE is the grain yield per unit of supplied N, also an integration of NUpE and NUtE. Another method to describe NUE is the utilization index (UI), which means the absolute amount of biomass produced per unit of N. NUE can also be described as NUEg, which is grain production per unit of N available, and HI, which is grain production of the total plant biomass. However, a crop plant could produce large amounts of biomass per unit N (high UI) without converting the acquired N to seed production and therefore have a low NUEg and HI. There are other NUE calculations taking various agronomic and physiological variations into account described elsewhere [14–16]. In summary, improving NUE could be achieved by improving either NUpE, NUtE, or both. However, owing to the fluctuations in the rhizosphere that influenced by microorganism, root exudates, and the volatile loss of gaseous N from the soil/plant canopy, it is difficult to quantify the "real" amount of N fertilizer available or actually acquired by plants.

Abbreviation	Term	Definitions
NUE	N use efficiency	NUpE × NUtE = yield/N available
NUpE	N uptake efficiency	NUp/Nav (soil + fertilizer) = acquired N/N available
NUtE	N utilization efficiency	Yield/NUp (assimilation of plant N to produce grain)
NUEg	N use efficiency of grain	Grain production/available N
ANR	Apparent N recovery rate	Net increased total N uptake by the plant with and without N fertilization/total amount of fertilizer N
AE	Agronomy N efficiency	Net increased yield of the plant with and without N fertilization/total amount of fertilizer N
NpUE	N physiological use efficiency	Net increased yield/net increased N uptake with and without application of fertilizer N
NTE	N transport efficiency	Total N transported into the aboveground parts/total N in the whole plant
UI	Utilization index	Total plant biomass/total plant N
FUE	Fertilizer use efficiency	(NUp/N applied) × 100
HI	Harvest index	Grain weight(vegetative organ weight + grain weight)
NHI	Nitrogen harvest index	Grain N accumulation/total N accumulation in aboveground biomass (e.g., grain + straw)
NRE	Nitrogen remobilization efficiency	N remobilization from source or senescent leaves/that of sink leaves or developing grains (seeds)

Table 1. Some definitions of NUE mostly used with respect to nitrogen.

3. Genes responsible for nitrogen use efficiency

Generally, NUE can be divided into two parts: assimilation efficiency involved in N uptake and assimilation, and utilization efficiency involved in N remobilization. Understanding the mechanisms regulating these processes is crucial for improving crop NUE. In soil, inorganic N is available for plants as nitrate (NO_3^-) in aerobic uplands and ammonium (NH_4^+) in flooded wetland or acidic soils. Rice roots in paddy soils release oxygen via their aerenchyma, generate rapid nitrification on their surface, and thus absorb N as NO_3^- at a rate comparable with that of NH_4^+ uptake [17, 18]. Direct molecular evidence for NO_3^- uptake in rice has been presented [19]. NH_4^+ or NO_3^- uptake by roots commonly results in acidification or alkalization of the rhizosphere, which in turn changes the soil N availability [14]. For many plants, some NO_3^- taken up by nitrate transporters (NAR2/NRTs) is assimilated in the roots, the other larger part transported to the shoots, where it is reduced to ammonium by a range of enzymes (**Figure 1**). The NH_4^+ derived from NO_3^- or directly from NH_4^+ uptake by ammonium transporters (AMTs) is assimilated into amino acids via the glutamine synthetase (GS)/glutamine-2-oxoglutarate aminotransferase (GOGAT) cycle and then is exported to sink organs [14]. Therefore, regulating gene function in N metabolism processes including N uptake, assimilation, compartmentation, translocation, and remobilization may be essential for improving NUE.

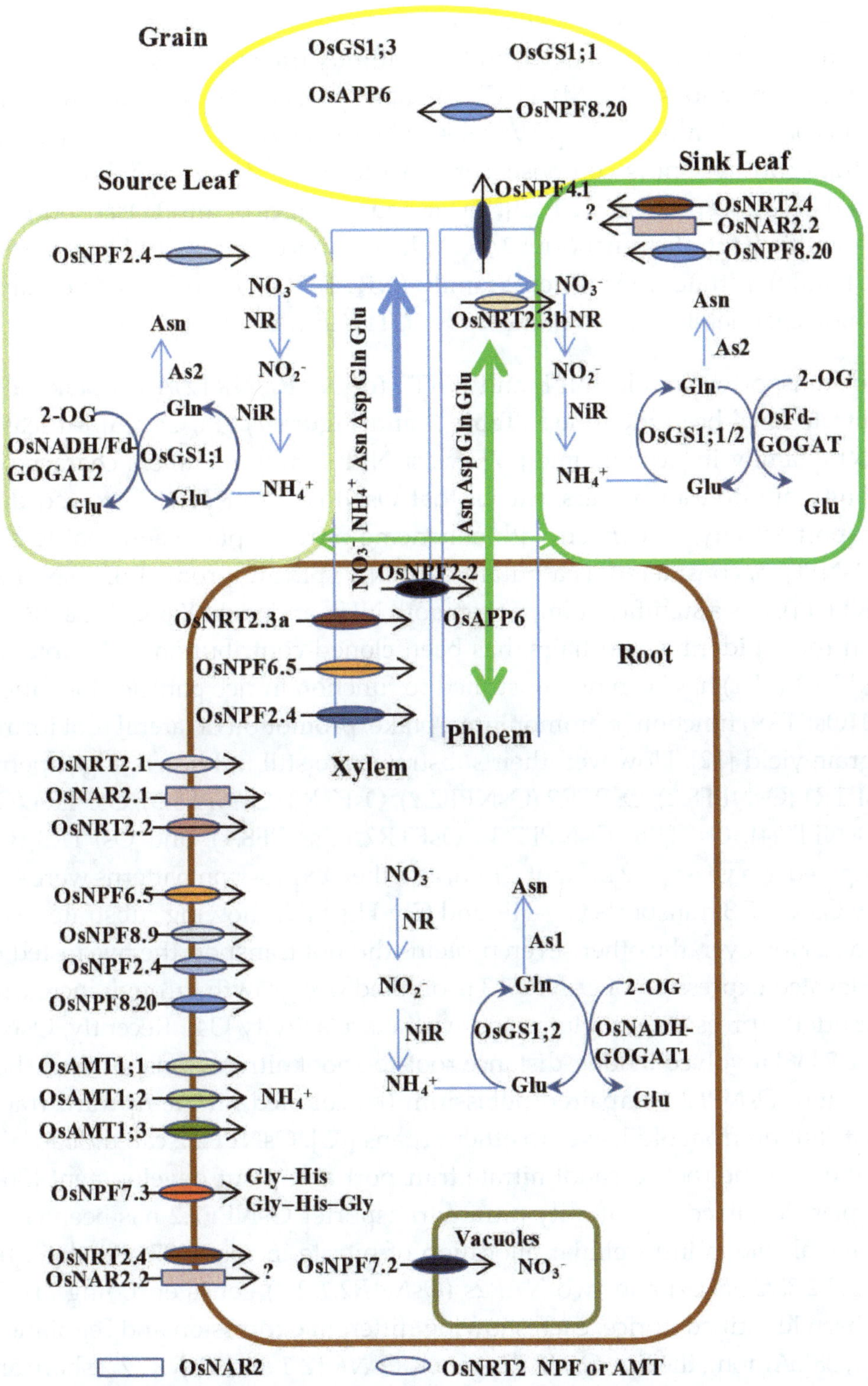

Figure 1. Schematic representation of characterized and predicted functions of the rice nitrate transporters of NRT and NPF families, ammonium transporters of AMT families, and nitrogen assimilation proteins of GS and GOGAT.

3.1. Nitrogen acquisition

Owing to the heterogeneity and dynamic variations of nitrate and ammonium concentrations, which range from lower than 100 μM to higher than 10 mM in soil solutions, plants have developed transporters for both nitrate and ammonium. These transporters are divided into high-affinity transporter system (HATS) and low-affinity transport system (LATS) [20]. Under low nitrogen concentrations (<1 mM), HATS mediates most of the N uptake, while under high concentrations of N (>1 mM), LATS plays roles in N uptake [21, 22]. Each high- and low-affinity nitrate transport system is composed of constitutive and nitrate-inducible components (cHATS and iHATS), respectively [20, 23]. So far, four families of nitrate transporters/channels have been identified: nitrate transporter 1/peptide transporter family (NPF, also known as the NRT1/PTR family), nitrate transporter 2 family (NRT2), the chloride channel family (CLC), and slow anion channel-associated homologues (SLAC/SLAH) [24].

In rice, two transporter families NPF and NRT2 (or NAR2/NRT2) for uptake and translocation of nitrate have been identified (**Table 2** and **Figure 1**) [14, 25, 26]. At least 80 genes belong to NPF family in rice genome [27]. Most NPF family members characterized so far are low-affinity nitrate transporters, except that OsNPF6.5 (NRT1.1b) showed dual-affinity nitrate transport activity, associated with enhancing nitrate uptake and root-to-shoot transport [28]. OsNPF6.5, considered as a putative mRNA splicing product of OsNPF8.9 (NRT1/NRT1.1/NRT1.1a), has a significant impact on both NUE and yield [26–29]. OsNPF8.9, mainly expressed in root epidermis and hairs, has been cloned contribution to N uptake [30]. The role of OsNPF4.1 (SP1) has been demonstrated to function in rice panicle elongation [31] and OsNPF8.20 (OsPTR9) function in ammonium uptake, promotion of lateral root formation, and increased grain yield [32]. However, their substrates are still unknown. Eight peptide transporters, OsPTR1 (OsNPF8.2), OsPTR2 (OsNPF2.2), OsPTR3 (OsNPF5.5), OsPTR4 (OsNPF7.1), OsPTR5 (OsNPF7.4), OsPTR6 (OsNPF7.3), OsPTR7 (OsNPF8.1), and OsPTR8 (OsNPF8.5), were investigated in a yeast ptr2 mutant strain, and their expression patterns were evaluated in plants. Only OsNPF7.3 transports Gly-His and Gly-His-Gly, showing substrate selectivity for di-/tripeptides. However, the other seven proteins did not transport the five tested di-/tripeptides [33]. Elevated expression of *OsNPF7.3* promoted rice growth through increasing ammonium transporter expression and glutamine synthetase activity [34]. Recently, OsNPF2.4 [35] and OsNPF2.2 [36] involved in long-distance root-to-shoot nitrate transport have been identified. Knockout of *OsNPF2.4* impaired potassium (K)-coupled nitrate upward transport and nitrate redistribution from old leaves to other organs [35]. OsNPF2.2 can unload nitrate from the xylem affecting the root-to-shoot nitrate transport and plant development [36]. In addition, a tonoplast-localized low-affinity nitrate transporter OsNPF7.2 has been characterized playing a pivotal role in intracellular allocation of nitrate in roots [37]. To date, five *NRT2s* (*OsNRT2.1/2.2/2.3a/2.3b/2.4*) and two *NAR2s* (*OsNAR2.1/2.2*) genes encoding HATS components have been identified in rice, each showing different expression and regulation patterns (**Table 2**) [19, 38]. Among the five *OsNRT2s* genes, *OsNRT2.1* and *OsNRT2.2* share an identical coding region sequence with different 5′- and 3′-untranscribed regions [38–40]. *OsNRT2.3a* and *OsNRT2.3b* are derived from the alternative splicing of *OsNRT2.3* [38]. *OsNRT2.3a* is mainly expressed in the xylem parenchyma of root participating in long-distance nitrate transport from root to shoot at low nitrate concentrations [41]. *OsNRT2.3b* is mainly expressed in the phloem of shoot, sensitive to pH. Elevated expression of *OsNRT2.3b* increased N, Fe, and P uptake and improved grain yield and NUE [42]. *OsNAR2.1*, *OsNRT2.1*, and *OsNRT2.2* were expressed abundantly throughout the primary and lateral roots. Overexpression of *OsNRT2.1*

Accession no.	Gene	Regulation	Expression pattern	Substrates	References
AF140606	*OsNPF8.9*	Unknown	Constitutively expressed in roots	NO_3^-	[29, 30]
AK066920	*OsNPF6.5*	NO_3^-	Root hairs, epidermis, and vascular tissues	NO_3^-	[28, 29]
AK099321.1	*OsNPF2.4*	NO_3^-	Root epidermis, xylem parenchyma, and phloem companion cells, leaf phloem cells	NO_3^-	[35]
AK068351	*OsNPF2.2*	NO_3^-, drought, salt	Parenchyma cells around the xylem	NO_3^-	[36]
XM_015767550	*OsNPF7.2*	NO_3^-	Root sclerenchyma, cortex, and stele cells	NO_3^-	[37]
AK101480	*OsNPF7.3*	NO_3^-	Root, seeds	Gly-His Gly-His-Gly	[33, 34]
AK064899	*OsNPF8.20*	N, light	Leaves, panicles, young root tips, cortical fiber cells of lateral roots, stems	Unknown	[32]
AK100802	*OsNPF4.1*	Unknown	Phloem of the branches of young panicles	Unknown	[31]
AK100112	*OsNPF8.2*	Drought, salt, cold	Seeds, leaf, panicle	Unknown	[33]
AK101055	*OsNPF5.5*	Unknown	Seeds, leaf	Unknown	[33]
AK101099	*OsNPF7.1*	Unknown	Constitutive expression	Unknown	[33]
AK070216	*OsNPF7.4*	Drought, salt	Root, panicle, node	Unknown	[33]
AK070036	*OsNPF8.1*	Drought, salt	Shoot, leaf, panicle, seeds	Unknown	[33]
AK072691	*OsNPF8.5*	Drought, salt	Constitutive expression	Unknown	[33]
AB008519	*OsNRT2.1*	NO_3^-, light, sucrose	Root tip, meristem	NO_3^-	[38–40]
AK109733	*OsNRT2.2*	NO_3^-, light, sucrose	Root tip, meristem	NO_3^-	[38–40]
AK109776	*OsNRT2.3a*	NO_3^-, light, sucrose	Root stele	NO_3^-	[38, 41]
AK072215	*OsNRT2.3b*	Light, sucrose, pH	Shoot phloem	NO_3^-	[38, 42]
NM_193361	*OsNRT2.4*	NO_3^-, light, sucrose, pH, NAA	Root, shoot	Unknown	[38–40]
NM_001053852.2	*OsNAR2.1*	NO_3^-, light, sucrose	Root epidermal cells	Unknown	[19, 38–40]
AK109571	*OsNAR2.2*	Light, sucrose	Root, shoot	None	[19, 38, 39]
AF289477	*OsAMT1;1*	NH_4^+, circadian rhythm	Constitutive expression	NH_4^+	[46, 48, 50, 52]

Accession no.	Gene	Regulation	Expression pattern	Substrates	References
AF289478	*OsAMT1;2*	NH_4^+	Root central cylinder and cell surface of root tips	NH_4^+	[46, 50]
AF289479	*OsAMT1;3*	Repressed, circadian rhythm	Root exodermis, sclerenchyma, endodermis, and pericycle cells of primary root	NH_4^+	[46, 47, 50, 53]
AB051864	*OsAMT2;1*	Unknown	Constitutive expression	NH_4^+	[46]
NM 190445	*OsAMT2;2*	NO_3^-, NH_4^+	Unknown	Unknown	[46, 55]
NM_001051237	*OsAMT2;3*	Unknown	Unknown	Unknown	[46]
AB083582	*OsAMT3;1*	Unknown	Roots, shoots	NH_4^+	[46]
AC104487	*OsAMT3;2*	Unknown	Unknown	Unknown	[46]
AP004775	*OsAMT3;3*	Unknown	Unknown	Unknown	[46]
AC091811	*OsAMT4*	Unknown	Unknown	Unknown	[46]
AB037664	*OsGS1;1*	NH_4^+	Leaves	NH_4^+, Glu	[58, 59]
AB180688	*OsGS1;2*	NH_4^+	Roots	NH_4^+, Glu	[58, 59, 64]
AB180689	*OsGS1;3*	Unknown	Spikelets	NH_4^+, Glu	[58, 59]
X14246	*OsGS2*	Unknown	Leaves	NH_4^+, Glu	[58, 60]
AB024716	*OsFd-GOGAT*	Light	Shoots	Gln, 2-OG	[60, 61]
AB008845	*OsNADH-GOGAT1*	NH_4^+, Gln	Developing tissues: root tip, premature leaf blade, spikelet at the early stage of ripening	Gln, 2-OG	[60, 61]
AB274818	*OsNADH-GOGAT2*	NH_4^+	Mature leaf blade and sheath: phloem companion and parenchyma cells	Gln, 2-OG	[60, 61, 68]

Table 2. Literature summary of the tissue expression and regulation of genes responsible for NUE.

gene alone did not increase nitrate uptake in rice [43], owing to that the nitrate uptake activity of *OsNRT2.1*, *OsNRT2.2*, and *OsNRT2.3a* requires a partner protein, OsNAR2.1 [19, 38, 44]. The transcripts of *OsNAR2.2* and *OsNRT2.4* were detected in roots and shoots, accumulation induced by nitrate [38–40]. However, their functions remain unknown.

Ammonium uptake is mainly mediated by proteins of the ammonia transport protein (AMT)/ transports methylammonium (MEP)/rhesus (RH) superfamily [45]. There are uncertainties regarding the exact chemical species transported by AMT, which can be in the form of either hydrophobic NH3 or charged ammonium [14, 45]. The activity of AMT members may play a more important role in NUpE in ammonium-preferring rice than in nitrate-utilizing crops. In rice, there are at least ten putative *OsAMT*-like genes grouped into four subfamilies (i.e., three each for *OsAMT1*, *OsAMT2*, and *OsAMT3*, respectively, and one for *OsAMT4*) (**Table 2**) [46]. So far, studies on expression regulation of *AMT* genes in rice are mainly focused on *OsAMT1* gene family, which displayed different spatiotemporal expression patterns in response to changes in N levels or daily irradiance (**Table 2**) [47, 48]. *OsAMT1;1* is constitutively expressed in rice roots and shoots showing a positive feedback regulation by endogenous glutamine [49]. It has been reported that *OsAMT1;1*, showing a higher expression level in roots under ammonium supply, contributes to NH_4^+ uptake and plays an important role in NK homeostasis [48, 50–52]. *OsAMT1;2* showed root-specific expression, is induced by ammonium, and may function as a nitrogen assimilator [49, 53]. Root-specific and nitrogen-derepressible expression for *OsAMT1;3* may function as a nitrogen sensor [49, 53]. Overexpression *OsAMT1;3* displayed significant decreases in growth but with poor nitrogen uptake ability, accompanied with a higher leaf C/N ratio [54]. *OsAMT2;1* showed constitutive expression in both roots and shoots, and *OsAMT3;1* showed very weak expression in roots and shoots [46]. *OsAMT2;2* is evenly expressed in roots and shoots and is induced by nitrogen [55].

3.2. Nitrogen assimilation

After taken up by the roots, nitrate is assimilated in the roots, the other larger part transported to the shoots, where it is first reduced to nitrite catalyzed by nitrate reductase (NR) in the cytoplasm and then further to ammonium by nitrite reductase (NiR) in the plastids. The ammonium derived from nitrate or directly from ammonium uptake by AMTs is finally assimilated into amino acids via the GS/GOGAT cycle (**Figure 1**) [14, 22]. GOGAT catalyzes the transfer of the amide group of glutamine (Gln) formed by GS to 2-oxoglutarate (2-OG) to yield two molecules of glutamate (Glu). One of the Gln molecules can be cycled back as a substrate for the GS reaction, and the other can be used for many synthetic reactions [56, 57].

Rice possesses three homologous but distinct genes for cytosolic glutamine synthetase (i.e., *OsGS1;1*, *OsGS1;2*, and *OsGS1;3*) and one chloroplastic gene (*OsGS2*). *OsGS1;1* and *OsGS1;2* both showed a high substrate affinity for ammonium and were induced by ammonium within the central cylinder of rice-elongating zone [58]. *OsGS1;1* was constitutively expressed , with higher expression profile in leaf blade and participated in rice normal growth and grain filling [59, 60]. *OsGS1;1* also functions in coordinating the global metabolic network in rice plants grown using ammonium as the nitrogen source [60] and is important for remobilization of nitrogen during natural senescence [61, 62]. *OsGS1;2* is constitutively expressed in surface cells of roots responsible for the primary assimilation of ammonium, and knockout of *OsGS1;2* showed severe reduction in active tiller number [63]. However, Ohashi et al. thought that the reduction in tiller number is an NH_4^+-specific event and the outgrowth of the axillary buds was severely suppressed caused by metabolic disorder in *OsGS1;2* mutants [64]. *OsGS1;3* is exclusively expressed in spikelet [59], indicating that it is probably important in grain ripening and/or germination. The OsGS2 subunit protein was present in leaves but was hardly detectable in roots [58]. There is also a small gene family for GOGAT: one ferredoxin (Fd)-dependent type and two NADH-dependent types [65]. *OsFd-GOGAT* is highly abundant in mesophyll cells and other chloroplast-containing cells regulated by light [56] and is important in reassimilation of ammonium generated by photorespiration in chloroplasts [65]. Recently, participating in nitrogen assimilation, C/N balance, [66], leaf senescence, and the nitrogen remobilization has been reported [67]. *OsNADH-GOGAT1* is mainly expressed in surface cells of rice roots in an NH_4^+-dependent manner and is important for primary ammonium assimilation in roots at the seedling stage and development of active tiller number until the harvest [62, 65]. *OsNADH-GOGAT2* is mainly expressed in vascular tissues of mature leaf blades and is important in the process of glutamine generation in senescing leaves for the remobilization of leaf nitrogen through phloem to the panicle during natural senescence. *OsNADH-GOGAT2* mutants had marked reduction in spikelet number per panicle [62, 68].

Although these observed phenotypes and those observed for GS enzymes have been identified, the interaction between isozymes of GOGAT and the GS isozymes, how they affect NUE, as well as posttranscriptional regulation of these enzymes needs to be further investigated.

3.3. Nitrogen remobilization and reassimilation

During the vegetative stage, the leaves are a sink for N; later, during senescence, this N is remobilized for reuse in the developing seeds, mainly as amino acids (**Figure 1**) [69]. Up to 95% of seed protein is derived from amino acids that are exported to the seed after the degradation of existing proteins in leaves [14], and the rest is supplemented from the soil and late top-dressed fertilizers [70]. Gln and asparagine (Asn) are major forms of total amino acids in phloem and xylem sap of rice plants [14, 71]. Increases of both Asn and Gln concentrations during senescence in the phloem sap suggest their key role in rendering N available for remobilization from the senescing leaves. Some isoforms of GS1, NADH-glutamate dehydrogenase (GDH), and asparagine synthetase (AS) are strongly activated during N remobilization [72]. The nature of the amino acid transporters, belonging to complex multigene families, is poorly understood in phloem loading for N redistribution during senescence [69].

The importance of GS/GOGAT activity in N remobilization, reassimilation, growth rate, yield, and grain filling has been emphasized previously. OsGS1;1 and OsNADH-GOGAT2 are important in remobilization of nitrogen during natural senescence [62]. GS1;2 is also important in the development of active tillers through the assimilation of NH_4^+ generated during lignin synthesis [64]. Together with GS, AS is believed to play a crucial role in primary N metabolism, catalyzing the formation of Asn and Glu from Gln and aspartate [14, 64]. There are two genes (i.e., *OsAS1* and *OsAS2*) identified encoding AS in rice. *OsAS1* is mainly expressed in root surface (epidermis, exodermis, and sclerenchyma) in an NH_4^+-dependent manner, which are very similar with *OsGS1;2* and *NADH-GOGAT1* in rice roots. Thus, AS1 is apparently coupled with the primary assimilation of NH_4^+ in rice roots. *OsAS2* detected in phloem companion and parenchyma cells [71, 73] is abundant in leaf blades and sheathes, along with the GS1;1 protein [61]. These suggest that AS2 in rice leaves is probably important in the long-distance transport of asparagine from rice leaves during natural senescence. In addition, the mitochondrial GDH plays a major role in reassimilation of photorespiratory ammonia and can alternatively incorporate ammonium into Glu in response to high levels of ammonium under stress [72]. Although there are a large number of amino acid permeases (AAPs) presented in rice [74, 75], no transporters have been functionally characterized with an exception for *OsAAP6*, which is mainly expressed in seeds for grain protein content [76]. Recently, the transport function of four rice AAP genes (*OsAAP1*, *OsAAP3*, *OsAAP7*, and *OsAAP16*) has been analyzed by expression in *Xenopus laevis oocytes*, electrophysiology, and cellular localization. *OsAAP1*, *OsAAP7*, and *OsAAP16* functioned as general AAPs and could transport all amino acids well except aspartate and β-alanine. While *OsAAP3* had a distinct substrate specificity transporting the basic amino acids lysine and arginine well but selected against aromatic amino acids [77].

4. Enhancing nitrogen use efficiency

As mentioned above, molecular studies have provided a general validation of the physiological conceptual framework of NUE in rice. However, besides genetics, there are other factors needed to consider such as the interactions between N uptake and water availability, the interaction between N utilization and carbon metabolism, and the interaction between different macronutrients and micronutrients [13]. Understanding the mechanisms regulating nitrogen movement in rice is crucial for improvement of NUE. Improvements in NUE result from NUpE, NUtE, or both. We describe approaches for increasing NUE with special consideration to genetics and agricultural management.

4.1. Increasing uptake capacity

Increased nitrogen uptake capacity may be achieved through better nitrogen transporters, more effective regulation of the transport systems, or better storage and assimilation. A simple example to improve NUpE would be to increase uptake by overexpressing more efficient transporters or all the transporters using transgenic methods [28, 42, 48, 78]. However, only increasing the uptake capacity of roots is not simple because of the tight regulation of N uptake, N taken up surplus to requirements increasing plant N status, which, in turn, leads to feedback regulation and reduction in uptake capacity [20].

Physiological traits that may also affect NUpE including root architecture and any other characteristic play a pivotal role in extracting available N from the soil [13, 79]. The capacity of the root for uptake depends on the degree to which the root extends and its absorption area, which is determined by complex root morphology. A common example is to target genes related to root morphology through a mapping approach, whereby traits are identified through genetic crosses using distinct populations, and then quantitative trait loci (QTLs) can be cloned by positional cloning [79–81]. To date, studies have been carried out to identify root morphological features such as root mass and depth, root axis length, and lateral branching related to NUE [82–84].

However, ammonium or nitrate uptake by rice roots commonly results in acidification or alkalization of the rhizosphere, which in turn changes the soil N availability for plants. In the rhizosphere, rice roots can also release oxygen and exudates that greatly influence local redox potential and the density and activity of microbial populations, which in turn can interconvert soil N forms, including those derived from fertilizer [14]. Thus, soil N availability fluctuating greatly in both space and time affects root morphology, which could make plants uptake N efficiently [14]. Studies in rice have been confirmed that compared to sole NH_4^+ nutrition, a mixture of NH_4^+ and NO_3^- promoted root growth as well as N absorption and assimilation [85, 86]. In the course of agricultural management, fertilizer type (i.e., controlled N release fertilizers, new potential N sources), methods of applying N fertilizers (e.g., the 4R nutrient stewardship framework: right source, right rate, right time, and right placement), soil types, tillage, transplanting density, cropping system, and microorganisms are governed to avoid nitrogen loss increasing fertilizer nitrogen use efficiency [87, 88].

Water is another key factor determining crop yield and NUE. Without sufficient water, plants cannot extract nutrients from the soil. Yield is constrained by moisture availability, not N availability, especially in maize [89]. In contrast to upland crops, alternate wetting and drying (AWD, flooding the soil and then allowing to dry down before being reflooded) to reduce total water for irrigation in rice has been developed for a number of decades. A number of studies have shown that AWD increases grain yield when compared to continuous flooding (CF) [90, 91].

On the base of current knowledge, scientists have developed a range of optimized crop management practices, such as site-specific nutrient management (SSNM) [92], real-time N management (RTNM) [93], and preliminary integrated precision rice management (PRM) system combining SSNM with alternate drying and wetting irrigation and optimized transplanting density [94]. Only integrated N management strategies are allowed for the achievement of production goals while minimizing the risk of environmental pollution. Sources of N and timing of application determine the most suitable method for application. The interest in implementing new knowledge about the methods of application is to develop sensors to diagnose the N status of crops in real time throughout large areas and decision support systems to help determine N fertilizer recommendations [88].

4.2. Increasing utilization efficiency

A number of physiological traits can affect the NUtE in crops, including the effect of N on carbohydrate partitioning, the storage of N, and the remobilization of N from senescent tissues, and these have been subdivided into a number of components by researchers [95, 96].

Increasing nitrogen utilization capacity can be achieved through overexpression of candidate genes in the pathways relating to N assimilation, translocation, remobilization, and reassimilation. As mentioned above, changes in the expression and activity of GS and GOGAT would have an effect on N assimilation, recycle, reassimilation, C/N balance, and senescence in rice, potentially affecting grain filling, yield, and NUE [62, 64, 66]. Identifying candidate genes cosegregate with NUE in genetic crosses is another efficient method. One of the first QTL studies conducted analyzing NUE in rice was carried out [97]. They looked at QTLs associated with NUE and determined whether they cosegregated with GS1 and NADH-GOGAT. The analysis identified seven loci that cosegregated with GS1 activity and six loci that cosegregated with NADH-GOGAT activity. A number of QTLs for agronomic traits related to N use and yield have been mapped to the chromosomal regions containing GS2 in rice [97, 98], suggesting that the genomic region surrounding GS2 may be valuable for breeding rice with improved agronomic performance and NUE. However, to date, no one has been able to introduce a GS gene into a NUE-inefficient background and show either enhanced NUE or yield.

C and N metabolisms are tightly linked with each other in plants. N assimilation requires carbon metabolism to provide adenosine triphosphate (ATP), reductants, and C skeletons through photosynthesis, photorespiration, and respiration. Large amounts of N are used in photosynthesis, particularly during ribulose 1,5-bisphosphate carboxylase-oxygenase

(Rubisco) and light-harvesting complexes to support the light-dependent use of CO_2, inorganic N, and water to produce sugars, amino acids, and organic acids [99]. Photorespiration, a side reaction of photosynthesis, has crucial implications in N reassimilation, which is catalyzed by the Rubisco. During photorespiration, NH_4^+ is produced during methylenetetrahydrofolate synthesis from glycine [100]. Respiration is a third fundamental process of energy metabolism in the dark and in nonphotosynthetic tissues, as well as in the light. In the respiratory pathways, the C skeletons for N assimilation are generated in different sectors, such as the oxidative pentose phosphate pathway (OPPP), glycolysis, and TCA cycle [101]. The operation of the TCA cycle in illuminated leaves is critical for the provision of 2-OG, which is necessary for glutamate and glutamine production [101–103]. Evidence has shown that the synthesis of 2-OG is induced by the activity of phosphoenolpyruvate carboxylase (PEPC), citrate synthase, isocitrate dehydrogenase, and aconitase, while the subsequent conversion of 2-OG to fumarate may be repressed in the light [101].

Thus, exploiting candidate genes involved in C/N metabolism is another approach to improve NUE. To date, there are two key genes identified to contribute to NUE in rice. Chloroplastic proteins are known to make up approximately 80% of the stored N in leaf tissues, with Rubisco accounting for up to 50% and 20% of the stored N in C_3 and C_4 plants, respectively [104]. Thus, Rubisco is an excellent N storage molecule, and its autophagic degradation in rice leaves may contribute to an efficient and rapid N remobilization by facilitating protein degradation for N mobilization in senescent leaves [70]. Rubisco is also involved in photorespiratory losses which can be as high as 20% of the total carbon fixation in C_3 plants and also liberates ammonia, which is required for reassimilation [105]. However, when rice plants overexpressing the Rubisco (rbcS) gene were analyzed, Rubisco-N to leaf-N increased, but there was no change in the rate of photosynthesis [106]. PEPC is a component of primary metabolism in plants and has a nonphotosynthetic role as one of its products is OAA, a component of the TCA cycle [107]. RNAi knockdown experiments of the chloroplastic isoform in rice have indicated that PEPC plays an important role in N assimilation, specifically when the main N source is NH_4^+ [108].

Growth and yield of rice plants are markedly affected by increased CO_2 concentration and temperature [109, 110]. Numerous studies have indicated that an increase in CO_2 generally stimulates photosynthesis, reduces stomatal conductance, and changes the rhizosphere conditions of plants, leading to increases in biomass and yield of crops [111–113], whereas an increase in temperature accelerates crop phenological development and shortens grain-filling period of crops, leading to decrease grain yield and reduce crop production in many regions of the world [114, 115]. Furthermore, high temperature, if occurring at critical stages of crop development (such as meiosis and flowering stages), reduces spikelet fertility [115]. Owing to elevated CO_2 under future climate change is associated with an increase in air temperature, many studies about plant response to the interaction of CO_2 and temperature have been reported [109, 110, 116]. Increases in CO_2 were unable to compensate for the negative impact of increases in temperature on biomass and yield in rice [109, 110]. Thus, selecting high-temperature-tolerant germplasm will be required to realize yield benefits in the future.

5. Conclusions

Plant NUE is a complex trait determined by quantitative trait loci and influenced by environmental changes and is the integration of NUpE and NUtE. There is a complex regulation of N uptake, assimilation, and remobilization.

Enhanced NUE can be achieved by genetically modifying plants and integrated agricultural management practices. The former is the most effective biotechnological method for increasing NUE. This can be achieved by overexpression of nitrate and ammonium transporters responsible for N uptake by roots and by manipulation of key genes controlling the balance of N and C metabolism.

Developing an integrated research program combining approaches, mainly based on whole-plant physiology, quantitative genetics, forward and reverse genetics, and agronomy approaches to improve NUE, is a major objective in the future.

Acknowledgements

We thank Dr. Chang Li for his comments on **Figure 1** and for his helpful discussions on this timely topic. We apologize to all colleagues whose work could not be cited owing for space limitations. This work is supported by grants from China's Agriculture Research System (#CARS-01-47) and the National Key Technology Support Program (2015BAD01B02).

Author details

Shuangjie Huang[1*], Chunfang Zhao[1], Yali Zhang[2] and Cailin Wang[1]

*Address all correspondence to: huangdeifan@163.com

1 Institute of Food Crops of Jiangsu Academy of Agricultural Sciences, Nanjing, China

2 State Key Laboratory of Crop Genetics and Germplasm Enhancement, Key Laboratory of Plant Nutrition and Fertilization in Low-Middle Reaches of the Yangtze River, Ministry of Agriculture, Nanjing Agricultural University, China

References

[1] Godfray HCJ, Beddington JR, Crute IR. Food security: The challenge of feeding 9 billion people. Science. 2010;**327**:812 818. DOI: 10.1126/science.1185383

[2] Tilman D, Balzer C, Hill J. Global food demand and the sustainable intensification of agriculture. Proceedings of the National Academy of Sciences of the United States of America. 2011;**108**:20260-20264. DOI: 10.1073/pnas.1116437108

[3] Cassman KG. Ecological intensification of cereal production systems: Yield potential, soil quality, and precision agriculture. Proceedings of the National Academy of Sciences. 1999;**96**: 5952-5959. DOI: 10.1073/pnas.96.11.5952

[4] Wu LL, Yuan S, Huang LY. Physiological mechanisms underlying the high-grain yield and high-nitrogen use efficiency of elite rice varieties under a low rate of nitrogen application in china. Frontiers in Plant Science. 2016;**7**:1024. DOI: 10.3389/fpls.2016.01024

[5] Robertson GP, Vitousek PM. Nitrogen in agriculture: Balancing the cost of an essential resource. Annual Review of Environment and Resources. 2009;**34**:97-125. DOI: 10.1146/annurev.environ.032108.105046

[6] Andrews M, Raven JA, Lea PJ. Do plants need nitrate? The mechanisms by which nitrogen form affects plants. Annals of Applied Biology. 2013;**163**:174-199. DOI: 10.1111/aab.12045

[7] Vitousek PM, Naylor R, Crews T. Nutrient imbalances in agricultural development. Science. 2009;**324**:1519-1520. DOI: 10.1126/science.1170261

[8] Good AG, Beatty PH. Fertilizing nature: A tragedy of excess in the commons. PLoS Biology. 2011;**9**:e1001124. DOI: 10.1371/journal.pbio.1001124

[9] Zhang FS, Cui ZL, Fan MS. Integrated soil-crop system management: Reducing environmental risk while increasing crop productivity and improving nutrient use efficiency in china. Journal of Environmental Quality. 2011;**40**:1051-1057. DOI: 10.2134/jeq2010.0292

[10] Smil V. Nitrogen in crop production: An account of global flows. Global Biogeochemical Cycles. 1999;**13**:647-662. DOI: 10.1029/1999gb900015

[11] Diaz RJ, Rosenberg R. Spreading dead zones and consequences for marine ecosystems. Science. 2008;**321**:926-929. DOI: 10.1126/science.1156401

[12] Guo JH, Liu XJ, Zhang Y. Significant acidification in major Chinese croplands. Science. 2010;**327**:1008-1010. DOI: 10.1126/science.1182570

[13] Han M, Okamoto M, Beatty PH. The genetics of nitrogen use efficiency in crop plants. Annual Review of Genetics. 2015;**49**:269-289. DOI: 10.1146/annurev-genet-112414-055037

[14] Xu GH, Fan XR, Miller AJ. Plant nitrogen assimilation and use efficiency. Annual Review of Plant Biology. 2012;**63**:153-182. DOI: 10.1146/annurev-arplant-042811-105532

[15] Good AG, Shrawat AK, Muench DG. Can less yield more? Is reducing nutrient input into the environment compatible with maintaining crop production?. Trends in Plant Science. 2004;**9**:597-605. DOI: 10.1016/j.tplants.2004.10.008

[16] Foulkes MJ, Hawkesford MJ, Barraclough PB. Identifying traits to improve the nitrogen economy of wheat: Recent advances and future prospects. Field Crops Research. 2009;**114**:329-342. DOI: 10.1016/j.fcr.2009.09.005

[17] Kirk GJD, Kronzucker HJ. The potential for nitrification and nitrate uptake in the rhizosphere of wetland plants: A modelling study. Annals of Botany. 2005;**96**:639-646. DOI: 10.1093/aob/mci216

[18] Li YL, Fan XR, Shen QR. The relationship between rhizosphere nitrification and nitrogen-use efficiency in rice plants. Plant Cell & Environment. 2008;**31**:73-85. DOI: 10.1111/j.1365-3040.2007.01737.x

[19] Yan M, Fan XR, Feng HM. Rice OsNAR2.1 interacts with OsNRT2.1, OsNRT2.2 and OsNRT2.3a nitrate transporters to provide uptake over high and low concentration ranges. Plant Cell & Environment. 2011;**34**:1360-1372. DOI: 10.1111/j.1365-3040.2011.02335.x

[20] Miller AJ, Fan XR, Orsel M. Nitrate transport and signalling. Journal of Experimental Botany. 2007;**58**:2297-2306. DOI: 10.1093/jxb/erm066

[21] Williams LE, Miller AJ. Transporters responsible for the uptake and partitioning of nitrogenous solutes. Annual Review of Plant Physiology & Plant Molecular Biology. 2011;**52**:659-688. DOI: 10.1146/annurev.arplant.52.1.659

[22] Neeraja CN, Subramanyam D, Surekha K. Advances in genetic basis of nitrogen use efficiency of rice. Indian Journal of Plant Physiology. 2016;**21**:504-513. DOI: 10.1007/s40502-016-0254-z

[23] O'Brien JA, Vega A, Bouguyon E. Nitrate transport, sensing, and responses in plants. Molecular Plant. 2016;**9**:837-856. DOI: 10.1016/j.molp.2016.05.004

[24] Krapp A, David LC, Chardin C. Nitrate transport and signalling in Arabidopsis. Journal of Experimental Botany. 2014;**65**:789-798. DOI: 10.1093/jxb/eru001

[25] Dechorgnat J, Nguyen CT, Armengaud P. From the soil to the seeds: The long journey of nitrate in plants. Journal of Experimental Botany. 2011;**62**:1349-1359. DOI: 10.1093/jxb/erq409

[26] Léran S, Varala K, Boyer J. A unified nomenclature of nitrate transporter 1/peptide transporter family members in plants. Trends in Plant Science. 2013;**19**:5-9. DOI: 10.1016/j.tplants.2013.08.008

[27] Tsay YF, Chiu CC, Tsai CB. Nitrate transporters and peptide transporters. FEBS Letters. 2007;**581**:2290-2300. DOI: 10.1016/j.febslet.2007.04.047

[28] Hu B, Wang W, Ou SJ. Variation in *NRT1.1B* contributes to nitrate-use divergence between rice subspecies. Nature Genetics. 2015;**47**:834-838. DOI: 10.1038/ng.3337

[29] Fan XR, Feng HM, Tan YW. A putative 6-transmembrane nitrate transporter *OsNRT1.1b* plays a key role in rice under low nitrogen. Journal of Integrative Plant Biology. 2016;**58**:590-599. DOI: 10.1111/jipb.12382

[30] Lin CM, Koh S, Stacey G. Cloning and functional characterization of a constitutively expressed nitrate transporter gene, *OsNRT1*, from rice. Plant Physiology. 2000;**122**:379-388. DOI: 10.1104/pp.122.2.379

[31] Li SB, Qian Q, Fu ZM. Short panicle1 encodes a putative PTR family transporter and determines rice panicle size. Plant Journal. 2009;**58**:592-605. DOI: 10.1111/j.1365-313X.2009.03799.x

[32] Fang ZM, Xia KF, Yang X. Altered expression of the PTR/NRT1 homologue *OsPTR9* affects nitrogen utilization efficiency, growth and grain yield in rice. Plant Biotechnology Journal. 2013;**11**:446-458. DOI: 10.1111/pbi.12031

[33] Ouyang J, Cai ZY, Xia KF. Identification and analysis of eight peptide transporter homologs in rice. Plant Science. 2010;**179**:374-382. DOI: 10.1016/j.plantsci.2010.06.013

[34] Fan XR, Xie D, Chen JG. Over-expression of *OsPTR6*, in rice increased plant growth at different nitrogen supplies but decreased nitrogen use efficiency at high ammonium supply. Plant Science. 2014;**227**:1-11. DOI: 10.1016/j.plantsci.2014.05.013

[35] Xia XD, Fan XR, Wei J. Rice nitrate transporter OsNPF2.4 functions in low-affinity acquisition and long-distance transport. Journal of Experimental Botany. 2015;**66**:317-331. DOI: 10.1093/jxb/eru425

[36] Li Y, Ouyang J, Wang YY. Disruption of the rice nitrate transporter OsNPF2.2 hinders root-to-shoot nitrate transport and vascular development. Scientific Reports. 2015;**5**:9635. DOI: 10.1038/srep09635

[37] Hu R, Qiu DY, Chen Y. Knock-down of a tonoplast localized low-affinity nitrate transporter *OsNPF7.2* affects rice growth under high nitrate supply. Frontiers in Plant Science. 2016;**7**:1529. DOI: 10.3389/fpls.2016.01529

[38] Feng HM, Yan M, Fan XR. Spatial expression and regulation of rice high-affinity nitrate transporters by nitrogen and carbon status. Journal of Experimental Botany. 2011;**62**:2319-2332. DOI: 10.1093/jxb/erq403

[39] Araki R, Hasegawa H. Expression of rice (*Oryza sativa* L.) genes involved in high-affinity nitrate transport during the period of nitrate induction. Breeding Science. 2006;**56**:295-302. DOI: 10.1270/jsbbs.56.295

[40] Cai C, Wang JY, Zhu YG. Gene structure and expression of the high-affinity nitrate transport system in rice roots. Journal of Integrative Plant Biology. 2008;**50**:443-451. DOI: 10.1111/j.1744-7909.2008.00642.x

[41] Tang Z, Fan XR, Li Q. Knockdown of a rice stelar nitrate transporter alters long-distance translocation but not root influx. Plant Physiology. 2012;**160**:2052-2063. DOI: 10.1104/pp.112.204461

[42] Fan XR, Tang Z, Tan, YW. Overexpression of a pH-sensitive nitrate transporter in rice increases crop yields. Proceedings of the National Academy of Sciences. 2016;**113**:7118-7123. DOI: 10.1073/pnas.1525184113

[43] Katayama H, Mori M, Kawamura Y. Production and characterization of transgenic rice plants carrying a high-affinity nitrate transporter gene (*OsNRT2.1*). Breeding Science. 2009;**59**:237-243. DOI: 10.1270/jsbbs.59.237

[44] Liu XQ, Huang DM, Tao JY. Identification and functional assay of the interaction motifs in the partner protein OsNAR2.1 of the two-component system for high-affinity nitrate transport. New Phytologist. 2014;**204**:74-80. DOI: 10.1111/nph.12986

[45] Khademi S, O'Connell J, Remis J. Mechanism of ammonia transport by amt/mep/rh: Structure of amtb at 1.35 A. Science. 2004;**305**:1587-1594. DOI: 10.1126/science.1101952

[46] Suenaga A, Moriya K, Sonoda Y. Constitutive expression of a novel-type ammonium transporter *OsAMT2* in rice plants. Plant & Cell Physiology. 2003;**44**:206-211. DOI: 10.1093/pcp/pcg017

[47] Kumar A, Silim SN, Okamoto M. Differential expression of three members of the AMT1, gene family encoding putative high-affinity NH_4^+, transporters in roots of *oryza sativa* subspecies indica. Plant Cell & Environment. 2003;**26**:907-914. DOI: 10.1046/j.1365-3040.2003.01023.x

[48] Ranathunge K, El-kereamy A, Gidda S. *AMT1;1* transgenic rice plants with enhanced NH_4^+ permeability show superior growth and higher yield under optimal and suboptimal NH_4^+conditions. Journal of Experimental Botany. 2014;**65**:965-979. DOI: 10.1093/jxb/ert458

[49] Sonoda Y, Ikeda A, Saiki S. Feedback regulation of the ammonium transporter gene family *AMT1* by glutamine in rice. Plant & Cell Physiology. 2003;**44**:1396-1402. DOI: 10.1093/pcp/pcg169

[50] Sonoda Y, Ikeda A, Saiki S. Distinct expression and function of three ammonium transporter genes (*OsAMT1;1-1;3*) in rice. Plant & Cell Physiology. 2003;**44**:726-734. DOI: 10.1093/pcp/pcg083

[51] Hoque MS, Masle J, Udvardi MK. Over-expression of the rice *OsAMT1-1* gene increases ammonium uptake and content, but impairs growth and development of plants under high ammonium nutrition. Functional Plant Biology. 2006;**33**:153-163. DOI: 10.1071/FP05165

[52] Li C, Tang Z, Wei J. The *OsAMT1.1* gene functions in ammonium uptake and ammonium-potassium homeostasis over low and high ammonium concentration ranges. Journal of Genetics & Genomics.2016;**43**:639-649. DOI: 10.1016/j.jgg.2016.11.001

[53] Yao SG, Sonoda Y, Tsutsui T. Promoter analysis of *OsAMT1;2* and *OsAMT1;3* implies their distinct roles in nitrogen utilization in rice. Breeding Science. 2008;**58**:201-207. DOI: 10.1270/jsbbs.58.201

[54] Bao A, Liang ZJ, Zhao ZQ. Overexpressing of *OsAMT1-3*, a high affinity ammonium transporter gene, modifies rice growth and carbon-nitrogen metabolic status. International Journal of Molecular Sciences. 2015;**16**:9037-9063. DOI: 10.3390/ijms16059037

[55] Li SM, Shi WM. Quantitative characterization of nitrogen regulation of *OsAMT1;1*, *OsAMT1;2*, and *OsAMT2;2*, expression in rice seedlings. Russian Journal of Plant Physiology. 2006;**53**:837-843. DOI: 10.1134/S102144370606015X

[56] Ishiyama K, Hayakawa T, Yamaya T. Expression of NADH-dependent glutamate synthase protein in the epidermis and exodermis of rice roots in response to the supply of ammonium ions. Planta. 1998;**204**:288-294. DOI: 10.1007/s004250050258

[57] Forde BG, Lea PJ. Glutamate in plants: Metabolism, regulation, and signalling. Journal of Experimental Botany. 2007;**58**:2339-2358. DOI: 10.1093/jxb/erm121

[58] Ishiyama K, Inoue E, Tabuchi M. Biochemical background and compartmentalized functions of cytosolic glutamine synthetase for active ammonium assimilation in rice roots. Plant & Cell Physiology. 2004;**45**:1640-1647. DOI: 10.1093/pcp/pch190

[59] Tabuchi M, Sugiyama K, Ishiyama K. Severe reduction in growth rate and grain filling of rice mutants lacking OsGS1;1, a cytosolic glutamine synthetase1;1. Plant Journal for Cell & Molecular Biology. 2005;**42**:641-651. DOI: 10.1111/j.1365-313X.2005.02406.x

[60] Kusano M, Tabuchi M, Fukushima A. Metabolomics data reveal a crucial role of cytosolic glutamine synthetase 1;1 in coordinating metabolic balance in rice. Plant Journal. 2011;**66**:456-466. DOI: 10.1111/j.1365-313X.2011.04506.x

[61] Tabuchi M, Abiko T, Yamaya T. Assimilation of ammonium ions and reutilization of nitrogen in rice (*Oryza sativa* L.). Journal of Experimental Botany. 2007;**58**:2319-2327. DOI: 10.1093/jxb/erm016

[62] Yamaya T, Kusano M. Evidence supporting distinct functions of three cytosolic glutamine synthetases and two NADH-glutamate synthases in rice. Journal of Experimental Botany. 2014;**65**:5519-5525. DOI: 10.1093/jxb/eru103

[63] Funayama K, Kojima S, Tabuchi-kobayashi M. Cytosolic glutamine synthetase1;2 is responsible for the primary assimilation of ammonium in rice roots. Plant & Cell Physiology. 2013;**54**:934-943. DOI: 10.1093/pcp/pct046

[64] Ohashi M, Ishiyama K, Kusano M. Lack of cytosolic glutamine synthetase1;2 in vascular tissues of axillary buds caused severe reduction in their outgrowth and disorder of metabolic balance in rice seedlings. Plant Journal. 2015;**81**:347-356. DOI: 10.1111/tpj.12731

[65] Tamura W, Hidaka Y, Tabuchi M. Reverse genetics approach to characterize a function of NADH-glutamate synthase1 in rice plants. Amino Acids. 2010;**39**:1003-1012. DOI: 10.1007/s00726-010-0531-5

[66] Yang XL, Nian JQ, Xie QJ. Rice ferredoxin-dependent glutamate synthase regulates nitrogen-carbon metabolomes and is genetically differentiated between *japonica* and *indica* subspecies. Molecular Plant. 2016;**9**:1520-1534. DOI: 10.1016/j.molp.2016.09.004

[67] Zeng DD, Qin R, Li M. The ferredoxin-dependent glutamate synthase (OsFd-GOGAT) participates in leaf senescence and the nitrogen remobilization in rice. Molecular Genetics & Genomics. 2017;**292**:385-395. DOI: 10.1007/s00438-016-1275-z

[68] Tamura W, Kojima S, Toyokawa A. Disruption of a novel NADH-glutamate synthase2 gene caused marked reduction in spikelet number of rice. Frontiers in Plant Science. 2011;**2**:57. DOI: 10.3389/fpls.2011.00057

[69] Okumoto S, Pilot G. Amino acid export in plants: A missing link in nitrogen cycling. Molecular Plant. 2011;**4**:453-463. DOI: 10.1093/mp/ssr003

[70] Yoneyama T, Tanno F, Tatsumi J. Whole-plant dynamic system of nitrogen use for vegetative growth and grain filling in rice plants (*Oryza sativa* L.) as revealed through the production of 350 grains from a germinated seed over 150 days: A review and synthesis. Frontiers in Plant Science. 2016;**7**:1151. DOI: 10.3389/fpls.2016.01151

[71] Ohashi M, Ishiyama K, Kojima S. Asparagine synthetase1, but not asparagine synthetase2, is responsible for the biosynthesis of asparagine following the supply of ammonium to rice roots. Plant & Cell Physiology. 2015;**56**:769-778. DOI: 10.1093/pcp/pcv005

[72] Masclaux-daubresse C, Danielvedele F, Dechorgnat J. Nitrogen uptake, assimilation and remobilization in plants: Challenges for sustainable and productive agriculture. Annals of Botany. 2010;**105**:1141-1157. DOI: 10.1093/aob/mcq028

[73] Nakano K, Suzuki T, Hayakawa T. Organ and cellular localization of asparagine synthetase in rice plants. Plant & Cell Physiology. 2000;**41**:874-880. DOI: 10.1093/pcp/pcd006

[74] Tegeder M, Ward JM. Molecular evolution of plant AAP and LHT amino acid transporters. Frontiers in Plant Science. 2012;**3**:21. DOI: 10.3389/fpls.2012.00021

[75] Zhao HM, Ma HL, Yu L. Genome-wide survey and expression analysis of amino acid transporter gene family in rice (*Oryza sativa* L.). PLoS ONE. 2012;**7**:e49210. DOI: 10.1371/journal.pone.0049210

[76] Peng B, Kong HL, Li YB. *OsAAP6* functions as an important regulator of grain protein content and nutritional quality in rice. Nature Communications. 2014;**5**:4847. DOI: 10.1038/ncomms5847

[77] Taylor MR, Reinders A, Ward JM. Transport function of rice amino acid permeases (AAPs). Plant & Cell Physiology. 2015;**56**:1355-1363. DOI: 10.1093/pcp/pcv053

[78] Chen JG, Zhang Y, Tan YW. Agronomic nitrogen-use efficiency of rice can be increased by driving *OsNRT2.1* expression with the *OsNAR2.1* promoter. Plant Biotechnology Journal. 2016;**14**:1705-1715. DOI: 10.1111/pbi.12531

[79] Garnett T, Conn V, Kaiser BN. Root based approaches to improving nitrogen use efficiency in plants. Plant Cell & Environment. 2009;**32**:1272-1283. DOI: 10.1111/j.1365-3040.2009.02011.x

[80] Zhang HM, Forde BG. An Arabidopsis mads box gene that controls nutrient-induced changes in root architecture. Science. 1998;**279**:407-409. DOI: 10.1126/science.279.5349.407

[81] McAllister CH, Beatty PH, Good AG. Engineering nitrogen use efficient crop plants: The current status. Plant Biotechnology Journal. 2012;**10**:1011-1025. DOI: 10.1111/j.1467-7652.2012.00700.x

[82] Yadav R, Courtois B, Huang N. Mapping genes controlling root morphology and root distribution in a doubled-haploid population of rice. Theoretical and Applied Genetics. 1997;**94**:619-632. DOI: 10.1007/s001220050459

[83] Shen L, Courtois B, Mcnally KL. Evaluation of near-isogenic lines of rice introgressed with QTLs for root depth through marker-aided selection. Theoretical and Applied Genetics. 2011;**103**:75-83. DOI: 10.1007/s001220100538

[84] Horii H, Nemoto K, Miyamoto N. Quantitative trait loci for adventitious and lateral roots in rice. Plant Breeding. 2006;**125**:198-200. DOI: 10.1111/j.1439-0523.2006.01124.x

[85] Duan YH, Zhang YL, Ye LT. Responses of rice cultivars with different nitrogen use efficiency to partial nitrate nutrition. Annals of Botany. 2007;**99**:1153-1160. DOI: 10.1093/aob/mcm051

[86] Song WJ, Makeen K, Wang DS. Nitrate supply affects root growth differentially in two rice cultivars differing in nitrogen use efficiency. Plant and Soil. 2011;**343**:357-368. DOI: 10.1007/s11104-011-0723-0

[87] Chen D, Suter H, Islam A. Prospects of improving efficiency of fertilizer nitrogen in Australian agriculture: A review of enhanced efficiency fertilizers. Australian Journal of Soil Research. 2008;**46**:289-301. DOI: 10.1071/SR07197

[88] Herrera JM, Rubio G, Häner LL. Emerging and established technologies to increase nitrogen use efficiency of cereals. Agronomy. 2016;**6**:25. DOI: 10.3390/agronomy6020025

[89] Bänziger M, Edmeades GO, Lafitte HR. Selection for drought tolerance increases maize yields across a range of nitrogen levels. Crop Science. 1999;**39**:1035-1040. DOI: 10.2135/cropsci1999.0011183x003900040012x

[90] Yang JC, Huang DF, Duan, H. Alternate wetting and moderate soil drying increases grain yield and reduces cadmium accumulation in rice grains. Journal of the Science of Food & Agriculture. 2009;**89**:1728-1736. DOI: 10.1002/jsfa.3648

[91] Zhang H, Xue YG, Wang ZQ. An alternate wetting and moderate soil drying regime improves root and shoot growth in rice. Crop Science. 2009;**49**:2246-2260. DOI: 10.2135/cropsci2009.02.0099

[92] Dobermann A, Witt C, Dawe D. Site-specific nutrient management for intensive rice cropping systems in Asia. Field Crops Research. 2002;**108**:37-66. DOI: 10.1016/S0378-4290(01)00197-6

[93] Peng SB, Buresh RJ, Huang JL. Strategies for overcoming low agronomic nitrogen use efficiency in irrigated rice systems in china. Field Crops Research. 2006;**96**:37-47. DOI: 10.1016/j.fcr.2005.05.004

[94] Zhao GM, Miao YX, Wang HY. A preliminary precision rice management system for increasing both grain yield and nitrogen use efficiency. Field Crops Research. 2013;**154**:23-30. DOI: 10.1016/j.fcr.2013.07.019

[95] Hirel B, Le Gouis J, Ney B. The challenge of improving nitrogen use efficiency in crop plants: Towards a more central role for genetic variability and quantitative genetics within integrated approaches. Journal of Experimental Botany. 2007;**58**:2369-2387. DOI: 10.1093/jxb/erm097

[96] Lea PJ, Azevedo RA. Nitrogen use efficiency. 2. Amino acid metabolism. Annals of Applied Biology. 2007;**151**:269-275. DOI: 10.1111/j.1744-7348.2007.00200.x

[97] Obara M, Kajiura M, Fukuta Y. Mapping of QTLs associated with cytosolic glutamine synthetase and NADH-glutamate synthase in rice (*Oryza sativa* L.). Journal of Experimental Botany. 2001;**52**:1209-1217. DOI: 10.1093/jexbot/52.359.1209

[98] Yamaya T, Obara M, Nakajima H. Genetic manipulation and quantitative-trait loci mapping for nitrogen recycling in rice. Journal of Experimental Botany. 2002;**53**:917-925. DOI: 10.1093/jexbot/53.370.917

[99] Zhu XG, Long SP, Ort DR. What is the maximum efficiency with which photosynthesis can convert solar energy into biomass?. Current Opinion in Biotechnology. 2008;**19**:153-159. DOI: 10.1016/j.copbio.2008.02.004

[100] Guo SW, Yi Z, Gao YX. New insights into the nitrogen form effect on photosynthesis and photorespiration. Pedosphere. 2007;**17**:601-610. DOI: 10.1016/S1002-0160(07)60071-X

[101] Nunesnesi A, Fernie AR, Stitt M. Metabolic and signaling aspects underpinning the regulation of plant carbon nitrogen interactions. Molecular Plant. 2010;**3**:973-996. DOI: 10.1093/mp/ssq049

[102] Sweetlove LJ, Beard KF, Nunes-Nesi A. Not just a circle: Flux modes in the plant TCA cycle. Trends in Plant Science. 2010;**15**:462-470. DOI: 10.1016/j.tplants.2010.05.006

[103] Foyer CH, Noctor G, Hodges M. Respiration and nitrogen assimilation: Targeting mitochondria-associated metabolism as a means to enhance nitrogen use efficiency. Journal of Experimental Botany. 2011;**62**:1467-1482. DOI: 10.1093/jxb/erq453

[104] Kant S, Bi YM, Rothstein SJ. Understanding plant response to nitrogen limitation for the improvement of crop nitrogen use efficiency. Journal of Experimental Botany. 2011;**62**:1499-1509. DOI: 10.1093/jxb/erq297

[105] Bauwe H, Hagemann M, Fernie AR. Photorespiration: Players, partners and origin. Trends in Plant Science. 2010;**15**:330-336. DOI: 10.1016/j.tplants.2010.03.006

[106] Suzuki YJ, Ohkubo M, Hatakeyama H. Increased Rubisco content in transgenic rice transformed with the 'sense' rbcS gene. Plant & Cell Physiology. 2007;**48**:626-637. DOI: 10.1093/pcp/pcm035

[107] Doubnerová V, Ryšlavá H. What can enzymes of C_4 photosynthesis do for C_3 plants under stress. Plant Science. 2011;**180**:575-583. DOI: 10.1016/j.plantsci.2010.12.005

[108] Masumoto C, Miyazawa SI, Ohkawa H. Phosphoenolpyruvate carboxylase intrinsically located in the chloroplast of rice plays a crucial role in ammonium assimilation. Proceedings of the National Academy of Sciences of the United States of America. 2010;**107**:5226-5231. DOI: 10.1073/pnas.0913127107

[109] Wang JY, Wang C, Chen NN. Response of rice production to elevated [CO_2] and its interaction with rising temperature or nitrogen supply: A meta-analysis. Climatic Change. 2015;**130**:529-543. DOI: 10.1007/s10584-015-1374-6

[110] Cai C, Yin XY, He SQ. Responses of wheat and rice to factorial combinations of ambient and elevated CO_2 and temperature in face experiments. Global Change Biology. 2016;**22**:856-874. DOI: 10.1111/gcb.13065

[111] Ainsworth EA, Rogers A. The response of photosynthesis and stomatal conductance to rising [CO_2]: Mechanisms and environmental interactions. Plant Cell & Environment. 2007;**30**:258-270. DOI: 10.1111/j.1365-3040.2007.01641.x

[112] Rajkumar M, Prasad MNV, Swaminathan S. Climate change driven plant-metal-microbe interactions. Environment International. 2013;**53**:74-86. DOI: 10.1016/j.envint.2012.12.009

[113] Xu ZZ, Jiang YL, Jia BR. Elevated-CO_2 response of stomata and its dependence on environmental factors. Frontiers in Plant Science 2016;**7**:657. DOI: 10.3389/fpls.2016.00657

[114] Wheeler TR, Craufurd PQ, Ellis RH. Temperature variability and the yield of annual crops. Agriculture Ecosystems & Environment. 2000;**82**:159-167. DOI: 10.1016/S0167-8809(00) 00224-3

[115] Jagadish SVK, Craufurd PQ, Wheeler TR. High temperature stress and spikelet fertility in rice (*Oryza sativa* L.). Journal of Experimental Botany. 2007;**58**:1627-1635. DOI: 10.1093/jxb/erm003

[116] Kimball BA. Crop responses to elevated CO_2 and interactions with H_2O, N, and temperature. Current Opinion in Plant Biology. 2016;**31**:36-43. DOI: 10.1016/j.pbi.2016.03.006

9

The Potential of Tree and Shrub Legumes in Agroforestry Systems

Ana I. Ribeiro-Barros, Maria J. Silva, Isabel Moura, José C. Ramalho, Cristina Máguas-Hanson and Natasha S. Ribeiro

Abstract

Climate variability and changes are utmost important primary drivers of biological processes. They are intimately associated with a wide array of abiotic stresses, highlighting the vulnerability of ecosystems and endangering biodiversity. Nitrogen-fixing trees and shrubs (NFTSs) constitute a unique group of plants for their wide range of applications at the environmental, social and economic levels. In this chapter, we review and analyse the potential of this group of legumes in agroforestry towards sustainable agriculture in Africa. In the first part, the intertwined pillar of sustainable agriculture is brought forward under the context of growing population and climate changes. The second part addresses general aspects of legumes, including botany and the symbiosis with rhizobia. The third part includes the application of NFTS as N-fertilizers in agroforestry, highlighting the importance of an accurate choice of the crop(s)/ NFTS combination(s) and cropping type (intercropping, multistrata or fallows). The implementation of agroforestry systems with NFTS should be supported by fundamental research strategies such as stable isotopes and systems biology and preceded by experimental assays, in order to identify the factors promoting N-losses and to design appropriate management strategies that synchronize legume-N availability with the crop demand.

Keywords: Africa, agroforestry, climate changes, sustainable agriculture, tree, shrub legumes

1. Introduction

Global agriculture is facing a series of challenges mainly related to growing population, increase more than 60% by the year 2050 to fulfil the needs of the world's population [1]. Secondly, drought and soil salinization are expected to result in losses of up to 50% of arable lands by the middle of this century [1, 2]. Thirdly, the spreading of agriculture to arid and semi-arid regions under intensive irrigation management will promote secondary soil salinization [3]. Thus, the future of agriculture must rely on the sustainable intensification of crop production to feed the increasingly growing population, as well as on the use of tolerant cultivars that are able to cope with extreme environmental conditions, i.e. low fertility and saline soils, increasing water shortage periods as well as raising air temperatures and CO_2 [4, 5].

These challenges will be particularly critical in the developing countries, which have the highest rates of population growth and where most of the farmland is managed by smallholders [6]. It is estimated that in these countries one of five persons still live on less than $1.25 a day [7]. In this context, intensive agriculture based on agro-chemicals and mechanization is not sustainable and the systems must rely on appropriate cropping and post-harvest practices, preferably based on local ecosystem-based resources. Such practices include, for example, the implementation of integrated agroforestry systems, crop-livestock integration and crop-aquaculture production, that concomitantly have the potential to promote the conservation and the rational use of biodiversity and other ecosystem services.

According to the Food and Agriculture Organization of the United Nations (FAO), sustainable agriculture lies at the core of the 2030 Agenda [7]. Indeed, 6 of the 17 sustainable development goals (SDGs) concentrate on this issue. These are as follows: (i) SDG 2—*End hunger, achieve food security and improved nutrition and promote sustainable agriculture*; (ii) SDG 6—*Ensure sustainable consumption and production patterns*; (iii) SDG 12—*Ensure sustainable consumption and production patterns*; (iv) SDG 13—*Take urgent action to combat climate change and its impacts*; (v) SDG 14—*Take urgent action to combat climate change and its impacts*; (vi) SDG 15—*Sustainably manage forests, combat desertification, halt and reverse land degradation, halt biodiversity loss*. Besides that, all the other 11 SDGs cross cut issues towards the end of hunger and poverty.

Since 2014, FAO has supported over 80 initiatives in Africa to promote sustainable agricultural production practices [8]. To achieve that, three intertwined pillars are considered essential: (i) *efficient use of resources*, i.e. agriculture intensification to produce more with less impact on natural resources; (ii) *environment protection and conservation*, i.e. better management of natural resources in order to protect biodiversity (and ecosystem's stability), water, soil fertility and reduce pollution and (iii) *resilient agriculture*, i.e. adopting approaches to adapt and mitigate the impact of climate change.

Legume fixing trees and shrubs play a crucial role in biodiversity dynamics. From the ecological point of view, their introduction in cropping systems may contribute to reduce the use of chemical fertilizers and to ecosystems stability.

2. Description and functioning

The Fabaceae or Leguminosae family is the third largest group of flowering plants and the second most important in agriculture [9]. According to recent molecular and morphological studies, Fabaceae is a single monophyletic family [9, 10], comprising more than 18,000 species distributed over ca. 800 genera and six sub-families (**Table 1**) [11]: (i) Cercidoideae and (ii) Detarioideae, both comprising mainly tropical species; (iii) Duparquetioideae, a sub-family from western and central Africa, with only one species identified; (iv) Dialioideae, widespread throughout the tropics; (v) the pantropical Caesalpinioideae, with more than 4000 species (including the former sub-family Mimosoideae) and (vi) the cosmopolitan and largest legume sub-family, Faboideae (Papilionoideae), with ca. 14,000 species, mainly herbs and small shrubs.

Varying in habit from annual herbs to large trees, legumes are conspicuous and well represented throughout temperate and tropical regions [9, 12, 13]. The family is particularly diverse in tropical forests and temperate shrub lands with a seasonally dry or arid climate. Such preference for semi-arid to arid habitats seems to be related to a nitrogen-demanding metabolism [9]. The vast majority of legume species (ca. 90%) is able to establish symbiosis with nitrogen-fixing diazotrophic bacteria of the genera *Rhizobium* or *Bradirhizobium* (collectively called rhizobia) at the root and, in some cases, at the shoot level [14]. The symbiosis results in the formation of a new plant organ, i.e. the root- or stem-nodule, where bacteria are hosted and fix atmospheric N_2, receiving in exchange energy and carbon to sustain their own metabolism as well as the symbiotic process [15]. This type of symbiosis has around 58 million years and arose from the genome duplication of the sub-family Papilionoideae [16].

Nitrogen is among the key elements for plant growth and production, being decisive to the adequate plant response to environmental stresses [17]. It is a major component of chlorophyll (photosynthesis), purines and pyrimidines (nucleic acids), amino acids (proteins) and ATP

Subfamily	Genera (number)	Species (number)	Distribution
Cercidoideae	12	ca. 335	Mainly tropical, e.g. *Bauhinia* spp., *Cercis* spp.
Detarioideae	84	ca. 760	Mainly tropical, e.g. *Amherstia* spp., *Detarium* spp., *Tamarindus* spp.
Duparquetioideae	1	1	West and Central Africa, *Duparquetia orchidaceae*
Dialioideae	17	ca. 85	Widespread throughout the tropics, e.g. *Dialium* spp.
Caesalpinioideae*	148	ca. 4400	Pantropical, e.g. *Caesalpinia* spp., *Senna* spp., *Mimosa* spp., *Acacia* spp.
Faboideae (Papilionoideae)	503	ca. 14,000	Cosmopolitan, e.g. *Astragalus* spp., *Lupinus* spp., *Pisum* spp.

*Includes the former sub-family Mimosoideae.

Table 1. Sub-families, number of genera and species, distribution and examples of tree and shrub legumes (adapted from LPWG [11]).

(energy). Although it is one of the most abundant elements in the Earth, its predominant form, i.e. N_2 (g), cannot be directly assimilated by the plants, which need reduced forms of this element (NH_4^+, NO_2^- and NO_3^-) [18, 19]. This conversion can be achieved chemically through the Harber-Bosch process, or biologically through bacterial nitrogen fixation [20]. While chemical nitrogen fixation is cost intensive and 40–50% of the nitrogen applied as fertilizer is lost via denitrification, runoff or leaching, only 10–20% of the biologically fixed nitrogen is lost that way [21]. Besides that, the use of chemical fertilizers has a series of ecological impacts, such as air, soil and water pollution [22]. Thus, there is a strong interest in symbiotic N_2 fixation between legumes and rhizobia towards the improvement of agricultural systems, i.e. better productivity with the least ecological impact [23, 24].

In Africa, tree and shrub legumes provide a wealth of goods and services (e.g. wood, food, medicines, energy and housing) to millions of rural and urban dwellers (**Table 2** and **Figure 1**) [25, 26]. The interest on this group of legumes has increased tremendously in the last decades, particularly regarding soil erosion control [27, 28] and farming systems (e.g. windbreaks, shade trees, nitrogen fertilizers, forage, fruits and vegetables) [29–32].

Species	Applications	Origin	$kg.N.ha^{-1}.yr^{-1}$
Acacia senegal (L.) Willd.	Poles, household, agriculture crafts, firewood, charcoal (stem and branches); tannin, ropes (bark); food (pods and seeds); forage (foliage and pods); honey (flowers); gum Arabic (Gum), medicine (various); erosion control; nitrogen fixation; fertilizer; fencing; intercropping [90, 91]	Drier tropical Africa, from Senegal and Mauritania (west) to Eritrea and Ethiopia (north-east) and South Africa (south); Oman, Pakistan and India [92]	28.7–46.7 [33] 7–12 [93] less than 20 [94]
Brachystegia boehmii Taub. (**Figure 1A**)	Small articles, firewood (stem and branches); ropes, twine, cloth and fishing nets, tanning, beehives (bark); food for edible larvae (leaves); medicinal (various) [25, 95–97]	Angola, Botswana, Malawi, Mozambique, Tanzania, Zaire, Zambia, Zimbabwe [13]	Not available
Brachystegia spiciformis Benth. (**Figure 1B**)	Construction, furniture, household items, firewood, charcoal (stem and branches); tannin, beehives, ropes, sacks (bark); forage (foliage and pods); honey (flowers); medicinal (various); nitrogen fixation, shading [91, 98, 99]	Angola, Kenya, Malawi, Mozambique, Tanzania, Zaire, Zambia, Zimbabwe [13]	Not available
Cajanus cajan (L.) Millsp. (**Figure 1C**)	Light construction, baskets, fuel (stems and branches); forage (vegetative parts); honey (flowers); food (seeds and pods); medicinal (various); erosion control; shading, sheltering, nitrogen fixation; fertilizer; intercropping [91, 100]	Unknown origin, probably Indian and African [13, 101]; India [100]	260 [91] 86 [94] 96 [102] 142 [103]
Gliricidia sepium (Jacq.) Walp.	Farm implements, furniture, posts, firewood, charcoal (stem and branches); forage (foliage and pods); honey and food (flowers); medicine and rodenticide (various); erosion control; shading; nitrogen fixation; fertilizer; fencing [91]	Central America, Caribbean, South America, Asia (Java and Peninsular Malaysia) [13]	210 [104] 35–38 [105] 108 [106]
Leucaena collinsii Britton & Rose1	Timber, firewood (stems and branches); forage (leaves); food (seeds); similar to gum Arabic (Gum); shading; nitrogen fixation; fertilizer; fencing; intercropping [91]	Mexico and Guatemala [13, 91]	Not available

Species	Applications	Origin	$kg.N.ha^{-1}.yr^{-1}$
Pterocarpus angolensis DC. (**Figure 1D**)	Building, furniture and handicrafts (stem); fish poison (bark); body anointment (root bark); tannin, dyestuff (Sap); forage (foliage); honey (flowers); medicine (various); erosion control; nitrogen fixation [91]	Angola, Mozambique, Namibia, South Africa, Swaziland, Tanzania, Zaire, Zambia [13]	Not available
Sesbania sesban (L.) Merr.	Firewood, charcoal (Stem and branches); tannin, ropes (Bark); forage (leaves and young branches); food (flowers), ropes, fishnets (fiber); gum (seeds and bark); medicine (various); shading; fencing; fertilizer; nitrogen fixation; intercropping [91]	Africa, Asia, Australia [13]	84 [102] 100 [103]
Tephrosia candida (Roxb.) DC.	Firewood (stem and branches); forage and insecticide (leaves); erosion control, shading; land reclamation; nitrogen fixation; fertilizer; fencing; intercropping [91]	India, SE Asia [13]	Not available
Tephrosia vogelii Hook.f	Fish poison, insecticide and molluscicide (leaves); medicine (various); shading; fencing; nitrogen fixation; fertilizer [91]	Tropical Africa, SE Asia [13, 91]	150 [103]

Table 2. Examples of tree and shrub legumes and their applications in formal and informal economies.

Figure 1. Details from *Brachystegia boehmii* leaves (A); *Brachystegia spiciformis* leaves and flowers (B); *Cajanus cajan* leaves, flowers and pods (C); *Pterocarpus angolensis* young leaves and mature fruit (D). Credits to Moura (A, D) and Catarino (B, C).

3. Importance and role in agroforestry tropical systems

Most tree and shrub legumes are resilient to extreme environments, e.g. erosion, low fertility, salinity, drought, fire and other adverse conditions [33–36]. Such abilities seem to be innate and enhanced by the symbiosis with N_2-fixing rhizobia [30, 37]. According to Diabate et al. [30] and Sprent [38], the use of nitrogen-fixing tree and shrubs (NFTSs) constitute a promising strategy to recover soil fertility, representing a sustainable agricultural approach to smallholder farmers. This is particularly important in sub-Saharan Africa where 80% of the farmland is managed by smallholders whose livelihoods depend strongly on the agricultural sector [6]. Most of these households live below the poverty line and therefore cannot afford the use of fertilizers. For example, smallholders from Niger, Namibia and Mozambique use less than 1 $kg.N.ha^{-1}.yr^{-1}$, i.e., 100 times less than the average fertilizer needs for most crops [1, 34].

The rates of N_2 fixation by NFTS depend on the species, climate and soil type, ranging from 0.1 to 700 $kg.N.ha^{-1}.yr^{-1}$ (**Table 2**) [33, 39, 40]. Despite the fact that many genera from the subfamilies Mimosoideae and Caesalpinioideae do not always establish root-nodule symbiosis, under proper environmental conditions, many species nodulate and fix atmospheric N at rates closer to those obtained with the traditional legumes belonging to the Papilionoideae [33]. Additionally there is also evidence that NTFS are also able to increase P availability in the soil, mostly due to mycorrhizal associations [41].

The use of fertilizer tree legumes (*Acacia anguistissima, Cajanus cajan, Gliricidia sepium, Leucaena collinsii, Sesbania sesban, Tephrosia candida* and *Tephrosia vogelii*) for sustainable maize (*Zea mays*) production has been analysed by Akinnifesi and collaborators [42] in East and Southern Africa (Zambia, Zimbabwe, Malawi and Tanzania). The authors reported a contribution of more than 60 $kg.N.ha^{-1}.yr^{-1}$ through biological nitrogen fixation (BNF), reducing the need of chemical N fertilizers in 75%. Besides that, N-fertilizer trees substantially increased crop yield, providing evidence that together with good management practices, maize yields can double as compared with traditional practices (without mineral fertilization). In Zambia, Mafongoya and Jiri [43] have analysed the use of *G. sepium* as green manure for cabbage (*Brassica oleracea*) and onion (*Allium cepa*) production. This practice produced higher crop yields than the unfertilized and full rate fertilized controls: ca. 16 (unfertilized), 43 (full rate fertilized), 48 (gliricidia) and 65 ton. ha^{-1} (half rate fertilizer and gliricidia) for cabbage and 22 (unfertilized), 43 (full rate fertilized), 65 (gliricidia) and 85 $ton.ha^{-1}$ (half rate fertilizer and gliricidia) for onion. In addition, gliricidia biomass replenished the soil with residual N, which was used by a subsequent crop, maize. In this case, the yields obtained (ca. 3–5 $ton.ha^{-1}$) were similar or slightly higher than those obtained with full rate fertilizer (ca. 2.5 to 4 $ton.ha^{-1}$) and unfertilized crop (ca. 1.5–3 $ton.ha^{-1}$). Nevertheless, caution should be paid to the potential environmental hazard of NO_3 leaching and resultant eutrophication [43–45], as well as to the choice of the best crop(s)/NFTS combination(s) and cropping type (intercropping, multistrata or fallows) [46–47].

Another interesting NFTS-based agroforestry system is the tree cropping system, like those used for coffee production (**Figure 2**). Such system is very popular in Latin America [48–51] and less exploited in Africa. In Mexico, the rates of N_2 fixation obtained by *Inga jinicuil* in

Figure 2. Agroforestry system with *Albizia* sp., coffee and maize in Gorongoza, Mozambique. Credits to Stalmans.

coffee plantations were above 40 kg.N.ha^{-1}.yr^{-1}, corresponding to 53% of the average amount of fertilizer applied annually. This observation reinforces the importance of the use of non-crop legumes in coffee agro-ecosystems [48–50]. According to the literature [49, 50], *Inga* spp. is the most popular choice from Mexico to Nicaragua. *G. sepium* and *Erythrina poeppigiana* are often the common choice in the low-lying areas of Honduras and Nicaragua and Costa Rica, respectively [51]. In Africa, similar systems may constitute a promising and sustainable solution to improve coffee (or fruits) productivity in the region.

4. Towards scientific knowledge

Legume research has been mainly focused on annual grain crops [34, 52–54]. Instead, a limited amount of knowledge has been produced in perennials. In this section, we will discuss the potential of the two promising strategies to analyse nitrogen mineralization and metabolism in tree legumes, i.e. stable isotopes and, briefly, systems biology.

4.1. Stable isotopes

The use of stable isotopes at natural abundance levels has brought a new dimension to our understanding of plant physiology and ecology. Analyses of the relative natural abundances of stable isotopes of carbon ($^{13}C/^{12}C$), oxygen ($^{18}O/^{16}O$), nitrogen ($^{15}N/^{14}N$) and deuterium (D/H) have been used across a wide range of scales, from cell to community and ecosystem level, contributing much to our understanding of the interactions between biosphere, pedosphere and atmosphere.

In general terms, processes such as diffusion and enzymatic incorporation favour the lighter isotope and lead to depletion of the heavier isotope as compared to source material. Natural abundance of ^{15}N can provide valuable information about N sources used by plants and fluxes of N in the ecosystems [e.g. 55–57]. There has been some debate on the interrelationship between nitrogen natural abundance ($^{15}N/^{14}N$) in soils and plants, and the use of a tracer or indicator of fractionation during N-uptake, assimilation and transport [58, 59]. Indeed, a variety of fractionations may occur during processes related to nitrogen transformation in soils

[60, 61] and plants [59], which may complicate source-sink relationships. For example, nitrification discriminates against ^{15}N more than N mineralization, which makes NH_4^+ isotopically heavier than the organic N from which it is derived [60]. Additionally, the $\delta^{15}N$ of a particular compound may change and, together with the complexity of the N geochemical cycle, the use of $\delta^{15}N$ should be carefully evaluated when applied to natural ecosystems.

However, there are substantial evidence that the natural abundance ratios for $^{15}N/^{14}N$ in soil and plants are useful integrators of the types and turnover rates of N cycling [62–64]. These ratios can indicate whether a range of plants have access to the same N source [59]. For instance, differences in leaf $\delta^{15}N$ can indicate differences in rooting depth or root characteristics, such as mycorrhizal or N-fixing root associations [59, 65]. Also, nitrification and plant uptake properties (such as timing and type of uptake) can be determined by the leaf $\delta^{15}N$ signatures [66, 67]. Robinson [59] developed a mixing model to account for contrasting N sources which provided useful insights on the quantification of biological N fixation in tree legumes [68, 69]. Since N_2-fixing species typically have $\delta^{15}N$ signatures close to the atmospheric value (0%), which strongly differ from the $\delta^{15}N$ signature of non-fixing species, $\delta^{15}N$ can be used as a sensitive tracer of N flow within an ecosystem. This approach was successfully used in the oligotrophic Portuguese primary dunes utilizing foliar $\delta^{15}N$ of the non-leguminous native shrub *Corema album* [70–72]. As the invasive *Acacia longifolia* and the native *Stauracanthus spectabilis* were the only legumes co-occurring with *C. album*, with no further sources of organic matter, this system represents an ideal model to quantify the impact of *A. longifolia* invasion. Similar to other ericoid $\delta^{15}N$ mycorrhizal plants [73], *C. album* exhibited depleted foliar values without legume influence which, together with its high abundance in this system, may function as a good monitoring plant for legume influence [71].

4.2. Systems biology

Systems biology is an emerging approach applied to biological scientific research that focuses on the complex interactions within biological systems, frequently associated with the environmental conditions. The best known example is the *Human Genome Project* which allowed major advances in human genetics and in the development of new medical therapies [74, 75]. Systems biology, commonly called '*Omics*' is associated with high-throughput analysis of e.g. genomes (DNA, *genomics*), transcriptomes (RNA, *transcriptomics*), proteomes (proteins, *proteomics*), metabolomes (metabolites, *metabolomics*), lipidomes (lipids, *lipidomics*) and interactomes (interactions between molecules, *interactomics*) coupled with bioinformatics, which integrates computational, statistical and mathematical modelling [76, 77]. In plants, systems biology has been essentially focused on models, such as arabidopsis and annual crops (e.g. rice, wheat, tomato, soybean, maize, sorghum, chickpea or groundnut) [78–85]. Systems biology research in perennial plants is still restricted to a small group of trees, namely eucalyptus, poplar, abies and pine (reviewed in Refs. [86, 87]). Among others, such studies led to significant advances on the global knowledge of plant biology (development and functioning), genomics-assisted breeding towards the production of crops tolerant to extreme temperatures, salinity, drought, pests and diseases, or the discovery of new bio-compounds with application in agriculture, medicine and in a wide range of industries [78, 79, 86, 87].

In our laboratory, we have recently initiated an integrated approach, combining eco-physiology and system's biology to understand the responses of two tree legumes (*Brachystegia boehmii* and

Colophospermum mopane) to abiotic stresses, namely high temperatures, drought and low soil fertility. Preliminary results indicate that these plants have an innate ability to cope with extreme environments and that such capacity is linked to an enhanced water and mineral use efficiency [88], reinforcement of the photosynthetic machinery and the antioxidant system as well as an elevated osmoprotection state [unpublished data; 89].

5. Concluding remarks

Agriculture has a primordial role to fight poverty and hunger and increase crop resilience to climate changes. The introduction of tree and shrub nitrogen-fixing trees into cropping systems is the most straightforward approach to reduce the use of chemical fertilizers, improving the soil ecosystem and the livelihoods of smallholder farmers in southern Africa. Additionally, agroforestry may improve ecosystem services such as, soil organic matter, biodiversity and N-retention. However, it is not devoid of environmental consequences, specifically N-leaching. Therefore, the implementation of agroforestry systems with NFTS should be preceded by experimental assays, in order to identify the factors promoting N-losses and design appropriate management strategies that synchronize legume-N availability with the crop demand.

Acknowledgements

The authors like to acknowledge Fundação para a Ciência e a Tecnologia through the research units UID/AGR/04129/2013 (LEAF), UID/GEO/04035/2013 (GeoBioTec), CE3C, Fundo para a Investigação Aplicada e Multissectorial. The authors thank Luís Catarino (CE3C), and Mark Stalmans (E.O. Wilson Biodiversity, Lab. Gorongoza National Park) for kindly providing the photographs.

Author details

Ana I. Ribeiro-Barros[1,2*], Maria J. Silva[1,2], Isabel Moura[1], José C. Ramalho[1,2], Cristina Máguas-Hanson[3] and Natasha S. Ribeiro[4]

*Address all correspondence to: aribeiro@itqb.unl.pt

1 Plant Stress & Biodiversity Lab, Linking Landscape, Environment, Agriculture and Food (LEAF), School of Agriculture (ISA), University of Lisbon (ULisboa), Portugal

2 GeoBioTec, Faculdade de Ciências e Tecnologia (FCT), Universidade NOVA de Lisboa (UNL), Portugal

3 Center for Ecology, Evolution and Environmental Changes (cE3c), Faculdade de Ciências, Universidade de Lisboa, Portugal

4 Faculty of Agronomy and Forest Engineering, Eduardo Mondlane University, Maputo, Mozambique

References

[1] FAO. World agriculture towards 2030/2050: The 2012 revision [Internet]. ESA E Working Paper No. 12-03. FAO. 2012. Available from: http://www.fao.org/fileadmin/userupload/esag/docs/AT2050_revision_summary.pdf

[2] Mahajan S, Tuteja N. Cold, salinity and drought stresses: An overview. Archives of Biochemistry and Biophysics. 2005;**444**:139-158. DOI: 10.1016/j.abb.2005.10.018

[3] Chaves MM, Flexas J, Pinheiro C. Photosynthesis under drought and salt stress: Regulation mechanisms from whole plant to cell. Annals of Botany. 2009;**103**:551-560. DOI: 10.1093/aob/mcn125

[4] CSSA. Position Statement on Crop Adaptation to Climate Change. [Internet]. Crop Science Society of America;. 2011. Available from: https://www.crops.org/files/science-policy/cssa-crop-adaptation-position-statement.pdf

[5] FAO. Save and Grow. A Policymaker's Guide to the Sustainable Intensification of Smallholder Crop Production [Internet]. 2011. Available from: http://www.fao.org/ag/save-and-grow/en/index.html

[6] FAO. Strategic Work of FAO to Increase the Resilience of Livelihoods [Internet]. 2017. Available from: http://www.fao.org/3/a-i6463e.pdf

[7] FAO. FAO and the 17 Sustainable Development Goals. [Internet]. 2015. Available from: http://www.fao.org/3/a-i4997e.pdf

[8] IFAD. Smallholders, Food Security, and the Environment. [Internet]. 2013. Available from: https://www.ifad.org/documents/10180/666cac24-14b6-43c2-876d-9c2d1f01d5dd

[9] Wojciechowski MF, Mahn J, Fabaceae JB. Legumes. The Tree of Life Web Project [Internet]. Available from: http://tolweb.org/

[10] Lewis G, Schrire B, MacKinder B, Lock M. Legumes of the World. Kew: Royal Botanical Garden; 2005. 577 p. ISSN: 0967-8018

[11] LPWG. A new subfamily classification of the Leguminosae based on a taxonomically comprehensive phylogeny. Taxon. 2017;**66**:44-77. Available from: http://www.ingenta-connect.com/contentone/iapt/tax/2017/00000066/00000001/art00004

[12] Rundel PW. Ecological success in relation to plant form and function in the woody legumes. In: Stirton CH, Zarucchi JL, editors. Advances in Legume Biology. Vol. 29. Monographs in Systematic Botany from the Missouri Botanical Garden. ST. Louis ; 1989. pp. 377-398

[13] ILDIS. International Legume Database & Information Service. [Internet]. 2015 Available from: http://www.ildis.org/leguminosae/

[14] De Faria SM, Lewis GP, Sprent JI, Sutherland JM. Occurrence of nodulation in the Leguminosae. New Phytologist. 1989;**111**:607-619. DOI: 10.1111/j.1469-8137.1989.tb02354.x

[15] Sprent J. Legume Nodulation: A Global Perspective. Wiley-Blackwell, Oxford; 2009. p. 200

[16] Young ND, Debelle F, Oldroyd GED, et al. The Medicago genome provides insight into the evolution of rhizobial symbioses. Nature. 2011;**480**:420-424. DOI: 10.1038/nature10625

[17] Carelli ML, Fahl JI, Ramalho JC. Aspects of nitrogen metabolism in coffee plants. Theoretical and Experimental Plant Physiology. 2006;**18**:9-21. DOI: 10.1590/S1677-04202006000100002

[18] Smil V. Nitrogen in crop production: an account of global flows. Global Biogeochemical Cycles. 1999;**13**:647-662. DOI: 10.1029/1999GB900015

[19] Vance CP. Symbiotic nitrogen fixation and phosphorous acquisition. Plant nutrition in a world of declining renewable resources. Plant Physiology. 2001;**127**:390-397. DOI: http://dx.doi.org/10.1104/pp.010331

[20] Burris RH, Roberts GP. Biological nitrogen fixation. Annual Review of Nutrition. 1993; **3**:317-335

[21] Galloway JN, Dentener FJ, Capone DG, et al. Nitrogen cycles: Past, present, and future generations. Biogeochemistry. 2004;**70**:153-226. DOI: 10.1007/s10533-004-0370-0

[22] Wagner SC. Biological nitrogen fixation. Nature Education Knowledge. 2011;**3**:15. Available from: http://www.nature.com/scitable/knowledge/library/biological-nitrogen-fixation-23570419

[23] Paul EA. Towards the year 2000: Directions for future nitrogen research. In: Wilson JR, editor. Advances in Nitrogen Cycling in Agricultural Ecosystems. CAB International, Wallingford; 1988. pp. 417-425. DOI: 10.1007/BF00032239

[24] Pawlowski K, Ribeiro A, Bisseling T. Nitrogen fixing root nodule symbioses: legume nodules and actinorhizal nodules. In: El-Gewely MR, editor. Biotechnology Annual Review. Vol. II. Elsevier Science, Rijeka; 1996. pp. 151-184. DOI: https://doi.org/10.1016/S1387-2656(08)70009-7

[25] Palmer E, Pitman N. Trees of Southern Africa. Balkema, Cape Town; 1972. 312 p.

[26] Campbell B, Frost P, Byron N. Miombo woodlands and their use: Overview and key issues. In: Campbell B editor. The Miombo in Transition: Woodlands and Welfare in Africa. Bogor: CIFOR; 1996. pp. 1-5. ISBN: 979-8764-07-2

[27] Ribeiro NS, Syampungani S, Nangoma D, Ribeiro-Barros A. Miombo woodlands research towards the sustainable use of ecosystem services in Southern Africa. In: Lo Y-H, Blanco JA, Roy S, editors. Biodiversity in Ecosystems – Linking Structure and Function. InTech, Rijeka; 2015. pp. 493-409. DOI: 10.5772/58494

[28] Moura I, Maquia I, Rija AA, et al. Biodiversity studies in key species from the African mopane and miombo woodlands. In: Bitz L, editor. Genetic Diversity. InTech, Rijeka; 2017. DOI: 10.5772/66845

[29] Smith OW. Fodder trees and shrubs in range and farming systems in tropical humid Africa. In: Speedy A, Pugliese PL, editors. Legume Trees and other Fodder Trees as Protein Sources for Livestock. FAO Animal Production and Health Paper 102; 1991. Available from: http://www.fao.org/docrep/003/T0632E/T0632E04.htm#ch4

[30] Diabate M, Munive A, De Faria SM, et al. Occurrence of nodulation in unexplored leguminous trees native to the West African tropical rainforest and inoculation response of native species useful in reforestation. New Phytologist. 2005;**166**:231-239. DOI: 10.1111/j.1469-8137.2005.01318.x

[31] Ribeiro A, Romeiras MM, Tavares J, et al. Ethnobotanical survey in Canhane village, district of Massingir, Mozambique: Medicinal plants and traditional knowledge. Journal of Ethnobiology and Ethnomedicine. 2010;**6**:33. DOI: 10.1186/1746-4269-6-33

[32] Dewees P, Campbell B, Katerere Y, et al. Managing the Miombo Woodlands of Southern Africa: Policies, incentives, and options for the rural poor. Journal of Natural Resources Policy Research. 2011;**2**(1): 57-73. DOI: 10.1080/19390450903350846

[33] Adams M, Simon J, Fautsch P. Woody legumes: A (re)view from the South. Tree Physiology. 2010;**30**:1072-1082. DOI: 10.1093/treephys/tpq061

[34] Bruning B, Rozema J. Symbiotic nitrogen fixation in legumes: Perspectives for saline agriculture. Environmental and Experimental Botany. 2013;**92**:134-143. DOI: http://dx.doi.org/10.1016/j.envexpbot.2012.09.001

[35] Maquia I, Ribeiro NS, Silva V, et al. Genetic diversity of *Brachystegia boehmii* Taub. and *Burkea africana* Hook.f. across a fire gradient in Niassa National Reserve, northern Mozambique. Biochemical Systematics and Ecology. 2013;**48**:238-247. DOI: https://doi.org/10.1016/j.bse.2012.12.020

[36] Ribeiro NS, Matos CN, Moura IR, et al. Monitoring vegetation dynamics and carbon stock density in miombo woodlands. Carbon Balance and Management. 2013;**8**:11. DOI: 10.1186/1750-0680-8-11

[37] Gutteridge RC, Shelton HM. Forage tree legumes in tropical agriculture. In: Gutteridge RC, Shelton HM, editors. Tropical Grassland Society of Australia, CSIRO Australia; 1998. Available from: http://www.fao.org/ag/agp/agpc/doc/publicat/gutt-shel/x5556e05.htm#1.1 the role of forage tree legumes in cropping and grazing systems

[38] Sprent J. West African legumes: The role of nodulation and nitrogen fixation. New Phytolologist. 2005;**166**:326-330. DOI: 10.1111/j.1469-8137.2005.01499.x

[39] Vanlauwe B, Giller KE. Popular myths around soil fertility management in sub-Saharan Africa. Agricultural Ecosystems and Environment. 2006;**116**:34-46. DOI: https://doi.org/10.1016/j.agee.2006.03.016

[40] Ajayi OC, Place F, Akinnifesi FK, et al. Agricultural success from Africa: The case of fertilizer tree systems in southern Africa (Malawi, Tanzania, Mozambique, Zambia and Zimbabwe). International Journal of Agriculture Sustainability. 2011;**9**:129-136. DOI: 10/3763/ijas.2010.0554

[41] Houlton BZ, Wang YP, Vitousek PM, Field CB. A unifying framework for dinitrogen fixation in the terrestrial biosphere. Nature. 2008;**454**:327-330. DOI: 10.1038/nature07028

[42] Akinnifesi KK, Sileshi AG, Chirwa PW, et al. Fertilizer trees for sustainable food security in the maize-based production systems of East and Southern Africa. A review. Agronomy for Sustainable Development. 2010;**30**:615-629. DOI: 10.1016/B978-0-444-52512-3.00022-X

[43] Mafongoya PL, Jiri O. Nutrient dynamics in wetland organic vegetable production systems in eastern Zambia. Sustainable Agriculture Research. 2016;**5**:78-85. DOI: 10.5539/sar.v5n1p78

[44] Hartemink A, Buresh R, Van Bodegom P, et al. Inorganic nitrogen dynamics in fallows and maize on an Oxisol and Alfisol in the highlands of Kenya. Geoderma. 2000;**98**:11-33. DOI: https://doi.org/10.1016/S0016-7061(00)00072-0

[45] Rosenstock TS, Tully KL, Arias-Navarro C, et al. Agroforestry with N_2-fixing trees: sustainable development's friend or foe? Current Opinion in Environmental Sustainability. 2014;**6**:15-21. DOI: http://doi.org/10.1016/j.cosust.2013.09.001

[46] Chikowo R, Mapfumo P, Leffelaar PA, et al. Integrating legumes to improve N cycling on smallholder farms in subhumid Zimbabwe: Resource quality, biophysical and environmental limitations. Nutrient Cycling in Agroecosystems. 2006;**76**:219-231. DOI: 10.1007/s10705-005-2651-y

[47] Mafongoya PL, Bationo A, Kihara J, et al. Appropriate technologies to replenish soil fertility in southern Africa. Nutrient Cycling in Agroecosystems. 2006;**76**:137-151. DOI 10.1007/s10705-006-9049-3

[48] Roskoski JP. Nitrogen fixation in a Mexican coffee plantation. Plant and Soil. 1982;**67**:283-291. DOI: 10.1007/BF02182775

[49] Soto-Pinto L, Perfecto Y, Castillo-Hernadez J, et al. Shade effect on coffee production at the northern Tzeltal zone of the state of Chiapas, Mexico. Agriculture, Ecosystems and Environment. 2000;**80**:61-69

[50] Galloway G, Beer J. Oportunidades para fomentar la silvicultura en cafetales de América Central. Serie Técnica 285: CATIE, Turrialba; 1997. 166 p.

[51] Bellefontaine R, Petit S, Pain-Orcet M, et al. Trees outside forests. Towards a better awareness. FAO Conservation Guide 35; 2002. ISBN 92-5-104656-5 Available from: http://www.fao.org/docrep/005/Y2328E/y2328e00.htm#toc

[52] Gupta S, Nadarajan N, Gupta DS. Legumes in the Omic Era. Springer, New York; 2014. 360 p. ISBN 978-1-4614-8370-0

[53] Lonati M, Probo M, Gorlier A, et al. Nitrogen fixation assessment in a legume-dominant alpine community: comparison of different reference species using the N-15 isotope dilution technique. Alpine Botany. 2015;**125**:51-58. DOI: 10.1007/s00035-014-0143-x

[54] Wahbi S, Maghraoui T, Hafidi M, et al. Enhanced transfer of biologically fixed N from faba bean to intercropped wheat through mycorrhizal symbiosis. Applied Soil Ecology. 2016;**107**:91-98. DOI: 10.1016/j.apsoil.2016.05.008

[55] Kohl DH, Shearer GB, Commoner B. Fertilizer nitrogen: Contribution to nitrate in surface water in a corn belt watershed. Science. 1971;**174**:1331-1334

[56] Amarger N, Mariotti A, Mariotti F. Essai d'estimation du taux d'azote fixé symbiotiquement chez le lupin par le traçage isotopique naturel (^{15}N). Comptes Rendus de l'Academie des Sciences. 1977;**284**:2179-2182

[57] Shearer G, Kohl DH. N_2-Fixation in field settings: estimations based on natural ^{15}N abundance. Australian Journal of Plant Physiology. 1986;**13**:699- 756. DOI: 10.1071/PP9860699

[58] Evans RD. Physiological mechanisms influencing plant nitrogen isotope composition. Trends in Plant Science. 2001;**6**:121-126. Available from: http://plants.trends.com

[59] Robinson D. $\delta^{15}N$ as an integrator of the nitrogen cycle. Trends in Ecology and Evolution. 2001;**16**:153-162. DOI: https://doi.org/10.1016/S0169-5347(00)02098-X

[60] Högberg P. ^{15}N natural abundance in soil-plant systems. New Phytologist. 1997;**137**:179-203. DOI: 10.1046/j.1469-8137.1997.00808.x

[61] Hopkins DW, Dowd RW, Shiel RS. Comparison of L- and D-amino acid metabolism in soils with differing microbial biomass and activity. Soil Biology and Biochemistry. 1997;**29**:23-29. DOI: https://doi.org/10.1016/S0038-0717(96)00266-0

[62] Nadelhoffer KJ, Fry B. Nitrogen isotopic studies in forest ecosystems. In: Lajtha K, Michener R, editors. Stable Isotopes in Ecology and Environmental Sciences. Blackwell Scientific Publications, Oxford; 1994. pp. 22-44. DOI: 10.1002/9780470691854

[63] Lorens P, Oliveras I, Poyatos R. Temporal variability of water fluxes in a *Pinus silvestris* forest patch in Mediterranean mountain conditions. Hydrology of Mediterranean and Semiarid Regions. 2003;**278**:101-105

[64] Fry B. Stable Isotope Ecology. Springer, New York; 2006. p. 300p.

[65] Cernusak LA, Pate JS, Farquhar GD. Diurnal variation in the stable composition of water and dry matter in fruiting *Lupinus angustifolius* under field conditions. Plant Cell and Environment. 2002;**25**:893-907. DOI: 10.1046/j.1365-3040.2002.00875.x

[66] Pate JS, Unkowich MJ, Erskine PD, et al. Australian mulga ecosystem $\delta^{13}C$ and $\delta^{15}N$ abundance of biota components and their ecophysiological significance. Plant Cell and Environment. 1998;**21**:1231-1242. DOI: 10.1046/j.1365-3040.1998.00359.x

[67] Pardo LH, Templer TH, Goodale CL, et al. Regional assessment of N saturation using foliar and root N-15. Biogeochemistry. 2006;**80**:143-171. DOI: 10.1007/s10533-006-9015-9

[68] Boddey RM, Silva LG, Reis V, et al. Assessment of bacterial nitrogen fixation in grass species. In: Triplett ED, editor. Prokaryotic Nitrogen Fixation: A Model System for Analysis of a Biological Process. Horizon Scientific Press, Wisconsin; 2000. ISBN: 978-1898486190

[69] Unkovich M, Blott K, Knight A, et al. Water use, competition and crop production in low rainfall, alley farming systems of south-eastern Australia. Crop Pasture Science. 2013;**54**:751-762. DOI: https://doi.org/10.1071/AR03049

[70] Hellmann C, Sutter R, Rascher KG, et al. Impact of an exotic N_2-fixing Acacia on composition and N status of a native Mediterranean community. Acta Oecologica. 2011;**37**:43-50. DOI: 10.1016/j.actao.2010.11.005

[71] Rascher K, Hellmann C, Werner C, et al. Community scale ^{15}N isoscapes: tracing the spatial impact of an exotic N_2-fixing invader. Ecology Letters. 2012;**15**:484-491 DOI: 10.1111/j.1461-0248.2012.01761.x

[72] Ulm F, Hellmann C, Cruz C, et al. N/P imbalance as a key driver for the invasion of oligotrophic dune systems by a woody legume. Oikos. 2016;**2**:231-240. DOI: 10.1111/oik.03810

[73] Craine J, Elmore A, Aidar M, et al. Global patterns of foliar nitrogen isotopes and their relationships with climate, mycorrhizal fungi, foliar nutrient concentrations, and nitrogen availability. The New Phytologist. 2009;**183**:980-992. DOI: 10.1111/j.1469-8137.2009.02917.x

[74] Snoep JL, Westerhoff HV. From isolation to integration, a systems biology approach for building the silicon cell. In: Alberghina L, Westerhoff V, editors. Systems Biology: Definitions and Perspectives. Topics in Current Genetics. Springer-Verlag, Berlin Heidelberg; 2005. pp. 13-30. ISBN: 978-3540229681

[75] Sauer U, Heinemann M, Zamboni N. Genetics: Getting closer to the whole picture. Science. 2007;**316**:550-551. DOI: 10.1126/science.1142502

[76] Kitano H. Computational systems biology. Nature. 2002;**420**:206-210. DOI: 10.1038/nature01254

[77] Kitano H. Systems biology: A brief overview. Science. 2002;**295**:1662-1664. DOI: 10.1126/science.1069492

[78] Nützman HW, Huang A, Osbourn A. Plant metabolic clusters – from genetics to genomics. New Phytologist. 2016;**211**:771-789. DOI: 10.1111/nph.13981

[79] Bode R, Ivanov AG, Hu NPA. Global transcriptome analyses provide evidence that chloroplast redox state contributes to intracellular as well as long-distance signalling in response to stress and acclimation in Arabidopsis. Photosynthetic Research. 2016;**128**:287-312. DOI: 10.1007/s11120-016-0245-y

[80] He B, Tao X, Gu Y, et al. Transcriptomic analysis and the expression of disease-resistant genes in *Oryza meyerian* under native conditions. PLoS One. 2015;7:e0144518. DOI: 10.1371/journal.pone.0144518.

[81] Pechanova O, Takáč T, Samaj J, et al. Maize proteomics: An insight into the biology of an important cereal crop. Proteomics. 2013;**13**:637-662. DOI: 10.1002/pmic.201200275

[82] Lee DG, Ahsan N, Lee SH, et al. Chilling stress-induced proteomic changes in rice roots. Journal of Plant Physiology. 2009;**166**:1-11. DOI: 10.1016/j.jplph.2008.02.001

[83] Peng Z, Wang M, Li F, et al. A proteomic study of the response to salinity and drought stress in an introgression strain of bread wheat. Molecular and Cellular Proteomics. 2009;**8**:2676-2686. DOI: 10.1074/mcp.M900052-MCP200

[84] Cheng L, Gao X, Li S, et al. Proteomic analysis of soybean [*Glycine max* (L.) Meer.] seeds during imbibition at chilling temperature. Molecular Breeding. 2010;**26**:1-17. DOI: 10.1007/978-3-319-41448-5

[85] Varshney RV, Kudapa H, Roorkiwal M, et al. Advances in genetics and molecular breeding of three legume crops of semi-arid tropics using next-generation sequencing and high-throughput genotyping technologies. Journal of Biosciences. 2012;**37**:811-820. DOI: 10.1007/s12038-012-9228-0

[86] Kole C, Muthamilarasan M, Henry R, et al. Application of genomics-assisted breeding for generation of climate resilient crops: progress and prospects. Frontiers in Plant Science. 2015;**6**:563. DOI: 10.3389/fpls.2015.00563

[87] Sundell D, Mannapperuma C, Netotea S, et al. The plant genome integrative explorer resource: PlantGenIE.org. New Phytologist. 2015;**208**:1149-1156. DOI: 10.1111/nph.13557

[88] Gaiser T, de Barros I, Lange FM, et al. Water use efficiency of a maize/cowpea intercrop on a highly acidic tropical soil as affected by liming and fertilizer application. Plant and Soil. 2004;**263**:165-171. DOI: 10.1023/B:PLSO.0000047733.98854.9f

[89] Antoniou C, Chatzimichail G, Xenofontos R, et al. Melatonin systemically ameliorates drought stress-induced damage in *Medicago sativa* plants by modulating nitro-oxidative homeostasis and proline metabolism. Nutrient Cycling in Agroecosystems. 2015;**101**:107-121. DOI: 10.1007/s 10705-014-9647-4

[90] Fagg CW, Allison GE. *Acacia senegal* and the gum arabic trade: monograph and annotated bibliography. Tropical Forestry Papers. Series 42: Oxford Forestry Institute; 2004. 260 p. 4 ISBN 10: 0850741572

[91] Orwa C, Mutua A, Kindt R, et al. Agroforestree Database: A Tree Reference and Selection Guide. World Agroforestry Centre, Kenya; 2009. Available from: http://www.worldagroforestry.org/output/agroforestree-database

[92] Boer E. Acacia senegal (L.) Willd. Record from PROTA4u. In: Oyen, LPA, Lemmens RHMJ, editors. Wageningen: PROTA; 2002. Available from: http://www.prota4u.org/search.asp

[93] Raddad AY, Salih AA, El Fadl MA, et al. Symbiotic nitrogen fixation in eight *Acacia senegal* provenances in dryland clays of the Blue Nile Sudan estimated by the ^{15}N natural abundance method. Plant and Soil. 2005;**275**:261-269. DOI: 10.1007/s11104-005-2152-4

[94] Sanginga N. Role of biological nitrogen fixation in legume based cropping systems; a case study of West Africa farming systems. Plant and Soil. 2003;**252**:25-39. DOI: 10.1023/A:1024192604607

[95] Augustino S, Hall JB, Makonda FBS, et al. Medicinal resources of the Miombo woodlands of Urumwa, Tanzania: Plants and its uses. Journal of Medicinal Plants Research. 2011;**5**:6352-6372. DOI: 10.5897/JMPR10.517

[96] Maroyi A. An ethnobotanical survey of medicinal plants used by the people in Nhema communal area, Zimbabwe. Journal of Ethnopharmacology 2011;**136**:347-354. DOI: 10.1016/j.jep.2011.05.003

[97] Bruschi P, Mancini M, Mattioli M, et al. Traditional uses of plants in a rural community of Mozambique and possible links with Miombo degradation and harvesting sustainability. Journal of Ethnobiology and Ethnomedicine. 2014;**10**:59. DOI: 10.1186/1746-4269-10-59

[98] Hines AD, Eckman K. Indigenous multipurpose trees of Tanzania. Uses and economic benefits for people. Working Paper. FAO, Rome; 1993. p. 276

[99] PROTA4U Web database. Available from: https://prota4u.org/database/

[100] van der Maesen, L.J.G., 2006. Cajanus cajan (L.) Millsp. Record from PROTA4U. In: Brink M, Belay G, editors. Wageningen: PROTA. Available from: http://www.prota4u.org/search.asp

[101] Le Houérou HN. *Cajanus cajan* (L.) Millsp. FAO. 2006. Available from: http://www.fao.org/ag/agp/agpc/doc/gbase/data/pf000150.htm

[102] Chikowo R, Mapfumo P, Nyamugafata P, et al. Woody legume fallow productivity, biological N2-fixation and residual benefits to two successive maize crops in Zimbabwe. Plant and Soil. 2004;**262**:303-315. DOI: 10.1023/B:PLSO.0000037053.05902.60

[103] Gathumbi SM, Ndufa JK, Giller KE, et al. Do species mixtures increase above- and belowground resource capture in woody and herbaceous tropical legumes? Agronomy Journal. 2002;**94**:518-526. DOI: 10.2134/agronj2002.5180

[104] Liyanage MS, Danso SKA, Jayasundara HPS. Biological nitrogen fixation in four *Gliricidia sepium* genotypes. Plant and Soil. 1994;**161**:267-274. DOI: 10.1007/BF00046398

[105] Hairiah K, Van Noordwijk M, Cadisch G. Quantification of biological N_2 fixation of hedgerow trees in Northern Lampung. Netherlands Journal of Agricultural Science. 2000;**48**:47-59. DOI: 10.1016/S1573-5214(00)80004-4

[106] Danso SKA, Bowen GD, Sanginga N. Biological nitrogen fixation in agro-ecosystems. Plant Soil. 1992;**141**:177-196. DOI: 10.1007/BF00011316

10

Optimization of Nitrogen in Durum Wheat in the Mediterranean Climate: The Agronomical Aspect and Greenhouse Gas (GHG) Emissions

Luigi Tedone, Salem Alhajj Ali and Giuseppe De Mastro

Abstract

Durum wheat (*Triticum turgidum* L. subsp. *durum*) is the most cultivated cereal crop in the Mediterranean basin, traditionally grown under rainfed conditions using conventional tillage. Agronomical practices, soil type and climate variables are known to influence crop productivity. Their interaction effect is very complex and the time in which they occur strongly affects yield and quality. The nitrogen supply, in combination with climatic conditions, is the main constraint determining the physiological performance, grain yield and quality response of wheat. In addition, the N formulation, fertilizer management, crop sequence, seasonal trends, and the supply of residual and mineralized N influence the response of wheat to N fertilizer. N fertilizer management must be optimized to prevent N deficiency in the critical crop growth period, to avoid yield and quality losses and also prevent the excessive application of N fertilizer, thus reducing the environmental impact. The split application of N fertilizer is a promising strategy that satisfies plant needs and reduces N losses through improved nitrogen use efficiency (NUE). Such a strategy can result in a remarkable reduction in greenhouse gas (GHG) emissions and the carbon footprint of Italian durum wheat, considering that the highest proportion of the total emissions deriving from N fertilizer production and its application.

Keywords: durum wheat, nitrogen, nitrogen use efficiency, greenhouse gas emission (GHG)

1. Introduction

Wheat is one of the main sources of carbohydrate foods worldwide, and as such its cultivation is considered strategic, along with other important cereals grown widely across other continents [1].

From 1950 to 1990, the production increased exponentially due to the combined effect of genetic improvement and new agronomic techniques, which allowed the crop yield to increase massively [2].

This increase in production is strongly related to the use of nitrogen, which is essential for the life of the crop and is a limiting input factor for the crop productivity [3]. Since the 1960s, there has been a ninefold increase in the use of nitrogen, and this is expected to grow by another 40–50% over the next 40 years [4].

The application of nitrogen fertilizer greatly improves agricultural productivity but increases the risk of aquifer pollution, estimated in the EU alone to be by 20% [5], also due to inefficient use of the fertilizer.

Sustainable agriculture is a very broad concept and includes ecological, economic, and social aspects.

"Sustainable agriculture" definition was defined by Congress in the 1990 "Farm Bill," [Public Law 101–624, Title XVI, Subtitle A, Section 1603 (Government Printing Office, Washington, DC, 1990) NAL Call # KF1692.A31 1990]. Under that law, the term "sustainable agriculture" means an integrated system of plant and animal production practices having a site-specific application that will, over the long term:

- satisfy human food and fiber needs;
- enhance environmental quality and the natural resource base upon which the agricultural economy depends;
- make the most efficient use of nonrenewable resources and on-farm resources and integrate, where appropriate, natural biological cycles and controls;
- sustain the economic viability of farm operations; and
- enhance the quality of life for farmers and society as a whole.

The Food and Agriculture Organization [2] has developed a common vision and an integrated approach to sustainability across agriculture, forestry, and fisheries. This unified perspective, valid across all agricultural sectors and taking into account social, economic, and environmental considerations, ensures the effectiveness of action on the ground and is underpinned by knowledge based on the best available science. It can be adapted at community and country levels to ensure local relevance and applicability.

From an environmental point of view, sustainability means a respectful farming system in terms of the use of natural resources such as water, soil fertility, and biodiversity, and a reduction in the use of chemicals. The study of energy systems has gained considerable importance in this regard, with the aim of reducing the use of fossil energy and, consequently, the emission of greenhouse gases into the atmosphere. The study of cropping systems that use improved crops and conservative systems becomes crucial for increasing the accumulation of organic matter in the soil.

The effect of climate change is of particular concern in agriculture, as it significantly affects crop yields and their inter-annual variability, and can potentially affect their traditional territorial distribution.

The phenomenon of climate change affects the whole planet but is very much observed at the Mediterranean basin level, where the climate is characterized by hot dry summers and mild wet winters. The main field crops are winter cereals, in particular durum wheat, which grows in the range of 200–600 mm per year [6]. Durum wheat and other cereals are typically sown in October to December and harvested in May to June.

The conditions for growth during this period are favorable for cereal crops, ensuring an optimal course of their cycle until the end of April. After this period, the phenomena of water and thermal stress during the final filling-ripening stage are likely.

Variable climatic conditions cause water stress with large fluctuations in both grain yield and in the physical property of the grain (hectoliter weight, 1000 seed weight), and also in grain protein content and composition, which has considerable effects on the rheological properties of semolina [7–9].

Nitrogen fertilization is a particularly significant parameter in the technique of wheat cultivation, in terms of its effects on specific agronomic characteristics, the production level, and the quality and quantity of grain proteins [10].

In the absence of other limiting factors, primarily the availability of water, wheat response to increasing nitrogen doses in terms of crop production and/or protein content is influenced by the level of chemical fertility of the soil [11]. The combined effect of nitrogen availability and optimum water conditions is evident in the tillering phase.

The effect of nitrogen and water deficiency is then evident in the subsequent phases of heading and spike emission/flowering, when stress conditions can cause an increase in flower abortions and a reduction in vegetation.

The effect of water stress, combined with high temperatures, contributes to determining the productivity and quality of wheat grain [10].

In Mediterranean environments, the crop response to nitrogen fertilization is conditioned by climate change (precipitation during the March–May period), agronomic practice (normal or conservative soil management), the quantity and number of fertilizer operations, and the type of fertilizer applied [12–14].

The effect of nitrogen availability, in optimum water conditions, is evident from the tillering phase. The combined effect of nitrogen and water deficiency is also evident in subsequent phases of heading and spikelet initiation/flowering, when stress conditions cause an increase in flower abortions and a reduction in vegetation [15, 16]. Management strategies, including the alteration of sowing time to avoid if possible the stressful period and using resistant wheat varieties would therefore be valuable to sustain productivity. In fact, alternative wheat management options have been proposed and tested worldwide to evaluate their performance with respect to yield [17]. These durum wheat cultivation management systems are necessary to maintain acceptable levels of grain yield with lower economic and environmental costs. One of the most effective and widely spread practices in wheat production is the use of a no-tillage system. This system has been increasingly adopted in dry areas as it enables water to be conserved in low rainfall regions. In addition to improving wheat productivity through improved soil quality [18], no-tillage represents an option to lower the environmental impact. It can save energy (a major direct expense for farmers in terms of fuel costs) through the reduction in mechanical operations for tillage and by lowering the emissions of GHGs, as compared to conventional tillage no-tillage acts as a sink source of soil emissions.

This chapter is a review of international studies that focus on the environmental implications of nitrogen fertilizer application in durum wheat, with respect to yield and quality response. Factors controlling wheat response to N fertilizer application are also discussed. The aim of this chapter is to identify options for enhancing NUE to achieve a more sustainable fertilization management of durum wheat in the Mediterranean environment.

2. Wheat diffusion and statistics

According to data collected between 2008 and 2013, cereal cultivation covered an area of more than 700 million hectare worldwide with a total production of 2.6 billion tonnes.

Wheat, as the main constituent of bread and durum, is the primary global cereal crop in terms of surface area invested, with more than 220 million hectares. By comparison, maize covers 184 million hectares and rice 162 million, according to FAO data.

Corn and rice are the top two crops in terms of productivity, while wheat is third with 729 million tonnes (**Table 1**).

Wheat production is mainly concentrated in Europe and Asia, where the crop's importance is increasing continuously, unlike on other continents where productivity has remained stable, or as in America, where the choice of alternatives crops (i.e., soybean) has led to a contraction in wheat production (**Figure 1**).

Improvements in cultivation, agricultural techniques, and mechanization have led to better wheat cultivars, and thus an increase in productivity.

The adoption of best agronomic practices has also been important in improving the sustainable production of cereals. Precision farming systems for the assessment of the physiological state of plants and conservative cultivation systems that ensure the optimization of inputs have enabled a higher level of soil water storage, resulting in an increase in crop production [18].

Species	Surface (ha)	Total production (tonnes)
Maize	184,800,969	1,037,791,518
Rice	162,716.862	741,477,711
Wheat	220,417,745	729,012,175

Table 1. Statistical data of the main cereal crops [2].

Conservative farming practices enabling the ability to retain water have been applied in countries such as Turkey, particularly in drought seasons, leading to an impressive yield increase in some regions [1].

FAO has consequently pushed key producer countries to change their farming practices so they meet sustainability criteria.

Durum wheat (*Triticum turgidum* L. subsp. *durum*) is a minor cereal crop representing just 5% of all wheat grown. It is thought to have originated in the Fertile Crescent (Mesopotamia) and is an allotetraploid (two genomes: AABB) with a total of 28 chromosomes (2n = 4x = 28), containing the full diploid complement of chromosomes from each of its progenitor species.

Durum wheat is mainly cultivated in three basins: the Mediterranean, the northern US and-Canada, and within the desert areas of the south-west US and northern Mexico. Other much smaller areas where durum wheat is cultivated include Russia and Kazakhstan, Australia, India, and Argentina (**Figure 2**).

In 2015, global production of durum wheat reached about 36 million tonnes, according to the International Grain Council [19]. Eurostat [20] reported that about 7.5 million tonnes of

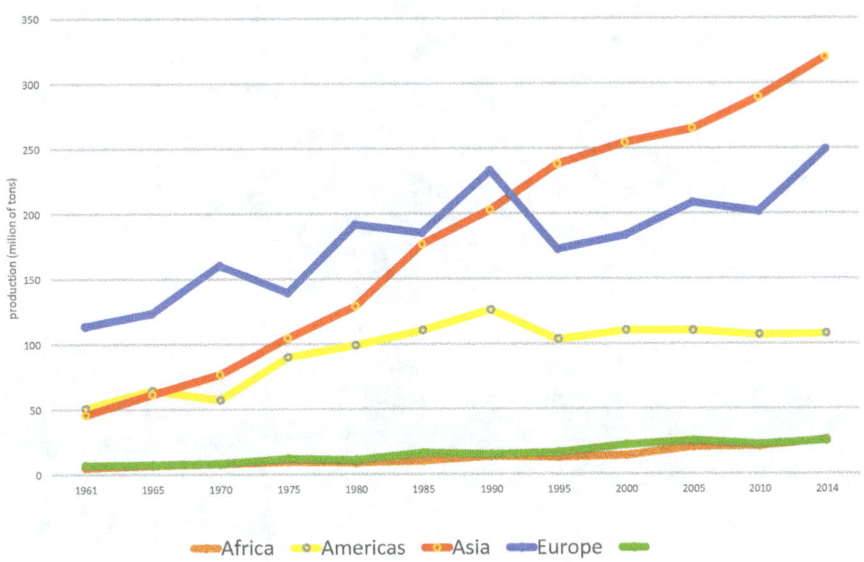

Figure 1. Trend of wheat production from 1961 to 2014.

durum wheat was produced in Europe in 2015/16, of which almost half (3.9 million tonnes) was produced in Italy (**Figure 3**), making it the primary durum wheat producer in Europe, followed by Turkey and France with average production of 2.7 and 1.7 million tonnes, respectively.

Smaller producers in the Mediterranean basin are concentrated in Morocco, Algeria, and Tunisia, mainly due to the effect on the crop cycle of the dry climate that often occurs there. The consumption and manufacturing of durum wheat is concentrated around the Mediterranean Sea, where the main food products of durum wheat are consumed: pasta, couscous, bulgur, and bread, which are obtained from four completely different technologies.

An additional 5 million tonnes guaranteed that local production is in fact required, which is mainly imported from North America.

The total production in the Mediterranean Basin under the winter cycle varies significantly, as the agronomic yields are highly influenced by rainfall during the spring period, and the total production can fluctuate between approximately 14 and 20 million tonnes in different years.

Currently, due to immigration to Europe and the growing popularity of pasta worldwide, increasing quantities of couscous are produced in Europe, and an increasing amount of pasta is produced in Northern Africa and in Tunisia.

Turkey is becoming an important exporter to the Middle Eastern and African pasta markets, where durum wheat is used for both bulgur and pasta production.

French breeding companies produce durum wheat varieties from commercial lots, which are very well managed from both agronomic and segregation points of view.

Semolina (milled durum wheat) is the raw material for four different food products: pasta, couscous, bulgur, and bread, all of which are typical of the Mediterranean diet. To ensure

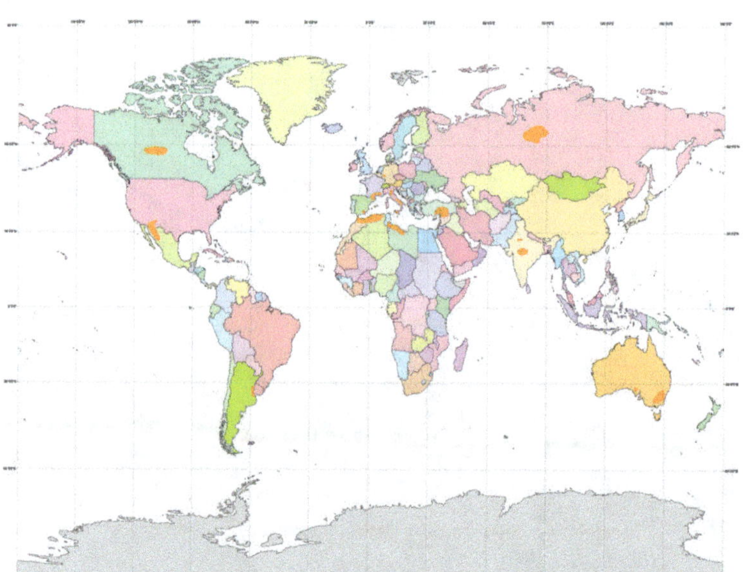

Figure 2. World map with the territories where durum wheat is grown highlighted in red (Eurostat).

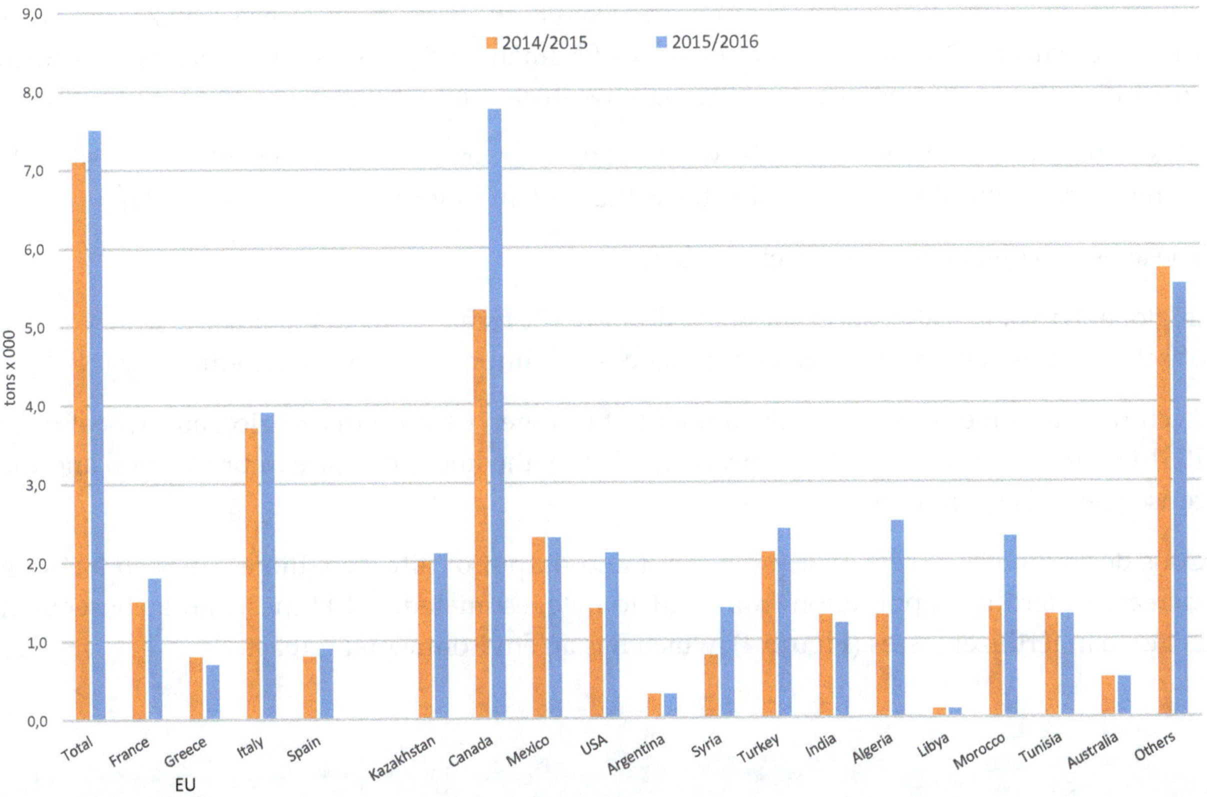

Figure 3. Main world producers of durum wheat (Eurostat).

a constant and sufficient supply in the Mediterranean basin, its cultivation and associated agronomic practices must be modified, to cope with the projected change in climate events. Meeting the increasing demand for durum wheat is a challenge even under current conditions, with the reductions from year-to-year variation in grain yield due to irregular and seasonal rainfall distribution. Under the winter cycle of the Mediterranean basin, rain-fed durum wheat production can experience remarkable yield losses due to insufficient rainfall, particularly during the spring.

3. Nitrogen consumption for wheat production

Nitrogen (N_2) is the main component of our atmosphere, at a percentage of 78.084%. It is essential for various metabolic reactions and vital processes, and is involved in various reactions that occur through air, soil, and water. It thus undergoes various chemical and biological transformations.

Nitrogen is a crucial mineral element and is involved in several fundamental compounds (amino acids, nucleic acids, chlorophyll, cytokines, polyamines, and secondary metabolites) essential for the biological plant cycle [3].

The agricultural world is heavily dependent on the nitrogen cycle, and sometimes the excessive use of nitrogen results in increased agricultural production [21]. More than 100 million Mt of N year^{-1} of reactive N is industrially produced by the Haber-Bosch process using fossil fuels as energy sources, and of this 50% is applied to the three main cereals (maize, 16%; rice, 16%; and wheat, 18%) that provide the bulk of human food calories and proteins consumed either directly as grain or indirectly through livestock products [22].

Most nitrogen is contained in seeds, which have an average protein content of 12%, slightly higher in hard wheat than tender. In straw, the content ranges from 3.5 to 4.8% [23].

The average content of N in protein is about 6%.

In global terms, it is almost certain that additional N fertilizer will be required, which can be offset to some extent by management practices that improve nitrogen use efficiency (NUE).

According to some studies [22], the amount of nitrogen used on maize, rice, and wheat from 1961 to 2010 increased, which has been justified by the linear increase in production and the consequent N asportation.

Considering the balance between the input and output of nitrogen, there has been a strong increase of fertilizer application from 1961 to date, estimated at 10 times, and a consequent increase in fertilizer losses (**Figure 4**), which has an environmental impact.

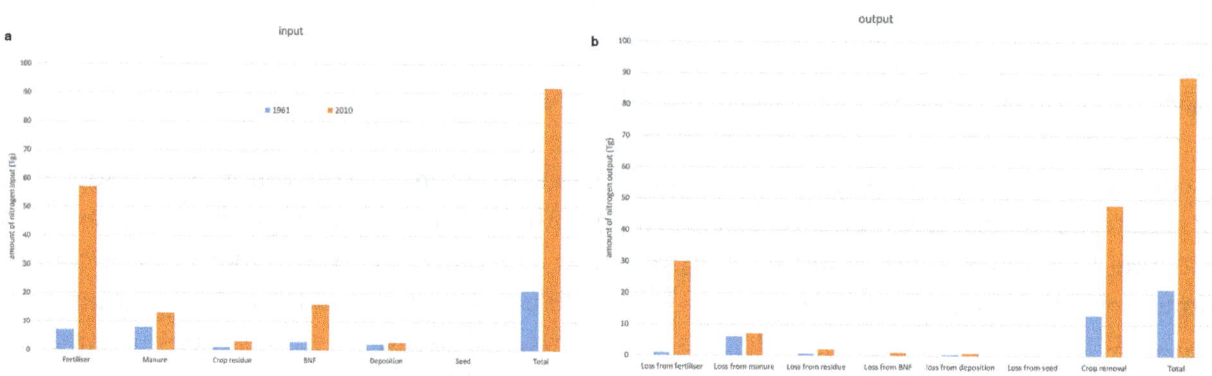

Figure 4. Estimated N inputs and outputs in 1961 and 2010 (from Ladha et al. [22]).

4. Agronomical implications of nitrogen use and the system of optimizing nitrogen use on durum wheat

Wheat is very sensitive to nitrogen deficiency and is highly reactive to the element. One of the most obvious responses to nitrogen deficiency is chlorosis, as a result of the lack of chlorophyll synthesis and reduced cell size and proliferation, leading to a stunted, reduced leaf surface and a yellowish (chlorotic) appearance of the crop.

Chlorosis is particularly apparent initially on mature leaves, and later on, the last growing leaves, as the N is shifted from old leaves to new to support growth. Thus, the older leaves dry out and cause poor plant growth and reduced yield.

Plants with an adequate amount of nitrogen grow rapidly and have dark green leaves and stems that grow very fast. Nitrogen is the mineral element most absorbed by plants and is placed in a plant's organs according to its physiological needs (seeds, leaves, and stems).

Nitrogen promotes digestion and increases vegetative growth, the number of heads per plant and the dimension of the spike, the weight of the kernels and the protein content [24].

Excess nitrogen may favor lodging, particularly in tall grain varieties, and retard the cycle with stress on grain filling if the season course is dry. The culture is also more susceptible to rust and septoria attacks.

Therefore, responses to increasing doses of nitrogen fertilizers on the yields are variable due to the sensitive influence of pedoclimatic conditions and the effect of crop precession. An optimum climatic performance in the filling phase of the kernels can favor both the accumulation of amidaceous and protein substances, so the availability of nitrogen in the soil at this time is critical to the qualitative improvement of the grain.

The calculation of N removal from seed durum wheat with 11% protein is [25]: SEED t ha^{-1} × 0.11 × 2.34 (correction factor) = X kg N ha^{-1} required.

This results in about 25.7 kg/ha of N being removed from 1 tonne of seed wheat. Studies have reported that in general between 24 and 28 kg/tonnes is removed from durum wheat grain.

Nitrogen in the plant is a basilar element, as it is essential to the plant's vital activity. The following distribution of nitrogen in the plant at different stages has been reported in several studies:

Recommendations of the correct dose of N fertilizer to be applied should take into account the different N source (**Figure 4**), which is calculated based on the following equation: (**Table 2**)

Plant part	Dry matter	Nitrogen
Anthesis		
Total (kg/ha)	7644	84.3
Shoot (%)	84	80
Root (%)	17	20
Harvest		
Total (kg/ha)	7422	71.7
Straw (%)	71.4	35.7
Grain (%)	28.6	64.3

Table 2. N distribution balance in wheat plant.

$$\text{N amount(D)} = \text{crop N demand(Fc)} - \text{environmental inputs(E)} + \text{environmental removal(U)} \quad (1)$$

and, in more detail: $D = Fc - (P + M + Cp) + (L + V + De + Um)$

where

- Fc is crop N demand
- E represents the amount of nitrogen that can be used from the culture but is not distributed with the fertilizer, and which is derived from: atmospheric precipitation (P); mineralization of soil organic matter (M); contributions arising from the previous crop (Cp); and U represents the amount of N that the environment removes due to the possible utilization by plants through:
- leaching (L)
- volatilization (V)
- denitrification (De)
- immobilization into soil organic matter (Um)
- **Atmospheric precipitation (P)**
- The amount of nitrogen from rain varies widely due to geographical position and fluctuations throughout the year, which makes it extremely difficult to generalize. However, in the Mediterranean climate, we can estimate a supply of nitrogen from atmospheric precipitation ranging from 20 kg/ha to 30 kg/ha for the entire crop cycle of wheat.
- **Mineralization of soil organic matter (M)**
- Soil nitrogen (N) content ranging between 0.05% and 0.2%. Considering a 30-cm layer of soil depth, this values correspond to an amount of total N between 1,950 to 7,800 kg/ha, higher than those necessary for the cultivation of wheat (**Figure 5**).
- Approximately 98%, however, is contained in the organic substance soil (humus) and is not absorbable by plants. The organic matter is gradually decomposed by microorganisms and atmospheric agents, and the nitrogen content is transformed into mineral form, which is then absorbable by plants.
- This complex set of processes is known as the mineralization of the organic matter. In warmer climates and sandy soils, 3% of nitrogen can be freed each year and 1% in more temperate climates and with clay soils. A 30-cm layer of soil, therefore, releases 40–80 kg/ha of nitrogen.

4.1. Contributions from the previous crop (Cp)

Previous crops such as legumes have certain characteristics that can result in nitrogen production. The legume species are able to establish a high amount of nitrogen for the symbiotic bacteria, which partly remain in the roots and enter into the cycle of organic soil

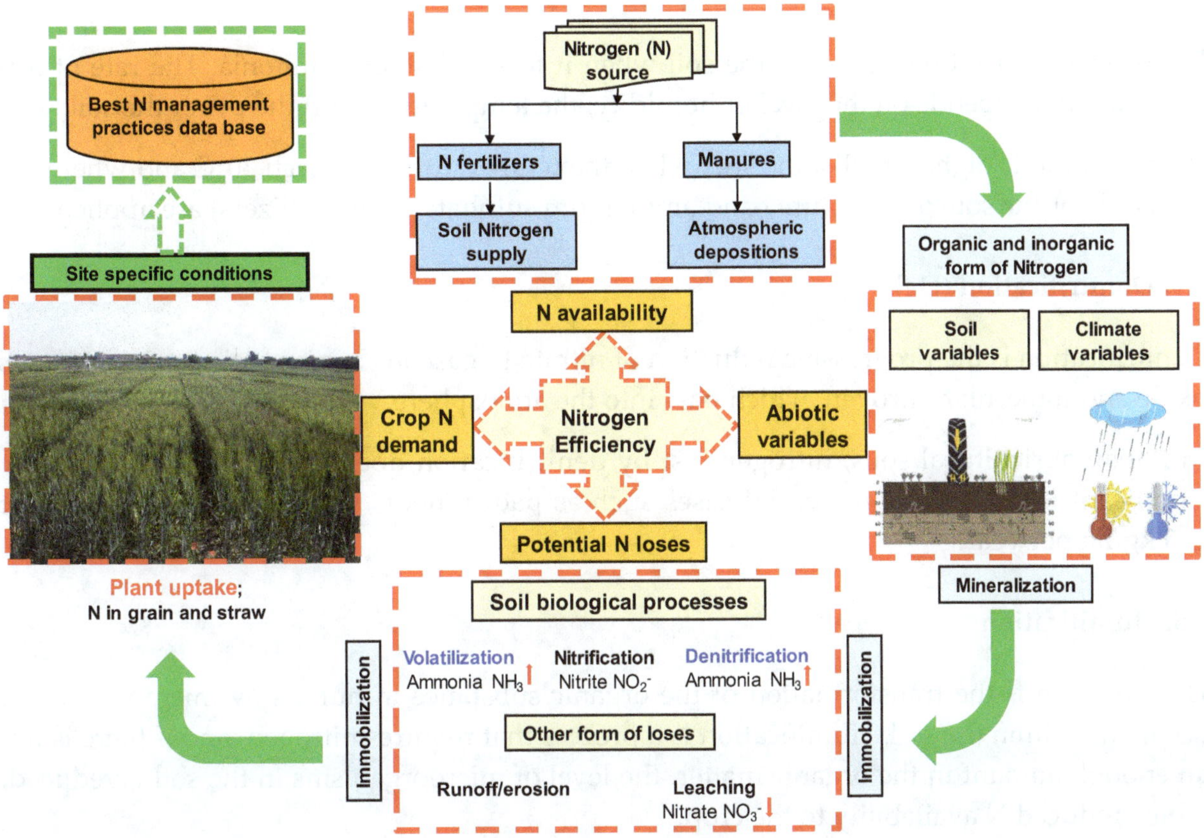

Figure 5. Potential fate of nitrogen in the agricultural system (in the case of durum wheat).

matter. Legumes leave nitrogen in the soil in available form for the crops that follow under a rotation system, but providing an accurate value of the quantity individual species is difficult. However, it can be estimated at about 30 kg/ha on average for perennial legume species and 20 kg/ha for annual species [26].

Nitrogen can also be derived from organic fertilization carried out in previous years.

4.2. Leaching (L)

Leaching is the movement of nitrogen in percolating water along the profile of the soil, when it exceeds that taken up by the root layer of plants.

As a result of leaching, nitrogen is not utilized by the crops and it thus represents both a nutritional and economic loss.

Normally leaching affects nitric form, even in sandy soils and it can also affect the ammonia form.

The water constitutes the vehicle by which the nitrate N in the soil moves, and nitrogen leaching is, therefore, a phenomenon that, for the parity of nitrates present in the soil, is closely dependent on its water balance.

Leaching is the main, if not the only, type of nitrogen removal affected by the environment.

4.3. Volatilization (V)

Nitrogen volatilization occurs in the soil when it forms gaseous ammonia. The rate of this phenomenon depends on the level of humidity, the temperature, and the pH of the soil.

Volatilization is higher in alkaline soil, when the temperature can reach 30°C and where the ammonia or the sources of it (urea and ammonium sulphate from fertilizers) are applied.

4.4. Denitrification (De)

Denitrification is the progressive reduction of nitrate to gaseous compounds, such as nitrous oxide and molecular nitrogen, which pass into the atmosphere.

In Italian agricultural soils, nitrogen lost by denitrification does not appear to reach very significant levels, except in special cases such as paddy fields, and has been estimated at 1–5 kg/ha per year.

4.5. Humification

Humification is the transformation of the organic substance in humus by microorganisms, occurring within the soil. Humification is a process that requires nitrogen, and if there is not an enough amount in the organic matter, the level of microorganisms in the soil is reduced, which reduced N availability to the crop.

To prevent fertility decreasing, the soil humus content must remain unchanged over time, and the humification value must equal that of the mineralization. Sufficient organic matter (crop residues or fertilizers) is, therefore, necessary to restore the proportion of humus and thus the organic nitrogen of the soil.

The response of wheat to N fertilizer is influenced by different factors: soil management, N fertilizer management, and timing of application, soil properties, crop sequence, seasonal trends and the supply of residual and mineralized N.

The agronomic efficiency of N fertilizer in a Mediterranean climate may be lower than that in temperate zones as the climatic conditions are largely outside the control of farmers, and it is difficult to predict the amount of N to apply to attain optimum growth.

4.6. Nitrogen use efficiency

This is the efficiency ratio of output (economic yield) to input (fertilizers) and increases in environmental and economic pressure make it a priority to optimize nitrogen use efficiency (NUE) in agriculture [27].

NUE has various definitions, but that of Moll et al. [28] is one of the most complete, as it does not only refer to the nitrogen of manure and fertilizers:

$$\text{NUE} = \frac{\text{N uptake}}{\text{N available}} \times \frac{\text{Seed yield}}{\text{N uptake}} \tag{2}$$

Here, the term "N available" also refers to the amount of N derived from mineralization of S.O. soil, atmospheric deposition, and bacterial nitrogen fixation. For simplicity of analysis, in many cases, the nitrogen arising from environmental inputs is defined as the absorption of a crop not fertilized.

The different viable approaches for the enhancement of NUE are explored in more detail in the introductions of each individual experiment. Several tools and strategies are available to improve the NUE, such as systems for genetic breeding that include genetic parameters and agronomical tools [29].

4.7. Breeding for NUE

Genetic improvements aimed at optimizing the NUE are crucial for achieving the economic benefits farmers seek, and result in a reduction of the environmental impact of nitrogen.

The increase in yield brought by a new cultivar results in an increase in nitrogen consumption, which apparently is the aspect that has been able to increase the yield from old to new varieties [30]. Indeed, studies have reported that Cimmyt varieties saw a gain in NUE of between 0.4 and 1.1% annually from 1962 to 1985 [31, 32].

Several components affect the level of NUE improvement. In particular, the N Harvest Index (NHI) component contributed to an estimated 0.15% improvement in the NUE per year, which was aimed at reducing the amount of straw produced during harvesting [33–35] (**Table 3**).

Other physiological, metabolic and physical-chemical components can help improve NUE by reducing the contribution of nitrogen fertilizer.

Aspects that act on both the cellular scale and the whole plant include root absorption, nitrate assimilation, N distribution, photosynthesis, senescence, nitrogen rebuilding, accumulation, and wheat composition. These have been taken into account in genetic improvement work [36] and are reported in **Table 4**.

Period	Genotypes	N level (kg N/ha)	NUE (% per year)	References
1962–1985	8	0	1.2	Ortiz-Monasterio et al. [31]
		75	0.4	
		150	0.6	
		300	0.9	
1977–2007	24	0	0.35	Sylvester-Bradley and Kindred [30]
		200	0.58	
1985–2010	195	150	0.37	Cormier et al. [34]
		250	0.3	

Table 3. Assessment of yearly percentage genetic gain in nitrogen use efficiency (NUE) from a direct comparison of old and modern cultivars [35].

Traits	Species	References
Root length density at depth	*T. aestivum, H. vulgare, O. sativa*	Gregory and Brown (1989), Steele et al. (2006), Reynolds et al. (2007) and Manschadi et al. (2006)
Glutamine synthetase (GS) activity	*T. aestivum, Z. mays*	Habash et al. (2001), Hirel et al. (2001), Masclaux et al. (2001) and Martin et al. (2006)
Alanine aminotransferase (AlaAT) activity	*O. sativa, B. napus*	Shrawat et al. (2008) and Good et al. (2007)
RuBisCo CO_2 specificity factor	*G. partitida*	Uemura et al. (1997)
Introduction of C4 "Krantz" anatomy into to C3 species	*Oryza* spp.	Hibberd et al. (2008)
Specific leaf N content	*Triticum* spp.	Austin et al. (1982) and Semenov et al. (2007)
Vertical N distribution with leaf layer	*T. aestivum, S. altissima*	Hirose and Werger (1987) and Critchley (2001)
Leaf posture	*T. aestivum*	Araus et al. (1993)
Leaf photosynthetic rate post-anthesis	*Z. mays, T. aestivum, O. sativa*	Reynolds et al. (2001, 2005), Wang et al. (2002) and Ding et al. (2007)
Stem N storage	*T. aestivum*	Critchley (2001)
Stay-green	*T. aestivum, T. durum, S. bicolor*	Borrell and Hammer (2000), Spano et al. (2003) and Verma et al. (2004)
N remobilization efficiency post-anthesis	*S. bicolour*	Borrell and Hammer (2000)
Post-anthesis N uptake	*T. aestivum*	Triboı et al. (2006)

Table 4. Physiological aspects studied for nitrogen use efficiency [36].

4.8. Effect of soil type

The soil-plant-environment system is very complex due to the interactions of several factors that affect plant growth, development, and the final yield. Agricultural soil represents the major source of nutrition for crop survival, and therefore, a good balance of macro- and micro-elements in the soil is fundamental to ensuring a better crop response.

Agriculture intensification has become common practice in different parts of the world to provide sufficient food for the increasing population, and while the high amounts of external inputs such as inorganic fertilizers have led to considerable increases in overall food production worldwide, they have resulted in continuous environmental degradation, particularly of the soil. This deterioration of soil quality and the reduction in agricultural productivity due to nutrient depletion, organic matter losses, and erosion have in turn led to a greater use of chemical inputs, particularly nitrogen. Today, N fertilizer is considered one of the main inputs for crop production, and any lack of N will lead to disruption of plant growth, which will result in economic losses. Nitrogen in the soil is present in soluble form as organic N (Org-N), as ammonium nitrogen (NH_4^+-N), and as nitrate nitrogen (NO_3^--N). The nitrogen cycle is extremely dynamic and complex, so climatic conditions and the physical and chemical properties of a particular soil can influence the microbiological processes responsible for mineralization, fixation, and denitrification of soil nitrogen. The N availability in the soil is known to be largely controlled by bacteria, so the rate of N cycling is dependent on factors such as soil type, soil moisture, temperature, and pH. Generally, the application of N fertilizer in wheat field depends on the availability of soil nitrogen and the potential losses. The effect of soil moisture on the response of wheat to N fertilizer application has been studied extensively, and the research has found that soil moisture must be considered in N fertilizer management for all soil types. The literature review shows that yield response varies with different nitrogen and moisture levels. In rain-fed areas, for example, soil moisture must be reserved when selecting fertilizer rates.

The application of N to wheat should be timed so an adequate amount is supplied when the crop needs it. As the nitrogen cycle is biologically influenced by prevailing climatic conditions, along with the physical and chemical properties of a particular soil, the choice of a suitable N fertilizer formula to be applied in a specific soil is fundamental. The electrical charges of soil particles vary according to the soil texture, so understanding the chemical properties and therefore, the chemical reactions within the soil will inform recommendations as to when fertilizers can be applied to soils. Soil scientists indicate that these reactions have a major influence on when fertilizer can be applied and how efficiently it is taken up by the crop. Clay particles in the soil have a negative electrical charge, so the amount of clay present in the soil surface is an important factor in these reactions. Clay particles with negative charges will react with components of the fertilizer that dissolve as positively (+) charged particles (cations) when added to soil. For example, the application of N fertilizers that include more ammonium (positive charges) and less nitrate (negative charges) forms of N can reduce the potential for losses in the short term. Ammonium (NH_4^+) forms of N bind to soil particles with negative charges and will not be subjected to losses through leaching or denitrification, therefore, increasing N availability for the plant. However, if not all of the applied ammonium is taken up by the plant, soil bacteria will convert the remaining amount to nitrate (NO_3^-), a form that is not bound to soil particles and can be lost when excessive rainfall leaches or saturates soils. Low soil temperatures can affect the transformation process as they minimize the activity of the nitrifying bacteria, which can naturally help prevent losses of ammonium forms of N.

4.9. Effect of soil management on nitrogen application in wheat

Soil is managed to ensure that sufficient nutrients are available for plant growth and development, and therefore, good soil management is critical for crop productivity. Poor soil management can lead to erosion, loss of fertility, deterioration of soil structure, and poor crop yields [37]. Understanding the nutritional status of the soil is important in any N fertilizer management plan, and involves the knowledge of different soil components such as microorganisms (densities, diversity, and activity), organic matter and the biological processes by which microorganisms make micro- and macroelements available to the plant, through the decomposition of organic matter. Soil microorganisms have many physical and chemical effects on plants. For example, their secretions help dissolve minerals, and they convert inorganic substances into other forms, making them available to plants. They also aerate the soil and help decompose organic matter [38].

Direct and indirect interventions can improve biological soil management. Direct methods attempt to alter the abundance or activity of specific groups of organisms [39]. Seeds or roots can, for example, be inoculated with rhizobia, mycorrhizae, fungi, and rhizobacteria to enhance soil fertility. Most agricultural practices are in fact indirect methods, such as incorporating plant residues into the soil, introducing legume crops and various tillage methods. They provide important sources of nutrition in the fields, and therefore, must be considered when applying N fertilizers [40]. Many studies have reported that the incorporation of straw, together with the N-rich aboveground biomass of the crops promotes the immobilization of soil mineral N, restores soil organic matter, and improves aeration. The N content in the residue is the input to the soil humus, and thus becomes a part of the supply for future crops. The levels of residual inorganic N in the root profile contribute to the total plant N and should be taken into account when formulating fertilizer recommendations for improving N utilization efficiency. The influence of tillage on the physical, chemical, and microbiological properties of the soil is becoming of great importance when studying issues related to soil fertility and productivity. Several studies suggest that tillage management can have beneficial effects on both grain yield and nutritional quality. Tillage practices such as no-tillage are believed to benefit the soil in terms of improving water retention and the conservation of organic matter in the soil [41].

Conservation agriculture (CA) is an ecologically interesting land management technique, aimed at reducing agronomical inputs related to the reduction of ploughing [42, 43].

The technique may have the potential to reduce the loss of important parameters of soil fertility.

Organic matter (OM) is one such fundamental parameter, and any increase in the soil has obvious benefits in terms of reducing greenhouse gas emissions and thus climate change [17, 37, 44]. An increase in OM, particularly in the more superficial layers, results in an improvement in the physical and chemical properties of the soil, its biological activity and, consequently, in N losses [45].

CA-related studies have also found a tendency to better water conservation and adsorption in the soil; however, under special conditions, CA adoption appears to result in lower NUE rates than conventional systems, largely due to N fertilizer immobilization through crop residues.

Fertilization strategies must therefore be appropriately adjusted by combining the rate, timing, splitting, and source of N, to optimize yield and quality [46].

Soil conditions such as temperature must also be considered, which is found to be lower in a CA system compared to conventional systems. Soil conditions are fundamentally important in N fertilizer management, as they strongly influence the nitrogen cycle. Studies have indicated that, even after CA has been applied for over 10 years, slightly higher doses of fertilizers are still needed to achieve significant yields in winter wheat, with the presence of crop residues on the surface layer.

4.10. Varietal assessment

Developments in wheat genetics since the 1960s have enabled a substantial increase in the yield potential of many crops, particularly cereals [47].

The large-scale genetic breeding of wheat to reduce the plant size (Rht) has improved efficiency in the redistribution of photosynthesis products (Harvest Index) and has had positive effects on the fertility of the spike [48].

As the upper physiological limit of HI (of approximately 60%) has almost been reached, genetic improvement has been directed toward increasing the capacity to produce aboveground biomass as part of the crop [49].

The maximum yield potential has undoubtedly led to an increase in the general nitrogen needs of non-nitrogen-fixing crops, which have in fact become dependent on substantial levels of mineral fertilizer.

The reduced height of the plants has materially allowed this increase, as the new varieties are much less sensitive to the risk of lodging than the old.

Breeding is an important factor in wheat improvement-related GM production. The overexpression of the French bean in GM wheat has led to an increased grain yield, and consequently NUE, estimated at about 20%. However, difficulties in carrying out experiments with GM materials in Europe have hindered further research into this.

Another gene studied in terms of the ability to modify NUE was the GS1 (Gln-1-3), whose overexpression was able to improve the rice crop index, N collection rate, and N utilization efficiency.

4.11. Time of application

To be effective, the amount of mineral nutrients applied must be consistent with the steps of development and the absorption capacity of the crop. This is known as the subdivision of the N extra dose fractionation best yield, with respect to a unique single intake. With a distributed total dose, nitrogen fertilization is reduced during the early phases (tillering) in favor of the later phases (stem elongation, earing), which encourages a greater absorption of nutrients and increases the efficiency of the crop [14].

Even if traditionally and environmentally the entire dose of N is supplied at the sowing, this approach can be considered valid: the correction of the fertilization dose, with contributions between late tillering and stem elongation, is an economically profitable technique (**Figure 6**).

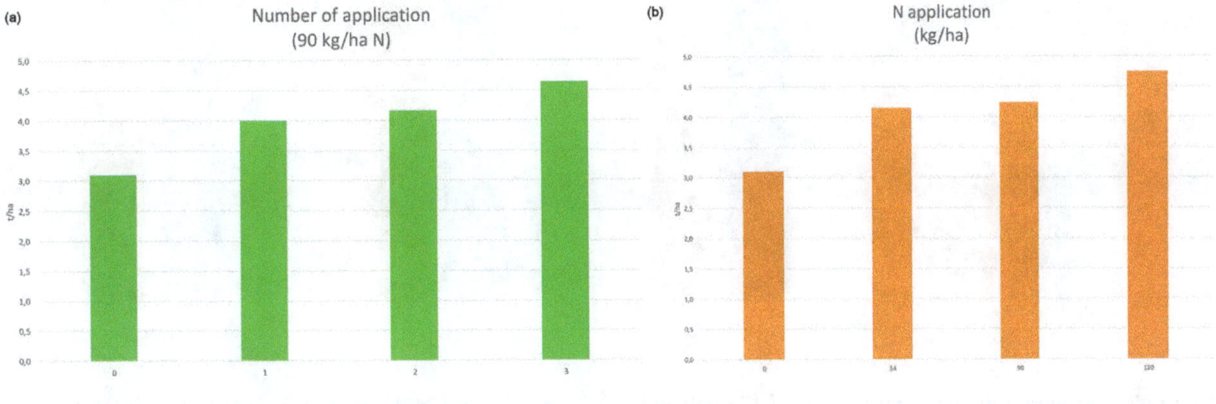

Figure 6. Effect of doses and number of N application on yield response in durum wheat [14].

The most effective nitrogen enhancement is administered during the phases of more intense vegetative activity (raised) and depends on the greater capacity of interception of plants. An experiment in the Mediterranean area (Spain) with a radioactive tracer (15N) has shown that only a small proportion of nitrogen supplied in presowing is effectively intercepted by wheat (14.1%). Nitrogen raised initially is instead absorbed in a greater proportion (54.8%) [50] (**Figure 7**).

The gross availability of nitrogen resulting from the mineralization of organic matter and from previous crop residues may adversely affect efficiency. Indications have emerged from the study of durum wheat in the same environment [50].

Very late fertilization around the time of earing/flowering, while not having significant effects on the yield, can increase the protein content of kernels and the production of protein per hectare [15, 51].

4.12. Use of precision farming for nitrogen application

Precision farming can represent a sustainable system, making nitrogen more sustainable while achieving the best economic results.

Although there are several definitions of precision agriculture (PF), Pierce and Nowak [52] have summarized it as "a system that provides the tools to do the right thing, the right place, at the right time," where "the right thing" is an agronomic operation. In fact, recent technological innovations have led to an increase of application opportunities and that definition can therefore be extended.

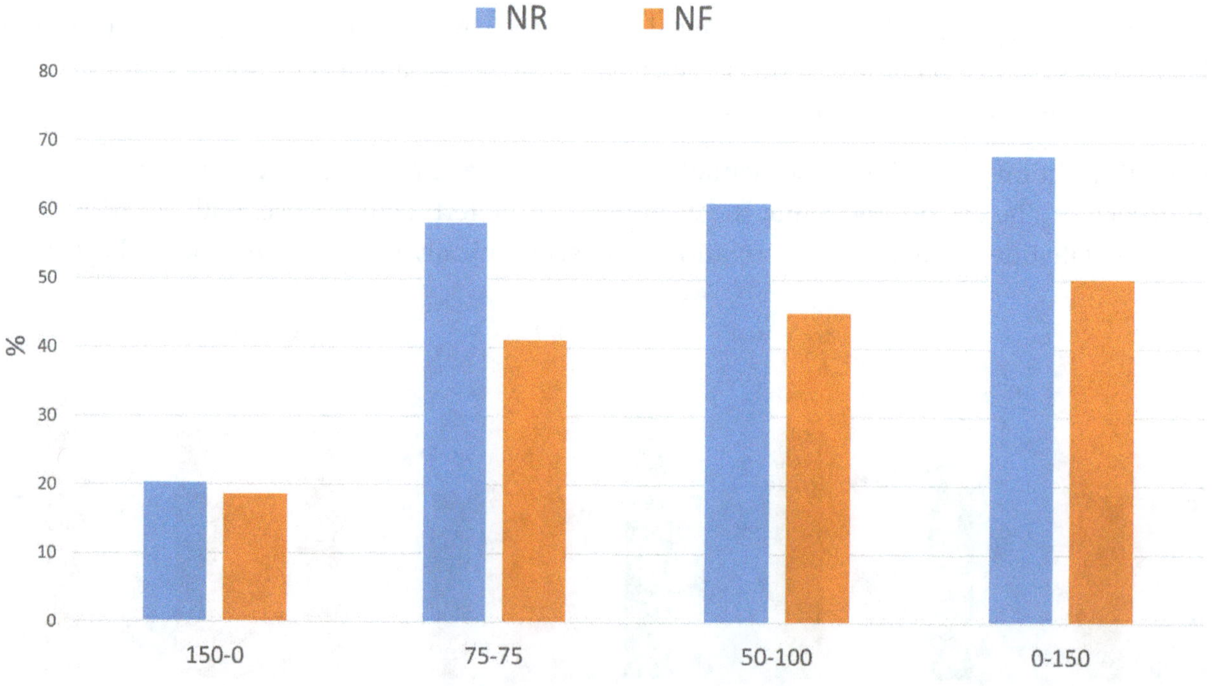

Figure 7. Nitrogen fertilizer recovery (%) of 15N-labeled fertilizer (NR) and N derived from 15N fertilizer (NF) in the whole plant at maturity for hard red spring wheat, as affected by N timing [49].

A more extensive definition [53] of PF is "farming management (agriculture, forestry and animal husbandry) based on the observation, measurement and response of the inter- and intra-field quantitative and qualitative variables that influence the production system of the farm." This aims to define, after site-specific data analysis, a complete decision support system for business management, with the aim of optimizing returns in terms of climate, environmental, economic, productive, and social sustainability [54].

The application of precision agriculture in extensive crops, particularly wheat, has been the subject of various studies [54–57], and involves the acquisition and integration of a large amount of information.

The first input to consider is soil variability. The physical and chemical characteristics of soil vary, which affects the production response [58]. In arid environments, in particular, the spatial variability of the soil factors, combined with climatic factors, greatly influences productivity [59].

Various techniques can be used to obtain soil data, although the approach of sensor-based gathering of crop and soil data is most common [60, 61]. They reported in their reviews that the widely used electrical and electromagnetic sensors provide valuable information about field heterogeneity.

These techniques are more intensive and generally cheaper than conventional sampling and analysis of soil or crop variables [56].

The use of sensors can be considered to spatially diagnose the seasonal pattern of crop conditions during the cycle. Sensors on planes or satellites can potentially collect the reflected electromagnetic radiation from the canopy at small scales of space and time. These remote sensors can evaluate the changes in growth environments from location to location and can potentially gather information about the field, predicting grain yield and nitrogen status [62].

Both passive and active-light proximal sensors can be used to collect the reflected radiation. Passive noncontact sensors depend on sunlight, while active sensors, with their own light sources, enable the assessment of the crop status irrespective of ambient light conditions [63, 64].

Commercial proximal sensors include:

- The passive Yara N-Sensor®/FieldScan (Yara International, ASA, Oslo, Norway).
- FieldSpec® Portable Spectroradiometer (ASD Inc., Boulder, CO, USA).
- The active sensor GreenSeeker® (Trimble Inc.).
- Crop Circle ™ (Dutch Scientific, Lincoln, NE).
- SPAD-502 Minolta analytical transmittance, which produces an estimate of leaf chlorophyll content.

Several studies have been conducted on wheat crops using these systems [65, 66], and the data collected must be converted into measures of vegetation, using vegetative indices.

The normalized difference vegetation index (NDVI) is the more precise vegetative normalization index and is based on differences in the red (670 nm) and near infrared (780 nm) spectrums.

NDVI has shown good correlation with N absorption in both bread and durum wheat. Correlations were often more stable when they involved single locations or varieties. An asymptotic tendency was found in the NDVI value for N uptake over 200 kg ha^{-1} [67–69].

The SPAD value showed a good correlation with the N status in bread wheat [70]. Of all observed indexes, the nitrogen nutrition index (NNI) was the most highly correlated. SPAD close to flowering was closely related to the final grain protein content.

5. Implication of nitrogen use on greenhouse gas (GHG) emissions

5.1. Problem statement

The rapid increase in the world population over recent decades has led to an expansion in agricultural land in order to provide sufficient food. Consequently, agricultural inputs have been extensively used to increase yield, resulting in serious environmental effects. The emission of anthropogenic gases into the atmosphere is a serious environmental burden, and it is generally accepted that agricultural activities are a major source of several of these gases. Since the industrialization of agriculture, or the so-called green revolution, significant increases in the atmospheric concentrations of these gases have been observed globally. These cause environmental change that will have an impact at both regional and global levels. Carbon dioxide (CO_2), methane (CH_4), and nitrous oxide (N_2O) are referred to as greenhouse gases (GHGs), as they are considered to contribute most to global warming. The effect of agricultural intensification, in particular the increased use of nitrogen fertilizer, on GHG emissions is likely to increase in the future. The Intergovernmental Panel on Climate Change (IPCC) has indicated that agricultural activities contributed around 10% of the total global GHG emissions [72]. The main agricultural contributors are soil emissions due to inorganic fertilizers and plant residues, in addition to biological processes. N is mainly lost from agriculture soils through nitrous oxide (N_2O) emissions, which has recently been the subject of considerable attention due to its greenhouse gas effect (300 times worse than carbon dioxide). The increase in population has made it global necessary to produce more foodly, so both the agricultural land area and N_2O emissions are likely to continue to rise in the coming decades (**Figure 8**). At a global level, agriculture land is estimated to contribute 60% of the N_2O emissions to the atmosphere annually. Pires et al. [73] derived interesting results from several field experiments. They found that the amount of N fertilizer applied in the agricultural fields is the strongest indicator of N_2O fluxes in major cropping systems. However, N fertilizer formulation and application timing, and the agronomic practices that determine N availability in the soil such as tillage and waste management can also influence the N_2O flow, providing an opportunity for mitigation options in agriculture through N management. Several studies [74, 75] have reported that an average of 2% of the N applied to cultivated soils is emitted into the atmosphere as N_2O. Inappropriate synchronization between N fertilization rate and crop demand would lead to a further increase in emission rates.

Although the N loss from agricultural soils in different gaseous forms is considered the dominant mechanism in agricultural production systems, the manufacturing processes of agricultural inputs such as N fertilizers are also responsible for direct emissions of CO_2 into the

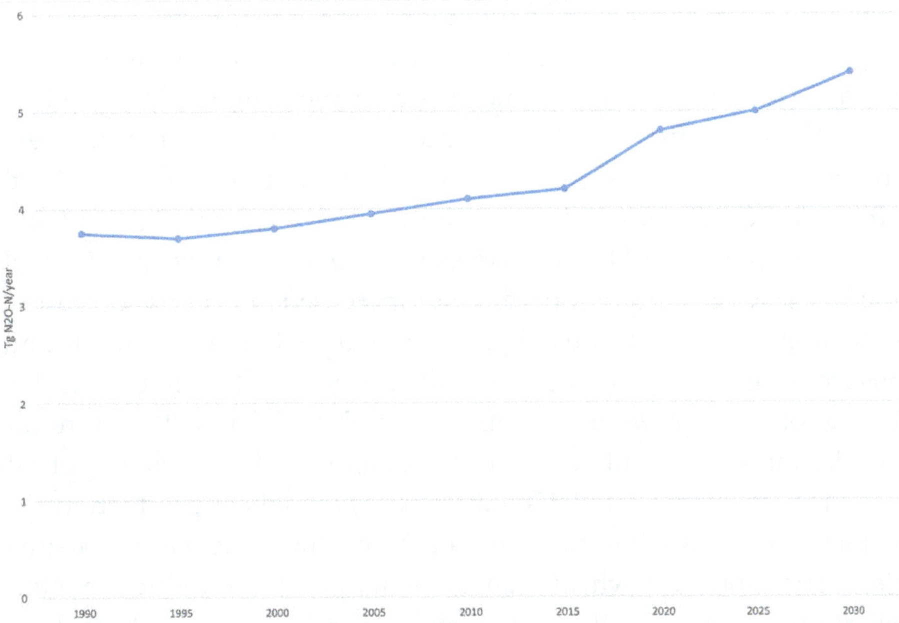

Figure 8. Global N_2O emissions (Tg N_2O-N yr^{-1}) from agricultural soils between 1990 and 2030 (adapted from Reay et al. [71]).

atmosphere [76]. According to the IPCC, nitrogen fertilizer application, crop residues, and N-fixing crops present a direct source of N_2O emissions in agricultural fields [72] due to the excessive levels of N applied, the higher amount fixed by N-fixing crops, and higher concentrations in plant residue. The remaining N not taken up by the plant can be lost indirectly through surface runoff, leached nitrate (NO_3^-) in groundwater, volatilization to the atmosphere or by biological soil processes such as nitrification and denitrification, all of which pose environmental concerns.

The environmental consequences of N fertilizer application in agriculture have become of great interest to many researchers worldwide. They aim to understand the mechanism of different N pools (both organic and inorganic forms of N in the soil), their interaction, and the processes by which they enter and leave the soil. N_2O gas is a natural product of soil microbial activities, so understanding the biological processes and the factors affecting them will help in the international efforts toward the development of mitigation strategies for greenhouse gas emissions from agricultural soil. In addition, to optimize the trade-off between economic development and the impact of agricultural inputs on GHG emissions, environmental issues must be included in any future political agendas, as they are extremely important in terms of agricultural productivity and food security.

5.2. Mitigation strategy

Nitrogen has been widely studied to find ways of improving its efficiency in terms of producing acceptable grain yield and quality with less environmental impact. Recent research efforts have been directed toward reducing N loss from the soil by improving the absorption mechanism and the metabolism of N in the plant. Most have in fact succeeded in reducing the excessive input of N fertilizers, while maintaining profitable grain yield with less environmental impact. The results indicate that efficiency and effectiveness of N use on crops can be achieved [77] through further synchronizing the timing of N fertilizer application and the N rate with plant nitrogen demand, and its capacity to uptake sufficient amounts of available nitrogen. This strategy is believed to be of paramount importance for the reduction of N_2O and therefore GHG emissions from the cropping system. As N uptake capacity is generally low at the beginning of the growing season, a large quantity of N fertilizer applied at sowing time will lead to a greater potential for increased N_2O emissions due to slow plant uptake. Bouwman et al. [77] indicated that not only does the N rate significantly influence N_2O emissions from fields but also the N fertilizer type, and the time when the N fertilizer is applied throughout the growing season. Hao et al. [78] found that autumn N fertilizer application resulted in significantly greater N_2O emissions than spring application. Despite being the key factor affecting total GHG emissions in the wheat production system, several studies in the literature found that the increased N fertilizer rate had no positive correlation with grain yield. Therefore, any wheat management option that reduces or eliminates the N fertilizer application (i.e., through the introduction of an N-fixing crop) would result in a significant reduction in total emissions. Rees et al. [79] highlighted the importance of N supply reduction in an emission mitigation strategy. Mosier et al. [80] indicated that improved soil management and better nutrient use efficiency could result in a reduction of 10–30% of total emissions from the soil.

To tackle the projected changes in climate events and the potential effects on crop yield, farmers should focus on the establishment of sustainable wheat production systems through efficient management of agricultural inputs [81]. As N inputs are the main cause of several environmental problems, a priority when establishing such a system would be to make them more efficient. The calculation of nitrogen use efficiency (NUE) as an agro-environmental indicator is important in the agro-policy context [82], as it informs the economic utility of a specific N fertilizer quantity applied to the field with reference to crop yield. However, to obtain the maximum from N fertilizer application in cereal production, Pires et al. [82] suggested that both the environmental and economic pillars of sustainability must be addressed along with agronomic efficiency. Alternative management strategies have recently been proposed to optimize the NUE and therefore increase crop yield, including N fertilizer timing, split applications, site-specific management, crop rotation, crop diversification, biological N fixation, and improved plant traits by genetic breeding, which is believed to help improve yields with a low input of N fertilizer, reducing its losses from the agriculture system [83]. Options such as the use of biological nitrification inhibition by *Brachiaria* roots exudates [84] and the adoption of alternative farming techniques based on no-till continuous cover cropping [85] are other strategies for improving NUE. However, the response to these management strategies can vary, depending on the complex interaction of soil, crop, and environmental factors.

The literature review shows that the yield response varies with the nitrogen and moisture level. The goal of N application to wheat and its timing is to supply adequate N when the crop needs it (**Figure 9**). As the nitrogen cycle is biologically influenced by prevailing climatic conditions, along with the physical and chemical properties of a particular

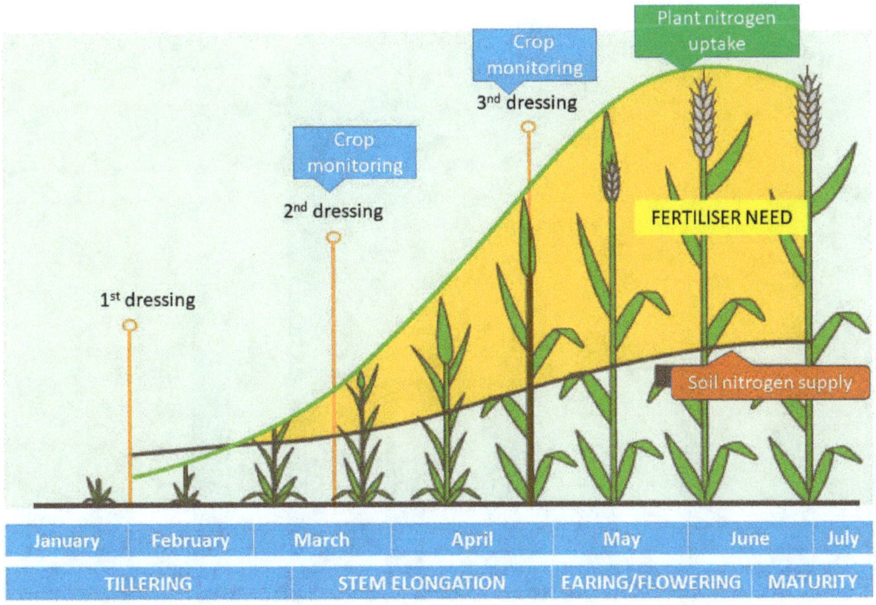

Figure 9. Timing and application of N fertilizer in winter durum wheat (from Fertilizers Europe).

soil, the choice of a suitable N fertilizer formula for a specific soil is fundamental [86]. The electrical charges of soil particles vary according to the soil texture, so studying the chemical properties and reactions within soil particles will help understanding when and which N fertilizers type can be applied to a specific soil. These reactions are believed to have a significant influence on when fertilizer can be applied and how efficiently the crop takes it up. As explained, clay particles in the soil have a negative electrical charge, so the amount of clay present in the soil surface, which the fertilizer first comes into contact with, is an important factor controlling plant response to the N type and the timing. The clay particles with negative charges will react with components of the fertilizer that dissolve as positively (+) charged particles (cations) when added to soil. For example, the application of N fertilizers that include more ammonium (positively charged) and less nitrate (negatively charged) forms of N can reduce their potential for losses in the short term. The application of the correct type of N fertilizer can help reduce N_2O emissions from the soil, according to the Fertilizers Europe Initiative. **Figure 10** shows comparative N_2O soil emissions resulting from the application of different nitrogen fertilizers, and **Figure 11** shows N losses via volatilization emissions. In both cases, N losses are significantly higher with urea than other N fertilizers.

Ammonium (NH_4^+) forms of N binding to soil particles with negative charges will not be subjected to losses through leaching or denitrification, and therefore, increase the N availability for the plant [87]. However, if the total amount of applied ammonium is not taken up by the plant, soil bacteria will convert the remainder to nitrate (NO_3^-), a form that cannot bind to soil particles and will be lost when excessive rainfall leaches or saturates the soil. As identified, soil conditions are important in the transformation process, and nitrifying bacteria activity is minimal in cold soil temperatures, which can naturally help protect ammonium forms of N from losses.

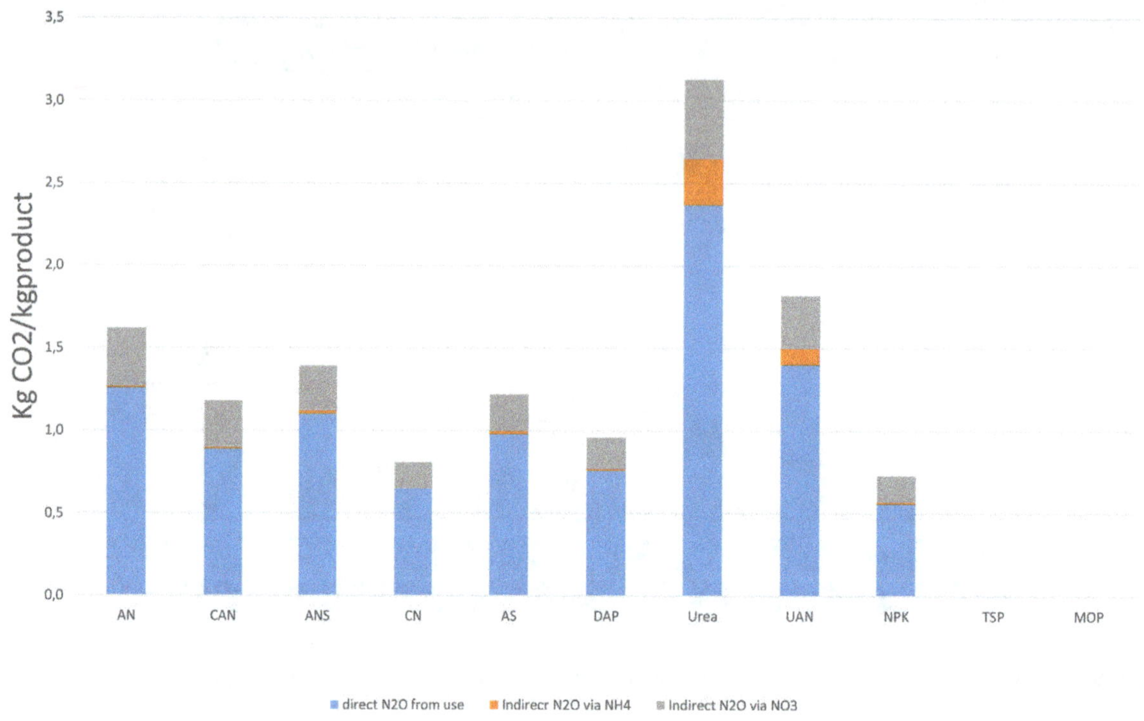

Figure 10. Comparative N_2O emissions from the application of different nitrogen fertilizers (adapted from Fertilizers Europe). AN, ammonium nitrate; CAN, calcium ammonium nitrate; ANS, ammonium nitrosulphate; CN, calcium nitrate; AS, ammonium sulphate; DAP, ammonium phosphates; Urea, urea; UAN, urea ammonium nitrate; NPK, NPK 15-15-15; TSP, triple superphosphate; MOP, muriate of potash.

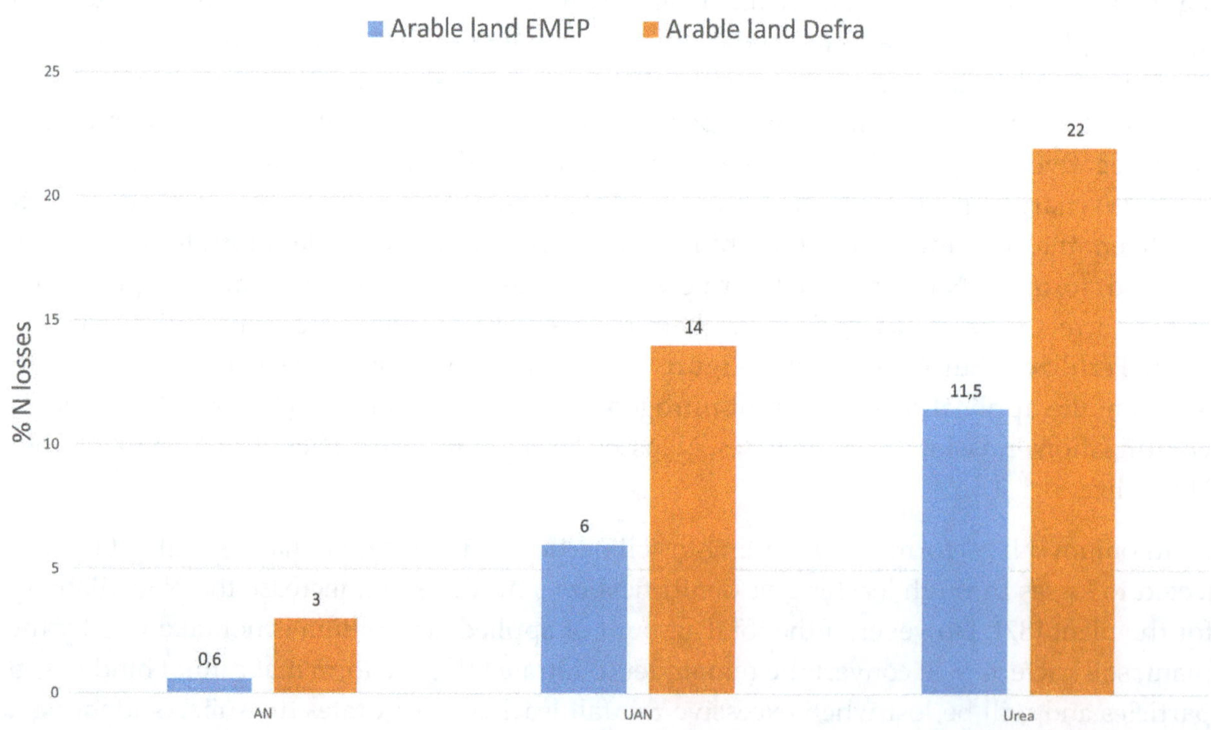

Figure 11. Volatilization losses (% N) from the application of different N fertilizers in arable land; Adapted from Fertilizers Europe; Data from the official European Emission Inventory (EMEP) as well as from a UK Government Department of Environment, Food and Rural Affairs (Defra) study.

6. Conclusion

Durum wheat is mainly cultivated in Mediterranean countries, but it is also of great importance in other European countries and in Canada and the United States, which are the world's leading exporters. The quality of durum wheat, often considered only in product market and technological terms, is in fact the result of a complex system that must take into account all the components of the supply chain. Climate change and the consequences of greenhouse gas emissions make crops very susceptible to climate variations, resulting in a variable yield response.

In addition to aspects related to the nutritional, safety, and technological characteristics of the product, it is also necessary to consider those related to the production environment, through the development of agronomic strategies that can improve precision farming, or techniques that can reduce the environmental impact of cultivation (conservative farming).

The possibility of using improved varieties with better efficiency and higher quality standards could offer an opportunity to increase Mediterranean production, making it less dependent on imports from foreign countries. The strategy should include changing the phenological model to avoid stressful times during plant growth. In addition, a comprehensive understanding of the responses of each plant growth phase to environmental variables such as high CO_2 concentration, temperature, and drought, alone or in combination, will help the crop plant adapt to these changes. Nitrogen fertilization is an indispensable agronomic technique that can achieve good levels of production, both quantitatively and qualitatively, and has been extensively studied in the Mediterranean environment.

The application of precision farming systems (drones, NDVI systems, N sensors, etc.) can provide useful indications for the precise application of nitrogen. Many experiments suggest that the best agronomic technique in nitrogen application is the splitting of nitrogen application (sowing, tillering, stem elongation, and if necessary grain filling), which appears to be more efficient than the application divided once, which gives a 15% increase, or twice, where the increase is 7%. This strategy appears to be effective in reducing the loss of soil nitrates through leaching runoff and denitrification, which is more of a risk during the winter period, when the rainfall in the Mediterranean climate is concentrated. In fact, several types of nitrogen efficiency, and efficiency in N fertilization, increase when the fertilizer is applied at the stem elongation phase, while higher levels of N at the sowing time and tillering result in poor efficiency. The splitting application of nitrogen appears to be an effective method of avoiding environmental problems associated with the potential loss of this element, but more importantly could maximize the efficiency of the fertilization of wheat, increasing the yield in the Mediterranean region and providing a rational management strategy for the crop.

Author details

Luigi Tedone*, Salem Alhajj Ali and Giuseppe De Mastro

*Address all correspondence to: luigi.tedone@uniba.it

Department of Agricultural and Environmental Science, Bari, Italy

References

[1] Curtis BC. Bread Wheat. Improvement and Production. Former International Maize and Wheat Improvement Center (CIMMYT); 2002

[2] FAOSTAT. Database internet FAOSTAT. http://faostat.fao.org/

[3] Yadav MR, Kumar R, Parihar CM, Yadav RK, Jat SL, Ram H, Meena RK, Singh M, Birbal M, Verma AP, Kumar U, Ghosh A, Jat ML. Strategies for improving nitrogen use efficiency: A review. Agricultural Reviews. 2017;**38**(1):29-40

[4] Tamang BG, Brasier KG, Thomason WE, Griffey CA, Fukao T. . Differential responses of grain yield, grain protein, and their associated traits to nitrogen supply in soft red winter wheat. Journal of Plant Nutrition and Soil Science. 2017:1-10. DOI: 10.1002/jpln.201600312

[5] European Environment Agency (EEA). https://europa.eu/european-union/about-eu/agencies/eea_en

[6] Olesen JE, Trnka M, Kersebaum KC, Skjelvåg AO, Seguin B, Peltonen-Sainio P, Rossi F, Kozyra J, Micale F. Impacts and adaptation of European crop production systems to climate change. European Journal of Agronomy. 2011;**34**:96-112

[7] Ercoli L, Masoni A, Mariotti M, Arduini I. Accumulation of dry matter and nitrogen in durum wheat during grain filling as affected by temperature and nitrogen rate. Italian Journal of Agronomy. 2009;**1**:3-13

[8] Ercoli L; Masoni A; Pampana S; Mariotti M, Arduini I. As durum wheat productivity is affected by nitrogen fertilization management in Central Italy. European Journal of Agronomy. 2017;**44**(1):38-45

[9] Giuliani MM, Giuzio L, De Caro A, Flagella Z. Relationships between nitrogen utilization and grain technological quality in durum wheat: II. Grain yield and quality. Agronomy Journal. 2011;**103**:1668-1675

[10] Flagella Z. Qualità nutrizionale e tecnologica del frumento duro. Italian Journal of Agronomy. 2006;**1**:203-239

[11] Desiderio E, D'Egidio MG, Ciriciofolo E, Codoni D, Lendini M, Palumbo M, Porfiri O, Fornara M, Cecchini C. Effetti della concimazione azotata su produzione e qualità della granella di frumento duro. Informatore agrario. 1998;**10**(98):31-38

[12] Basso B, Cammarano D, Grace PR, Cafiero G, Sartori L, Pisante M, Landi G, De Franchi S, Basso F. Criteria for selecting optimal nitrogen fertilizer rates for precision agriculture. Italian Journal of Agronomy. 2009;**4**:147-158

[13] Lopez-Bellido L, Munoz-Romero V, Benítez-Vega J, Fernández-García P, Redondo R, López-Bellido RJ. Wheat response to nitrogen splitting applied to a Vertisols in different tillage systems and cropping rotations under typical Mediterranean climatic conditions. European Journal of Agronomy. 2012;**43**:24-32

[14] Tedone L, Verdini L, Grassano N, Tarraf W, De Mastro G. Optimising nitrogen in order to improve the efficiency, eco-physiology, yield and quality on one cultivar of durum wheat. Italian Journal of Agronomy. 2014;**9**(2):49-54

[15] Flagella Z, Giuliani MM, Giuzio L, De Caro A, Di Fonzo N, Masci S. Influence of water deficit on durum wheat grain protein composition and technological quality. The 2nd International Conference on Integrated Approaches to Sustain and Improve Plant Plant Production Under Drought Stress; 2005. pp. 2-31

[16] Alhajj Ali S, Tedone, L, De Mastro G. Climate variability impact on wheat production in Europe: Adaptation and mitigation strategies. In: Quantification of Climate Variability, Adaptation and Mitigation for Agricultural Sustainability, 1st ed. Chapter: 12. Springer International Publishing; 2017. pp. 251-321. DOI: 10.1007/978-3-319-32059-5_12

[17] Alhajj Ali S, Tedone L, De Mastro G. A comparison of the energy consumption of rain-fed durum wheat under different management scenarios in southern Italy. Energy. 2013;**61**:308-318

[18] De Vita P, Di Paolo E, Fecondo G, Di Fonzo N, Pisante M. No-tillage and conventional tillage effects on durum wheat yield, grain quality and soil moisture content in southern Italy. Soil & Tillage Research. 2007;**92**:69-78

[19] International Grains Council. http://www.igc.int/en/

[20] Eurostat.http://ec.europa.eu/eurostat/statistics-xplained/index.php/Main_annual_crop_statistics

[21] Taiz L. Agriculture, plant physiology, and human population growth: Past, present, and future. Theoretical and Experimental Plant Physiology. 2013;**25**(3):167-181

[22] Ladha A, T-Padre CK, Reddy K, Cassman G, Sudhir Verma D, Powlson S, van Kessel C, Richter Dde B, Chakraborty D, Himanshu P. Global nitrogen budgets in cereals: A 50-year assessment for maize, rice, and wheat production systems J. K. Scientific Reports. 2016;**6**:19355. DOI: 10.1038/srep1935

[23] Antongiovanni M, Sargentin C. Variability in chemical composition of straws. Options Mediterranee. 1991;**16**:49-53

[24] Blumenthal JM, Baltensperger DD, Cassman KG, Mason SC, Pavlista AD.Importance and effect of nitrogen on crop quality and health. Nitrogen in the Environment: Sources, Problems, and Management, 2nd ed. Oxford: Elsevier; 2008. pp. 51-70

[25] AA.VV. N-Farm Trials for Adapting and Adopting Good Agricultural Practices. FAO; 2003

[26] Amato G, Ruisi P, Frenda AS, Di Miceli G, Saia S, Plaia A, Giambalvo D. Long-term tillage and crop sequence effects on wheat grain yield and quality. Agronomy Journal. 2013;**105**:1317-1327

[27] Ladha JK, Pathak H, Krupnik TJ, Six J, Kessel CV. Efficiency of fertilizer nitrogen in cereal production: Retrospects and prospects. Advanced in Agronomy. 2005;**87**:85-156

[28] Moll, RH, Kamprath EJ, Jackson WA. Analysis and interpretation of factors which contribute to efficiency to nitrogen utilization. Agronomy Journal. 1982;**74**:562-564.

[29] Barraclough P, Howarth RJ, Jones J, Lopez-Bellido R, Parmar S, Shepherd CE, Hawkesford, MJ. Nitrogen efficiency of wheat: Genotypic and environmental variation and prospects for improvement. European Journal of Agronomy. 2010;**33**:1-11

[30] Bradley RS, Kindred DR. Analysing nitrogen responses of cereals to prioritize routes to the improvement of nitrogen use efficiency. Journal of Experimental Botany. 2009;**60**(7):1939-1951. DOI: 10.1093/jxb/erp116

[31] Ortiz-Monasterio I, Sayre KD, Rajaram S, McMahon M. Genetic progress in wheat yield and nitrogen use efficiency under four N rates. Crop Science. 1997;**37**:898-904

[32] Arregui LM, Quemada, M. Strategies to improve nitrogen use efficiency in winter cereal crops under rainfed conditions. Agronomy Journal. 2008;**100**:277-284

[33] Gooding MJ, Addisu M, Uppal RK, Snape JW, Jones HE. Effect of wheat dwarfing genes on nitrogen-use efficiency. Journal of Agricultural Science. 2012;**150**:3-22

[34] Cormier F, Faure S, Dubreuil P, Heumez E, Beauchene K, Lafarge S, Praud S, Le Gouis J. A multi-environmental study of recent breeding progress on nitrogen use efficiency in wheat (*Triticum aestivum* L.). Theoretical and Applied Genetics. 2013;**126**:3035-3048. (García-Torres et al., 2003, Gilbert, 2012)

[35] Cormier F, Foulkes J, Hirel B, Gouache D, Moënne-Loccoz Y, Le Gouis J. Breeding for increased nitrogen-use efficiency: A review for wheat (*T. aestivum* L.). Plant Breeding. 2016;**135**:255-278

[36] Foulkes MJ, Hawkesford MJ, Barraclough PB, Holdsworth MJ, Kerr S, Kightley S, Shewry PR. Identifying traits to improve the nitrogen economy of wheat: Recent advances and future prospects. Field Crops Research. 2009;**114**:329-342

[37] Pisante M. Agricoltura sostenibile. Principi, sistemi e tecnologie applicate all'agricoltura produttiva per la salvaguardia dell'ambiente e la tutela climatica. Il Sole 24 Ore. Bologna, Italy: Edagricole; 2013

[38] Franzluebbers AJ, Hons FM, Zubberer DA. Tillage induced seasonal changes in soil physical properties affecting soil CO_2 evolution under intensive cropping. Soil & Tillage Research. 1995;**34**:41-60

[39] 39 Hendrix, PF, Crossley Jr DA, Blair JM, Coleman DC. Soil biota as components of sustainable agroecosystems. In: Sustainable agricultural systems, Edwards CA, Lal R, Madden P, Miler RH, House G, editors. Ankeny, USA: SWCS; 1990. pp. 637-654

[40] Amato G, Giambalvo D, Frenda A S, Di Miceli G, Ruisi P, Badagliacca G. Long-term effects of no tillage treatment on soil N availability, N uptake, and 15N-fertilizer recovery of durum wheat differ in relation to crop sequence. Field Crops Research. 2016;**189**:51-58

[41] Doran JW, and Parkin TB. Defining and assessing soil quality. In Doran JW, Coleman DC, Bezdicek DF, Stewart BA, editors. Defining Soil Quality for a Sustainable Environment. Madison, Wisconsin, USA: SSSA, Inc.; 1994

[42] Campbell CA, Zentner RP, Selles F, McConkey BG, Dyck FB. Nitrogen management for spring wheat grown annually on zero-tillage: Yield and nitrogen use efficiency. Agronomy Journal. 1993;**85**:107-114

[43] Stagnari F, Onofri A, Codianni P, Pisante M. Durum wheat varieties in N deficient environments and organic farming: A comparison of yield, quality and stability performances. Plant Breeding. 2013;**132**:266-275

[44] Zhang L, Schwärzel K. Applying Multifunctionality to address the challenges and benefits of land use management. In: Multifunctional Land-Use Systems for Managing the Nexus of Environmental Resources; 2017. Springer ed. ISBN: 978-3-319-54956-9. DOI 10.1007/978-3-319-54957-6_1

[45] Abedi T, Alemzadeh A, Kazemeini SA. Effect of organic and inorganic fertilizers on grain yield and protein banding pattern of wheat. Australian Journal of Crop Science. 2010;**4**:384-389

[46] Austin RB, Bingham J, Blackwell RD, Evan LT, Ford MA, Morgan CL, Taylor M. Genetic improvements in winter wheat yields since 1900 and associated physiological changes. Journal of Agricultural Science. 1980;**94**:675-689

[47] Flintham JE, Börner A, Worland AJ, Gale MD. Optimizing wheat grain yield: Effects of Rht (gibberellin-insensitive) dwarfing genes. The Journal of Agricultural Science. 1997; 128(1):11-25

[48] Shearman, VJ, Sylvester-Bradley R, Scott RK, Foulkes MJ. Physiological processes associated with wheat yield progress in the UK. Crop Science. 2005:45:175-185

[49] Lopez-Bellido L, López-Bellido RJ, Redondo R. Nitrogen efficiency in wheat under rainfed Mediterranean conditions as affected by split nitrogen application. Field Crops Research. 2005;**94**:86-97

[50] Miceli F, Martin M, Zerbi G. Yield, quality and nitrogen efficiency in winter wheat fertilized with increasing N levels at different times. Journal of Agronomy and Crop Science. 1992;**168**:337-344

[51] Pierce FJ, Nowak P. Aspects of precision agriculture. In: Sparks DL, editor. Advances in Agriculture. Academic Press; 1999. pp. 1-85

[52] Blasi G, Pisante M. Linee guida per lo sviluppo dell'agricoltura di precisione in Italia; 2016. https://www.politicheagricole.it

[53] Casa R. Agricoltura di precisione. Edagricole Ed; 2016

[54] Aparicio N, Villegas D, Casadesus J, Araus L, Royo C. Spectral vegetation index as non destructive tools for determining durum wheat yield. Agronomy Journal. 2000;**92**:83-91

[55] Adamchuk VI, Viscarra Rossel RA, Sudduth KA, Lammers PS. Sensor fusion for precision agriculture. In: Thomas C, editor. Sensor Fusion—Foundation and Applications. InTech; 2011. ISBN:978-953-307-446-7

[56] Diacono M, Rubino P, Montemurro F. Precision nitrogen management of wheat. A review. Agronomy for Sustainable Development. Springer Verlag/EDP Sciences/INRA. 2012;**33**(1): 219-241

[57] Fitzgerald G, Rodriguez D, O'Leary G. Measuring and predicting canopy nitrogen nutrition in wheat using a spectral index—The canopy chlorophyll content index (CCCI). Field Crops Research. 2010;**116**:318-324. DOI: 10.1016/j.fcr.2010.01.010

[58] Whelan BM, McBratney AB. The "null hypothesis" of precision agriculture management. Precision Agriculture. 2000;**2**:265-279

[59] Wong MTF, Asseng S. Determining the causes of spatial and temporal variability of wheat yields at sub-field scale using a new method of upscaling a crop model. Plant and Soil. 2006;**283**:203-215. DOI: 10.1007/s11104-006-0012-5

[60] Thomason WE, Phillips SB, Davis PH, Warren JG, Alley MM, Reiter MS. Variable nitrogen rate determination from plant spectral reflectance in soft red winter wheat. Precision Agriculture. 2011;**12**:666-681. DOI: 10.1007/s11119-010-9210-5

[61] Song X, Wang J, Huang W, Liu L, Yan G, Pu R. The delineation of agricultural management zones with high resolution remotely sensed data. Precision Agriculture. 2009;**10**:471-487. DOI: 10.1007/s11119-0099108-2

[62] Rodriguez D, Sadras VO, Christensen LK, Belford R. Spatial assessment of the physiological status of wheat crops as affected by water and nitrogen supply using infrared thermal imagery. Australian Journal of Agricultural Research. 2005;**56**:983-993

[63] Samborski SM, Tremblay N, Fallon E. Strategies to make use of plant sensors-based diagnostic information for nitrogen recommendations. Agronomy Journal. 2009;**101**:800-816. DOI: 10.2134/agronj2008. 0162Rx

[64] Erdle K, Mistele B, Schmidhalter U. Comparison of active and passive spectral sensors in discriminating biomass parameters and nitrogen status in wheat cultivars. Field Crop Research. 2011;**124**:74-84. DOI: 10.1016/j.fcr.2011.06.007

[65] Singh B, Sharma RK, Kaur J, Jat ML, Martin KL, Singh Y, Singh V, Chandna P, Choudhary OP, Gupta RK, Thind HS, Singh-J UHS, Khurana HS, Kumar A, Uppal RK, Vashistha M, Raun WR, Gupta R. Assessment of the nitrogen management strategy using an optical sensor for irrigated wheat. Agronomy for Sustainable Development. 2011;**31**(3):589-603. DOI: 10.1007/s13593-011-0005-5

[66] Eitel JUH, Long DS, Gessler PE, Smith AMS. Using in-situ measurements to evaluate the new RapidEye™ satellite series for prediction of wheat nitrogen status. International Journal of Remote Sensing. 2007;**28**:4183-4190. DOI: 10.1080/01431160701422213

[67] Blondlot A., Gate P. & Poilvé H., 2005. Providing operational nitrogen recommendations to farmers using satellite imagery. In Stafford J., editor. Precision Agriculture '05: Proceedings of the 5th European Conference on Precision Agriculture, p 345-352. Uppsala, Sweden. Wageningen Academic Publishers

[68] Basso B, Cammarano D, Fiorentino D, Ritchie JT. Wheat yield response to spatially variable nitrogen fertilizer in Mediterranean environment. European Journal of Agronomy. 2013;**51**:65-70

[69] Arregui LM, Lasa B, Lafarga A, Irañeta I, Baroja E, Quemada M. Evaluation of chlorophyll meters as tools for N fertilisation in winter wheat humid Mediterranean environment. European Journal of Agronomy. 2006;**24**:140-148

[70] Pires JCM, Martins FG, Sousa SIV, Alvim-Ferraz MCM, Pereira MC. Prediction of the daily mean PM10 concentrations using linear models. American Journal of Environmental Sciences. 2008;**4**(5):445-453

[71] Reay DS, Davidson EA, Smith KA, Smith P, Melillo JM, Dentener F, Crutzen PJ. Global agriculture and nitrous oxide emissions. Nature Climate Change. 2012;**2**(6):410-416

[72] Stehfest E, Bouwman L. N_2O and NO emission from agricultural fields and soils under natural vegetation: Summarizing available measurement data and modeling of global annual emissions. Nutrient Cycling in Agroecosystems. 2006;**74**:207-228

[73] Linquist, B, van Groenigen KJ, Adviento-Borbe MA, Pittelkow C, and van Kessel C. An agronomic assessment of greenhouse gas emissions from major cereal crops. Global Change Biology. 2012;**18**(1):194-209. DOI: 10.1111/j.1365-2486.2011.02502.x

[74] Iqbal J, Ronggui H, Lijun D, Lan L, Shan L, et al. Differences in soil CO_2 flux between different land use types in mid-subtropical China. Soil Biology and Biochemistry. 2008;**40**:2324-2333

[75] IPCC—Intergovernmental Panel on Climate Change. http://www.ipcc.ch

[76] Snyder CS, Bruulsema TW, Jensen TL, Fixen PE. Review of greenhouse gas emissions from crop production systems and fertilizer management effects. Agriculture, Ecosystems & Environment. October 2009;**133**(3-4):247-266

[77] Bouwman, AF, Boumans LJM, Batjes NH. Modeling global annual N_2O and NO emissions from fertilized fields, Global Biogeochem Cycles. 2002;**16**(4):1080. DOI: 10.1029/2001GB001812

[78] Hao X, Chang C, Carefoot J, Janzen HH, Ellert BH. Nitrous oxide emissions from an irrigated soil as affected by fertilizer and straw management. Nutrient Cycling in Agroecosystems. 2001;**60**:1. DOI: 10.1023/A:1012603732435

[79] Rees RM, Augustin J, Alberti G, Ball BC, Boeckx P, Cantarel A, Castaldi S, Chirinda N, Chojnicki B, Giebels M, Gordon H, Grosz B, Horváth L, Juszczak R, Klemedtsson ÅK, Klemedtsson L, Medinets S, Machon A, Mapanda F, Nyamangara J, Olesen Jørgen E, Reay DS, Sanchez L, Cobeña AS, Smith KA, Sowerby A, Sommer M, Soussana JF, Stenberg M, Topp CFE, Van Cleemput O, Vallejo A, Watson CA, Wuta M. Nitrous oxide emissions from European agriculture—An analysis of variability and drivers of emissions from field experiments. Biogeosciences. 2013;**10**:2671-2682

[80] Mosier A, Kroeze C, Nevison C, Oenema O, Seitzinger S, Cleemput O. Mitigation strategies for greenhouse gas emissions from animal production systems: Synergy between measuring and modelling at different scales. Nutrient Cycling in Agroecosystems. 1998;**52**:225. DOI: 10.1023/A:1009740530221

[81] Brentrup F, Palliere C. Nitrogen Use Efficiency as an Agro-Environmental Indicator. Leysin: OECD; 2010. Available: http://www.oecd.org/tad/sustainable-agriculture/44810433.pdf

[82] Pires MV, da Cunha DA, de Matos Carlos S, Costa MH. Nitrogen-use efficiency, nitrous oxide emissions, and cereal production in Brazil: Current trends and gorecasts. PLoS One. 2015;**10**(8):e0135234. DOI: 10.1371/journalpone.0135234

[83] Subbarao GV, Sahrawat KL, Nakahara K, Ishikawa T, Kishii M, Rao IM, Hash CT, George TS, Srinivasa Rao P, Nardi P, Bonnett D, Berry W, Suenaga K, Lata JC. Biological nitrification inhibition—A novel strategy to regulate nitrification in agricultural systems. Advances in Agronomy. 2012;**114**. DOI: 10.1016/B978-0-12-394275-3.00001-8

[84] Hirel B, Tétu T, Lea Peter J, Dubois F. Improving nitrogen use efficiency in crops for sustainable agriculture. Sustainability. 2011;**3**:1452-1485. DOI: 10.3390/su3091452

[85] Nelson KA, Motavalli PP, Nathan M. Nitrogen fertilizer sources and application timing affects wheat and inter-seeded red clover yields on claypan soils. Agronomy. 2014;**4**:497-513. DOI: 10.3390/agronomy4040497

[86] Osman KT. Soils: Principles, Properties and Management; 2013. Spriger ed. SBN 978-94-007-5663-2 pp. 261. DOI: 10.1007/978-94-007-5663-2_5

[87] Fageria NK. The Use of Nutrients in Crop Plants. CRC Press/Taylor and Francis Group; 2016

11

Nitrogen Fixation and Transfer in Agricultural Production Systems

M. Anowarul Islam and Albert Tetteh Adjesiwor

Abstract

There is a consensus within the scientific community that nitrogenous fertilizers are almost indispensable in today's agriculture. However, the geometric increase in nitrogenous fertilizer applications and the associated environmental concerns call for focus on more sustainable alternatives. Biological dinitrogen (N_2) fixation (BNF) is one of the most sustainable approaches to meeting crop nitrogen (N) demands. The BNF is, especially, important in low value crops (e.g., forages) and in developing economies. However, just like synthetic N fertilizers, BNF has issues of its own. Among the issues of great importance is the low and highly variable proportion of fixed N_2 transferred to non-N_2-fixing plants. The proportion of transfer ranges from as low as 0% to as high as 70%, depending on a myriad of factors. Most of the factors (e.g., N fertilizer application, species, and cultivar selection) are management related and can, therefore, be controlled for improved N_2 fixation and transfer. In this chapter, we discuss current trends in BNF in selected legume crops, the global economics of BNF, and recent reports on N_2 transfer in agricultural production systems. Additionally, factors affecting N_2 transfer and management considerations for improving N_2 fixation and transfer are discussed.

Keywords: biological nitrogen fixation, nitrogen transfer, fertilizers, legumes, grass-legume mixtures

1. Introduction

Plants require N in relatively large quantities to grow and reproduce. In fact, N is the third most important factor in the growth and development of crop plants [1]. This made N one of the most important nutrients in agricultural production systems. The important role N plays in global food production is evident in the ever-increasing amounts of N fertilizers applied annually. It has been estimated that approximately 100 Tg of synthetic N fertilizers were applied in 2009 [2]. The geometric increase in N fertilizer use worldwide is in part, attributable to the need to produce enough food to feed the over 7 billion people currently living on earth. Although there is a consensus within the scientific community that N fertilizers are almost indispensable in today's agriculture, there are great concerns with the use of N fertilizers. Some of these include pollution of surface and underground waters, greenhouse gas (e.g., nitrous oxide: N_2O) emissions, and low N use efficiency (NUE). There is, therefore, a multi-pronged approach to N management in global food production. While N fertilizers are being increasingly applied to crops to increase crop productivity, there are calls for more sustainable approaches to meeting N demand of crops such as climate-smart agriculture and sustainable intensification.

The BNF, the process whereby micro-organisms use nitrogenase enzyme to convert atmospheric inert N_2 to plant usable forms [3, 4], was the main source of N prior to the industrial revolution [5]. It is generally agreed that BNF is one of the most sustainable approaches to meeting crop N demands. For example, it has been estimated that NUE increases exponentially with increasing levels of biologically fixed N_2 in soils while NUE decreases linearly with increasing levels of applied synthetic N fertilizers [2]. There are concerns about the best approach for quantifying inputs of fixed N_2. Conservative estimates based on harvested areas and yields from 2005 Food and Agricultural Organization (FAO) database on world crop production (FAOSTAT) showed that 2.95 and 18.5 Tg N was fixed annually by pulses and oilseed crops, respectively [6]. Soybean (*Glycine max* (L.)) fixed 16.4 Tg N, representing 77% of total N_2 fixation by legume crops in 2005 [6]. Although BNF contributes ~25 Tg N which is dwarfed by the ~100 Tg contributed by synthetic N fertilizers [2], the importance of BNF to the global N budget is substantial.

Just like synthetic fertilizers, BNF has issues of its own. Among the issues of great importance is the transfer of fixed N_2 to non-N_2-fixing plants. The proportion of biologically fixed N_2 transferred to neighboring plants can range from as low as 0% to as high as 73%, depending on a myriad of factors [1]. The biology, chemistry, and processes involved in BNF have been extensively described in the literature [7–12]. Therefore, in this chapter, we discuss briefly the organisms involved in BNF and then proceed to current trends in global N_2 fixation and value of BNF transfer in agricultural production systems with special emphasis on N_2 fixation from *Rhizobia*-legume symbiosis. Finally, we summarize current findings on N transfer in agricultural systems, discuss the factors responsible for low and variable transfer of biologically fixed N_2, and provide some suggestions for improved transfer of fixed N_2.

2. Biological dinitrogen fixation: importance and economics

Several micro-organisms can convert inert atmospheric N_2 to plant usable forms. These organisms may exist in association and symbiosis with host plants or independent of a host plant (**Table 1**). Organisms relying solely on atmospheric N_2 as their N source for growth are referred to as diazotrophs [7]. Biological N_2 fixation is a significant source of N in agricultural and natural ecosystems. The N input from BNF is particularly important in low value crops (e.g., forages)

Micro-organism	Properties and importance
Rhizobia	Symbiosis with roots of legumes (nodules); important source of N for legumes; proper Rhizobia strains required for effective nodulation and N_2 fixation
Frankia (Actinomycetes)	Symbiosis with non-legume angiosperms (e.g., *Alnus, Myrica, Alder, Casuarina*); important source of N in agroforestry
Anabaena	Autotrophic; mostly aquatic but can be terrestrial; symbiosis with non-legumes (e.g., *Azolla* sp.); important in paddy rice (*Oryza sativa* L.) production; can be utilized as green manure
Bradyrhizobium	Aerobic, heterotrophic, free-living N_2-fixer
Azospirillum	Microaerophilic; heterotrophic; free-living N_2-fixer or in association with grass roots; can be important source of N for non-legumes
Acetobacter	Heterotrophic; endophytic, can be important source of N for sugarcane (*Saccharum officinarum* L.) and some tropical grasses
Azotobacter	Aerobic; heterotrophic; free-living N_2-fixer
Cyanobacteria	Autotrophic; free-living N_2-fixer (e.g., *Escherichia coli*) or symbiotic; symbiosis with lichens (fungi), cycads, etc.

†Modified from [3, 7, 18].

Table 1. Properties of selected micro-organisms involved in biological N_2 fixation in agriculture and natural ecosystems†.

and developing economies, where farmers either have limited access to synthetic N fertilizers or are unable to afford N fertilizers. In fact, forage accumulation and profitability from grass-legume mixtures have been reported to be equal or greater than N-fertilized grass monocultures [13–15]. Aside direct N input from BNF, N from BNF reduces the amount of synthetic N fertilizers applied in agriculture and natural ecosystems. This, in turn, reduces cost of production, greenhouse gas (GHG) emissions, and pollution of surface and underground waters. Low NUE and N recovery are major issues associated with use of N fertilizers [16, 17]. In a comprehensive analysis, Lassaletta et al. [2] showed that the efficiency of N use of biologically fixed N_2 is greater than synthetic N. Among the micro-organisms involved in BNF, N_2 fixation from *Rhizobia*-legume symbiosis is a significant source of N in agriculture. Needless to say, BNF from associative and free-living bacteria and diazotrophs are important in natural ecosystems and water-logged production areas (e.g., paddy fields) [6].

2.1. Amount and value of N_2 fixed by legumes

The amount of N_2 fixed from *Rhizobia*-legume symbiosis varies greatly depending on many factors. These include, but not limited to, plant species and cultivar, residual soil N, *Rhizobia* strains, and environmental conditions. Generally, perennial forages fix greater amounts of N_2 compared to annual forages since they live longer in the field [18]. For example, estimated total BNF from alfalfa (*Medicago sativa* L.), red clover (*Trifolium pratense* L.), and white clover (*Trifolium repens* L.) are 465, 252, and 102 kg N ha^{-1} $year^{-1}$ while from faba bean (*Vicia faba* L.), field pea (*Pisum sativum* L.), and lentil (*Lens culinaris* Medik.) are 165, 111, and 52 kg N ha^{-1} $year^{-1}$, respectively [19]. Estimates of N_2 fixation from selected crops has shown that in 2014, up to 29

Tg N was fixed by eight crops (**Figure 1**). Soybean (*Glycine max* (L.) Merr.) alone contributed 23.4 Tg, representing 81% of total N_2 fixed by these crops (**Figure 1**). While these might not be precise estimates, there is a clear indication that the contribution N_2 fixation to the global N budget is enormous. Though N_2 fixation from peas, lentils, common bean (*Phaseolus vulgaris* L.), faba bean, cowpea (*Vigna unguiculata* (L.) Walp.), chickpeas (*Cicer arietinum* L.), and groundnut

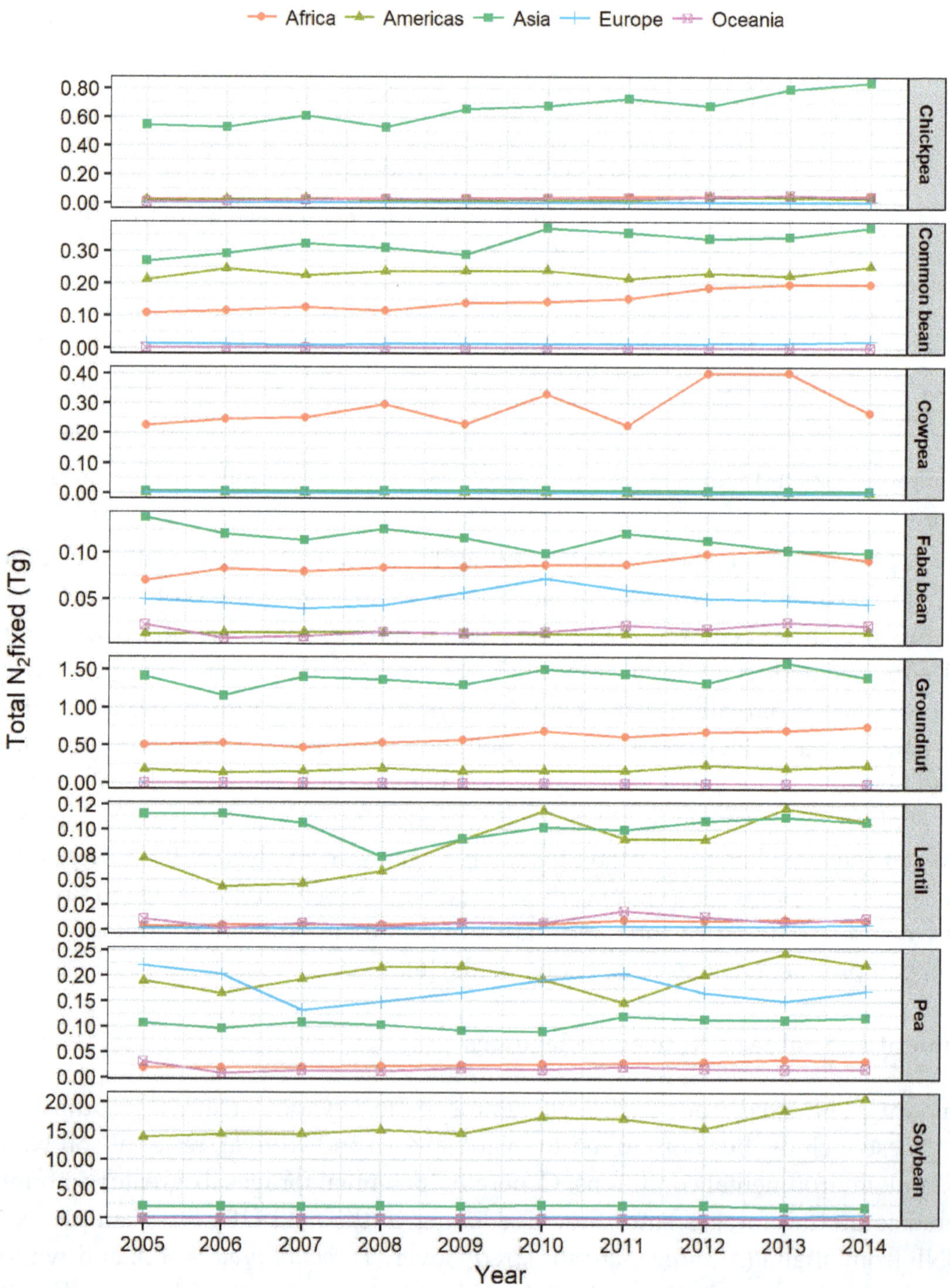

Figure 1. Estimates of global trends in biological N_2 fixation for selected legume crops. The N_2 was estimated based on harvested areas and yield data from Food and Agricultural Organization (FAO) database on world crop production (FAOSTAT) [21]. This follows the procedure described by [6].

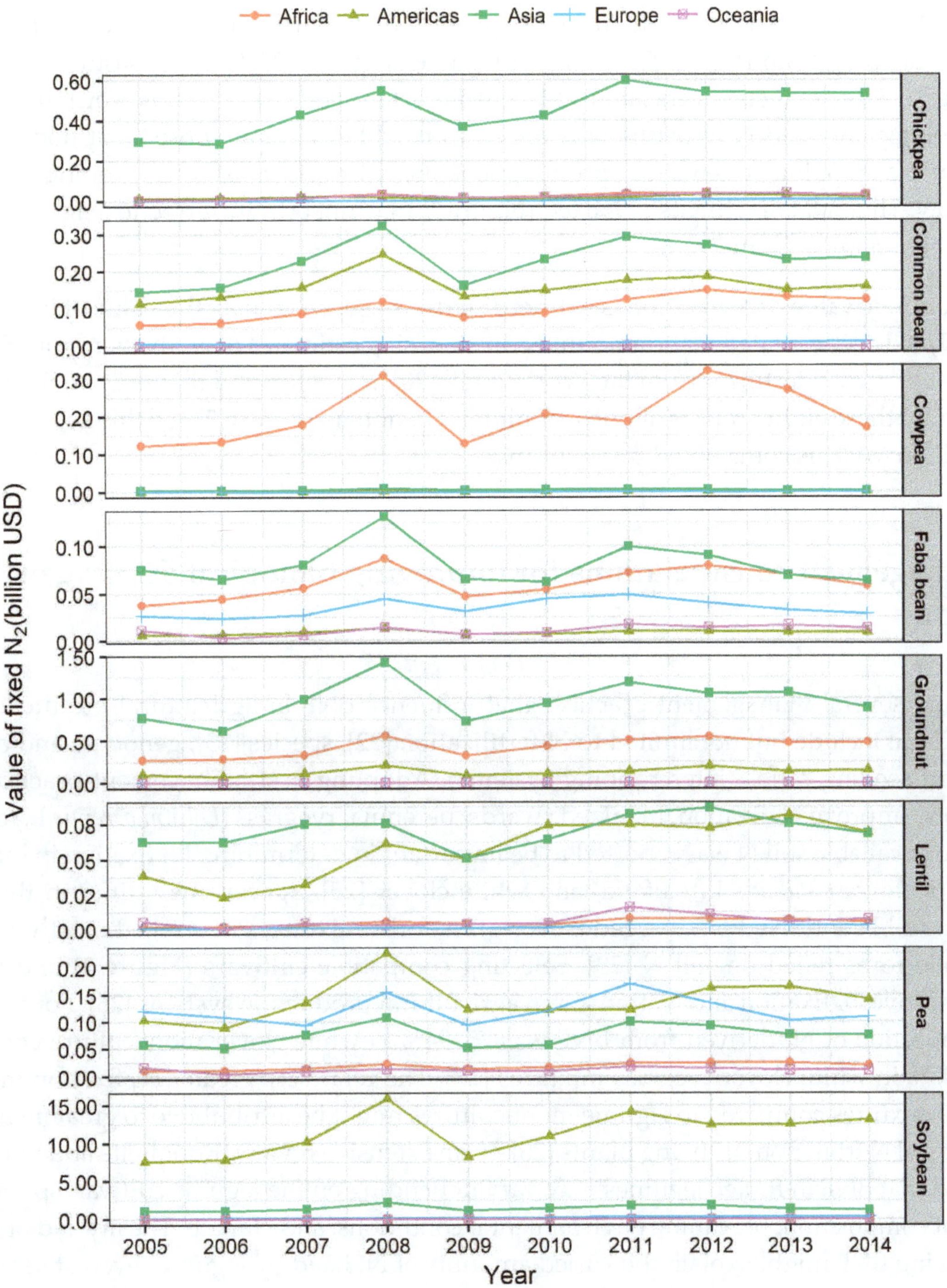

Figure 2. Trends in global economics of biological N_2 fixation. Value of fixed N_2 was calculated based on estimated N_2 fixation (**Figure 1**) and price of urea fertilizer from 2005 to 2014 reported by the World Bank [20].

(*Arachis hypogaea* L.) is dwarfed by soybean (because of the larger area planted to soybean) based on these estimates, the contribution of N_2 fixation from these crops (e.g., cowpea) to farmers in developing countries is substantial. Unlike forages, grains from grain legumes are harvested and removed from the field. Thus, grain legumes usually remove more soil N than forages [18]. The uncertainties associated with estimating N_2 fixation from forages, extensively grazed savannas, sugarcane (*Saccharum officinarum* L.), and rice (*Oryza sativa* L.) production systems have been acknowledged [6]. Nonetheless, the estimated annual N_2 fixation from these systems are 5 Tg from rice, <4 Tg from non-legume crops, 12–25 Tg from pasture and fodder legumes, 0.5 Tg from sugarcane, and <14 Tg from extensive savannas. It is worth mentioning that biologically fixed N_2 must be transferred to neighboring and subsequent non-N_2-fixing crops in the cropping systems for optimum benefits. Nitrogen transfer in cropping systems is often low. Thus, all the estimated N_2 fixed (**Figure 1**) may not be transferred to neighboring and subsequent non-N_2-fixing crops.

The economic value of N_2 fixation is extraordinarily large. Of course, the value of biologically fixed N_2 is directly related to the amount N_2 fixed. Using estimates of N_2 fixation from **Figure 1** and cost of urea N fertilizer from the World Bank [20], it is estimated that in 2014, the value of N fixed by these eight crops is about 18.5 billion US dollars (**Figure 2**). Of this amount, about 14.9 billion (81%) is contributed by soybeans.

3. Management considerations for improving biological dinitrogen fixation

There are several management practices that influence BNF in agricultural production systems. These include but not limited to N-fertilization [22], species [23], genotype and cultivar [24], and seeding ratios (intercropping systems). Adopting best management practices can, therefore, improve N_2 fixation. In mixed swards, perennial ryegrass (*Lolium perenne* L.) competition for available soil N was reported to be important in determining N_2 fixation in birdsfoot trefoil (*Lotus corniculatus* L.), alfalfa, and white clover [25]. Species may differ in their reliance on soil N and fixed N_2. In a red clover-grass-forbs mixture, grass relied mostly on fixed N_2, while forbs relied on soil N [23]. Selecting compatible cultivars (**Figure 3**) and species may improve N_2 fixation and N_2 transfer in agricultural production systems [26]. For example, the proportion of N_2 derived from BNF was 75–94% in white clover monoculture compared to 85–97% in white clover-ryegrass mixtures [27]. The relatively greater N_2-fixation in grass-legume mixtures compared to legume monocultures might be attributable to greater competition for soil N from non-N_2-fixing plants [28]. In an extensive review, Rouquette and Smith [29] asserted that BNF in forage legumes may vary depending on the legume cultivar, species, soil nutrient composition, prevailing environmental conditions, and climate. The myriad of factors influencing BNF might explain the varied amounts of N_2 fixed by legumes even at same locations reported by many researchers [30–35]. For example, at the same location, the proportion of plant total N derived from BNF was reported to range from 12 to 96% on grazed plots [36]. Application of N fertilizers has been found to suppress BNF in legumes [22]. For example, the application of N fertilizer decreased atmospheric derived N_2 of clover from 77 to 43% [37].

The strain of *Rhizobia* also determines the level of N_2 fixation [38]. Most of these *Rhizobia* strains are highly specialized and due to this specialization and the intricacy of interaction between N_2-fixing plant species and bacteria involved in N_2 fixation, any disturbance or manipulation may be detrimental to the amount of N_2 fixed [39]. Thus, inoculation with the right strains of *Rhizobia* would improve N_2 fixation. There are three major constraints to BNF in grass-legume

Figure 3. Established stands of meadow bromegrass (*Bromus biebersteinii* Roem. & Schult.) monoculture (a) and 50% meadow bromegrass: 50% sainfoin (*Onobrychis viciifolia* Scop.) (b), 50% meadow bromegrass:50% birdsfoot trefoil (c), 50% meadow bromegrass:16.7% sainfoin:16.7% birdsfoot trefoil:16.7% alfalfa (d) seeding ratios in 2014 at the University of Wyoming Sheridan Research and Extension Center, WY, USA.

mixtures and these include: low forage yield, low proportion of legumes in mixtures, and low reliance of the legume on N_2 fixation [40]. To maintain optimal N_2 fixation, sufficient legume populations must be maintained in grass-legume stands. This might be difficult to achieve because of the selective grazing of legumes by livestock (in grazing systems), poor soil conditions, and pest and disease problems [28]. However, using optimal seed mass ratios and good grazing and haying practices may help maintain optimal legume proportions [15].

4. Transfer of biologically fixed nitrogen in agricultural production systems

Biologically fixed N_2 satisfies the immediate N needs of the host plants. However, the fixed N_2 can be transferred to other crops in the cropping system, especially non-N_2-fixing plants. The transfer is accomplished through three main routes, viz.: decomposition of nodules and secondary roots that are not thickened, exudates of soluble N compounds, and transfer mediated by mycorrhizal fungi [1, 41–43]. The transfer of N through nodule and root decomposition and exudation of N compounds is termed as rhizodeposition [44]. The proportion of biologically fixed N_2 transferred to neighboring or succeeding crop plants is highly variable [45]. This can range from as low as 0% to as high as 73%, depending on a myriad of factors [1]. In an extensive review, rhizodeposition was reported to vary from 4 to 71% [44]. Review of literature from 2015 to 2017 on transfer of N in selected crops has shown that N transfer ranged from 0 to 70% (**Table 2**). Among the three main N transfer routes, rhizodeposition through decomposition of the nodules and roots represents the main pathway of N transfer.

Nitrogen transfer from signal grass (*Brachiaria decumbens* Stapf.) to stylo (*Stylosanthes guianensis* (Aublet) Sw.) was reported to be mainly through decomposition of roots compared to root exudates and transfer mediated by mycorrhizae [46]. This might be particularly true for forage species since aboveground biomass is the economic part of the plant. Additionally, non-tree legumes have relatively greater proportion of fine roots that have faster turnover rate. It must be noted that despite the greater contribution of decomposition of the nodules and roots to N transfer, this transfer route is relatively slower compared to exudates of soluble N compounds and transfer mediated by mycorrhizae [1]. Nitrogen transfer from the tropical legume, gliricidia (*Gliricidia sepium* (Jacq.) Kunth ex Walp.) to yellow-blue stem (*Dichanthium aristatum* (Poir.) C.E. Hubb.) was reported to be mainly via root exudates [47]. In a short-term rhizodeposition study, 3.5 and 5.3% N was rhizodeposited through root exudates in white clover monocrop and white clover-perennial ryegrass mixture, respectively, over a 3-day period [42]. This significant N transfer within a short period is an indication of the importance of exudation of N compounds in meeting N needs of crops, especially during early growing stages [42]. It is well documented that mycorrhizae can facilitate the transfer of biologically fixed N_2

Crop(s)	**Amount of N transferred (% of fixed N)**	**Reference(s)**
Caragana (*Caragana arborescens* Lam.)-oat (*Avena sativa* L.)	38–45 kg ha^{-1} (60–70)§	[62]
Alfalfa-tall fescue (*Schedonorus arundinaceus* (Schreb.) Dumort.)	0–650 kg ha^{-1} (0–12)†	[66]
White clover-perennial ryegrass	0–340 kg ha^{-1} (0–47)†	[66]
Mung bean-oat	12.8 mg plant^{-1} (9.7)	[68]
Soybean-maize	7.84 mg pot^{-1} (7.57)	[53]
Soybean-maize	10.77–13.72 mg pot^{-1} (1.26–2.17)	[55]
Faba bean-wheat	0.17 mg plant shoot^{-1} (14.9)	[52]
Red clover-bluegrass (*Poa pratensis* L.)	35.85 mg plant^{-1} (1.5)	[24]
Pigeon pea (*Cajanus cajan* (L.) Millsp.-coffee (*Coffea arabica* L.)	21.8 g kg^{-1} (na)	[63]
Crotalaria-coffee	13.5 g kg^{-1} (na)	[63]
Velvet bean (*Mucuna pruriens* (L.) DC.)-coffee	19.7 g kg^{-1} (na)	[63]
Red clover-perennial ryegrass and forbs	25–58 kg ha^{-1} (9.5–15)	[23]

na, could not be estimated from data.
§4 m distance from caragana shelterbelt.
†Cumulative over 3-year period.

Table 2. Amount of nitrogen (N) fixed and proportion transferred to soil or neighboring plants in agricultural systems.

from N_2-fixers to non-N_2-fixing plants [48–51]. In a rice and mung bean (*Vigna radiata* L.) intercropping study, arbuscular mycorrhizal fungi (AMF) inoculation increased N transfer from 5.4 to 15.7% [49]. Proportion of fixed N_2 transferred from faba bean to wheat (*Triticum aestivum* L.) was 50% when inoculated with AMF compared 15% in uninoculated stands [52]. Similar results were also reported in garden pea-barley (*Hordeum vulgare* L.) and soybean-maize (*Zea mays* L.) intercropping studies [48, 53]. The AMF-mediated transfer of N can be both unidirectional and bidirectional [48, 54] and often along with a concentration gradient [47]. Thus, transfer of N from N_2-fixing plants to non-N_2-fixers is often expected to be greater than from non-N_2-fixing plants to N_2-fixers [55].

5. Factors affecting nitrogen transfer

It has long been acknowledged that since plant N composition is partitioned into various plant organs or parts, not all the N_2 fixed by plants will be transferred to neighboring plants or succeeding plants in cropping systems [56]. However, there are a number of biotic and abiotic factors influencing N transfer in agricultural production systems [1]. Environmental factors such as water, temperature, and light have direct and indirect effects on N transfer in cropping systems. Soil moisture has a great influence on decomposition and it is required for the uptake of N. Thus, moisture stress affects both the mineralization of fixed N_2 and uptake of mineralized N by plants. However, moisture stress promotes nodule senescence, implying that more nodule biomass will be available for mineralization during moisture stress conditions [57]. Nitrogen is highly soluble. Thus, excess water can result in N leaching out of the rooting zone of plants making it unavailable for uptake. Flooding (e.g., low land rice production systems) results in anaerobic conditions, and thus could result in gaseous N losses in the form of N_2O [18]. Optimum light conditions (quality, quantity, and duration) and temperature have a direct effect on photosynthesis and hence, promote both N_2 fixation and transfer. For example, nodule activity and N exudation from roots of soybean and sesbania (*Sesbania cannabina* (Retz.) Poir.) were the greatest at 30 and 35°C day and night temperatures, respectively [58]. Prolonged dark treatment affected nodule functioning in barrel medic (*Medicago truncatula* Gaertn.) and induced nodule senescence [59]. This condition is common in intercropping systems (e.g., grass-legume mixtures) [1], especially in species with varied canopy heights.

A common practice in agricultural production systems is intercropping N_2-fixing legumes with non-N_2-fixing crops (**Figure 3**) [15]. This is particularly important in low value crops (e.g., forages) and in developing countries. In intercropping systems, the proximity of the N_2-fixing crop to the non-N_2-fixing determines the amount of N transferred. The concentration of N in the rhizosphere is the greatest closer to the root surface [60]. Therefore, N transfer predominantly occurs in upper soil layers [23]. Since N uptake is along with concentration gradients [47], close proximity between N_2-fixing legumes and non-N_2-fixing crops reduces the distance of travel for dissolved N compounds [1]. Close proximity is achieved either through direct root contact or mycorrhizal hyphae connections [61]. However, Issah et al. [62] reported that maximum oat productivity was obtained when grown 4 m from caragana shelterbelt compared to 2 m from the shelterbelt.

Aside proximity, species (**Table 2**) of N_2-fixing legumes as well as the non-N_2-fixing crops (when grown in mixtures) influence the amount of N_2 fixed and transferred to neighboring crops. The amount of N transferred to Arabian coffee (*Coffea arabica* L.) ranged from 13.5 to 21.8 g kg^{-1} depending on the N_2-fixing legume (**Table 2**) [63]. There was no observable N transfer from berseem clover (*Trifolium alexandrinum* L.) to annual ryegrass (*L. perenne* L. subsp. *multiflorum* [Lam.] Husnot) when gown in mixtures [64]. This was attributed to the greater efficiency of annual ryegrass in the uptake of available soil N which resulted in berseem clover becoming reliant on fixed N_2 [64]. In an alfalfa-Bermudagrass (*Cynodon dactylon* (L.) Pers.) intercrop, alfalfa fixed 80 to 222 kg N ha^{-1} $year^{-1}$ and transferred about 18 kg N ha^{-1} $year^{-1}$ to Bermudagrass [65]. Alfalfa fixed twice as much N as white clover but transferred only 59 kg N ha^{-1} compared to 147 kg N ha^{-1} transferred by white clover over a 3-year period [66]. Although decomposed alfalfa roots released greater N than that of birdsfoot trefoil, the opposite was true for decomposed nodules [41]. There was no transfer of N from any of seven legumes [snail medick (*M. scutellata* L.), common vetch (*V. sativa* L.), squarrosum clover (*T. squarrosum* L.), hairy vetch (*V. villosa* Roth), sulla (*Hedysarum coronarium* L.), and fenugreek (*Trigonella foenum-graecum* L.)] to annual ryegrass under Mediterranean conditions [67]. N transfer is also influenced by crop cultivars. For example, red clover cultivars differed in amount of N transferred to Kentucky bluegrass [24]. Compatibility of species grown in mixed swards affects the amount of N_2 fixed and the proportion transferred. A recent study has shown that grass N demand in grass-legume mixtures might be more important than legume N supply in determining N transfer efficiency [26].

Other factors such as age or stage of growth [68], season or year [69–71], proportion of N-fixing species [71], compatibility [45], and stand persistence [35] affect N transfer in cropping systems. For example, N in naked oats (*Avena nuda* L.) derived from N_2 fixed by mung bean was 7.6% at pod setting and increased to 9.7% at maturity [68]. The proportion of N transferred from red clover to Kentucky bluegrass was reported to have increased over time [24]. This is particularly true for perennial forages because of relatively low N_2 fixation in establishment year compared to well-established stands [1]. It is generally expected that as the proportion of legumes in mixed swards increases, N_2 fixation and transfer increases [1]. However, in a continental-scale field study with two perennial N_2-fixing legumes (red clover and white clover) and four perennial grasses (perennial ryegrass, Timothy (*Phleum pratense* L.), Kentucky bluegrass, and orchardgrass (*Dactylis glomerata* L.)), it was reported that N gained in mixed swards increased with increasing legume proportion up to 30% [71]. This supports the assertion by [26] that grass N demand in grass-legume mixtures might be more important than legume N supply in determining N transfer efficiency. In an annual garden pea-barley intercropping system, greatest N transfer was obtained in 1:1 garden pea: barley compared to 2:1 system [72].

6. Conclusions

It is generally agreed that BNF is one of the most sustainable sources of N in agricultural production systems. The BNF is especially important in low value crops (e.g., forages) and in developing economies. Estimated N_2 fixation from selected crops showed that the contribution of N_2 fixation to the global N budget is enormous. Though N_2fixation from peas, lentils, common bean, faba bean, cowpea, chickpeas, and groundnut is dwarfed by soybean (because of the larger area planted to soybean) based on these estimates, the contribution of N_2 fixation from these crops (e.g., cowpea) to farmers in developing countries is substantial. Unlike forages, grains from grain legumes are harvested and removed from the field. Thus, grain legumes usually remove more soil N than forages. There are, however, several issues related to BNF that are of concern to the scientific community. Among the issues of great importance is the low and highly variable proportion of fixed N_2 transferred to non-N_2-fixing plants. Proportion of fixed N_2 transferred to non-N_2-fixing plants ranges from as low as 0% to as high as 70%, depending on a myriad of factors. This was not different than the range of values reported from previous reviews. However, most of the factors (e.g., N fertilizer application, species, and cultivar selection) are management related and can, therefore, be controlled for improved N_2 fixation and transfer. Most *Rhizobia* strains are highly specialized and due to this specialization, inoculation with the right strains of *Rhizobia* would improve N_2 fixation. One of the constraints to BNF in grass-legume mixtures is low proportion of legumes in the mixtures. It is, therefore, important to maintain sufficient legume populations in the grass-legume systems for optimal N_2 fixation. This might, however, be difficult to achieve because of the selective grazing of legumes by livestock (in grazing systems). Nonetheless, using optimal seed mass ratios and good grazing and haying practices may help maintain optimal legume proportions.

Author details

M. Anowarul Islam* and Albert Tetteh Adjesiwor

*Address all correspondence to: mislam@uwyo.edu

Department of Plant Sciences, University of Wyoming, Laramie, WY, USA

References

[1] Thilakarathna MS, McElroy MS, Chapagain T, Papadopoulos YA, Raizada MN. Below-ground nitrogen transfer from legumes to non-legumes under managed herbaceous cropping systems. A review. Agronomy for Sustainable Development. 2016;**36**:1-16

[2] Lassaletta L, Billen G, Grizzetti B, Anglade J, Garnier J. 50 year trends in nitrogen use efficiency of world cropping systems: The relationship between yield and nitrogen input to cropland. Environmental Research Letters. 2014;**9**:1-9

[3] Unkovich M, Herridge D, Peoples M, Cadisch G, Boddey B, Giller K, Alves B, Chalk P. Measuring Plant-Associated Nitrogen Fixation in Agricultural Systems. Canberra, Australia: Australian Centre for International Agricultural Research (ACIAR); 2008. p. 258

[4] Varley JB, Wang Y, Chan K, Studt F, Norskov JK. Mechanistic insights into nitrogen fixation by nitrogenase enzymes. Physical Chemistry Chemical Physics. 2015;**17**:29541-29547

[5] Vitousek PM, Menge DNL, Reed SC, Cleveland CC. Biological nitrogen fixation: Rates, patterns and ecological controls in terrestrial ecosystems. Philosophical Transactions of the Royal Society of London B: Biological Sciences. 2013;**368**:1-9

[6] Herridge DF, Peoples MB, Boddey RM. Global inputs of biological nitrogen fixation in agricultural systems. Plant and Soil. 2008;**311**:1-18

[7] Sylvia DM, Fuhrmann JJ, Hartel PG, Zuberer DA. Principles and Applications of Soil Microbiology. Upper Saddle River: Pearson Prentice Hall; 2005. p. 640

[8] de Bruijn FJ, editor. Biological Nitrogen Fixation. Vol. 1&2. Hoboken, NJ: John Wiley & Sons, Inc.; 2015. p. 1196

[9] Darnajoux R, Zhang X, McRose DL, Miadlikowska J, Lutzoni F, Kraepiel AM, Bellenger JP. Biological nitrogen fixation by alternative nitrogenases in boreal cyanolichens: Importance of molybdenum availability and implications for current biological nitrogen fixation estimates. New Phytologist. 2017;**213**:680-689

[10] Jayakumar A, Chang BX, Widner B, Bernhardt P, Mulholland MR, Ward BB. Biological nitrogen fixation in the oxygen-minimum region of the eastern tropical North Pacific Ocean. ISME Journal. 2017;**11**:2356-2367

[11] Puschel D, Janouskova M, Voriskova A, Gryndlerova H, Vosatka M, Jansa J. Arbuscular *Mycorrhiza* stimulates biological nitrogen fixation in two *Medicago* spp. through improved phosphorus acquisition. Frontiers in Plant Science. 2017;**8**:1-12

[12] Sulieman S, Tran L, editors. Legume Nitrogen Fixation in a Changing Environment. New York: Springer; 2016. p. 133

[13] Sturludóttir E, Brophy C, Bélanger G, Gustavsson A-M, Jørgensen M, Lunnan T, Helgadóttir Á. Benefits of mixing grasses and legumes for herbage yield and nutritive value in Northern Europe and Canada. Grass and Forage Science. 2014;**69**:229-240

[14] Sanderson MA, Brink G, Stout R, Ruth L. Grass–legume proportions in forage seed mixtures and effects on herbage yield and weed abundance. Agronomy Journal. 2013; **105**:1289-1297

[15] Adjesiwor AT, Islam MA, Zheljazkov VD, Ritten JP, Garcia y Garcia A. Grass-legume seed mass ratios and nitrogen rates affect forage accumulation, nutritive value, and profitability. Crop Science. 2017;**57**:1-13

[16] Abbasi MK, Tahir MM, Sadiq A, Iqbal M, Zafar M. Yield and nitrogen use efficiency of rainfed maize response to splitting and nitrogen rates in Kashmir, Pakistan. Agronomy Journal. 2012;**104**:448-457

[17] Conant RT, Berdanier AB, Grace PR. Patterns and trends in nitrogen use and nitrogen recovery efficiency in world agriculture. Global Biogeochemical Cycles. 2013;**27**:558-566

[18] Havlin JL, Tisdale SL, Nelson WL, Beaton JD. Soil Fertility and Fertilizers: An Introduction to Nutrient Management. 8th ed. Vol. 516. Upper Saddle River, NJ: Pearson Prentice Hall; 2014. 516 p

[19] Anglade J, Billen G, Garnier J. New relationships for estimating N_2 fixation in legumes: Incidence for N balance of low-input cropping systems in Europe. Ecological Applications. 2015;**6**:1-24

[20] The World Bank. Commodity Markets. Latest Commodity Prices. Available from http://www.worldbank.org/en/research/commodity-markets. [Accessed: August 11, 2017]

[21] FAOSTAT. Food and Agricultural Organization (FAO) Database on World Crop Production (FAOSTAT). Available from http://www.fao.org/faostat/en/#data/QC. [Accessed: August 11, 2017]

[22] Enriquez-Hidalgo D, Gilliland TJ, Hennessy D. Herbage and nitrogen yields, fixation and transfer by white clover to companion grasses in grazed swards under different rates of nitrogen fertilization. Grass and Forage Science. 2016;**71**:559-574

[23] Dhamala NR, Rasmussen J, Carlsson G, Soegaard K, Eriksen JN. Transfer in three-species grass-clover mixtures with chicory, ribwort plantain or caraway. Plant and Soil. 2017;**413**:217-230

[24] Thilakarathna MS, Papadopoulos YA, Rodd AV, Grimmett M, Fillmore SAE, Crouse M, Prithiviraj B. Nitrogen fixation and transfer of red clover genotypes under legume-grass forage based production systems. Nutrient Cycling in Agroecosystems. 2016;**106**:233-247

[25] Rasmussen J, Soegaard K, Pirhofer-Walzl K, Eriksen J. N_2-fixation and residual N effect of four legume species and four companion grass species. European Journal of Agronomy. 2012;**36**:66-74

[26] McElroy MS, Papadopoulos YA, Glover KE, Dong Z, Fillmore SAE, Johnston MO. Interactions between cultivars of legume species (*Trifolium pratense* L., *Medicago sativa* L.) and grasses (*Phleum pratense* L., *Lolium perenne* L.) under different nitrogen levels. Canadian Journal of Plant Science. 2017;**97**:214-225

[27] Jørgensen FV, Jensen ES, Schjoerring JK. Dinitrogen fixation in white clover grown in pure stand and mixture with ryegrass estimated by the immobilized ^{15}N isotope dilution method. Plant and Soil. 1999;**208**:293-305

[28] Barnes R, Nelson C, Collins M, Moore K. Forages: The Science of Grassland Agriculture. 6th ed. Vol. 2. Ames, Iowa: Blackwell Publishing Ltd.; 2007. p. 556

[29] Rouquette F, Smith G. Effects of biological nitrogen fixation and nutrient cycling on stocking strategies for cow-calf and stocker programs. The Professional Animal Scientist. 2010;**26**:131-141

[30] Querné A, Battie-laclau P, Dufour L, Wery J, Dupraz C. Effects of walnut trees on biological nitrogen fixation and yield of intercropped alfalfa in a Mediterranean agroforestry system. European Journal of Agronomy. 2017;**84**:35-46

[31] Ritchie ME, Raina R. Effects of herbivores on nitrogen fixation by grass endophytes, legume symbionts and free-living soil surface bacteria in the Serengeti. Pedobiologia. 2016;**59**:233-241

[32] Scheifele M, Hobi A, Buegger F, Gattinger A, Schulin R, Boller T, Mäder P. Impact of pyrochar and hydrochar on soybean (*Glycine max* L.) root nodulation and biological nitrogen fixation. Journal of Plant Nutrition and Soil Science. 2017;**180**:199-211

[33] Denton MD, Phillips LA, Peoples MB, Pearce DJ, Swan AD, Mele PM, Brockwell J. Legume inoculant application methods: Effects on nodulation patterns, nitrogen fixation, crop growth and yield in narrow-leaf lupin and faba bean. Plant and Soil. 2017;**419**:25-39

[34] Dey SK, Chakrabarti B, Prasanna R, Pratap D, Singh SD, Purakayastha TJ, Pathak H. Elevated carbon dioxide level along with phosphorus application and cyanobacterial inoculation enhances nitrogen fixation and uptake in cowpea crop. Archives of Agronomy and Soil Science. 2017;**63**:1927-1937

[35] Zheng M, Li D, Lu X, Zhu X, Zhang W, Huang J, Fu S, Lu X, Mo J. Effects of phosphorus addition with and without nitrogen addition on biological nitrogen fixation in tropical legume and non-legume tree plantations. Biogeochemistry. 2016;**131**:65-76

[36] Eriksen J, Høgh-Jensen H. Variations in the natural abundance of ^{15}N in ryegrass/white clover shoot material as influenced by cattle grazing. Plant and Soil. 1998;**205**:67-76

[37] Ledgard S, Sprosen M, Penno J, Rajendram G. Nitrogen fixation by white clover in pastures grazed by dairy cows: Temporal variation and effects of nitrogen fertilization. Plant and Soil. 2001;**229**:177-187

[38] Fujita H, Aoki S, Kawaguchi M. Evolutionary dynamics of nitrogen fixation in the legume–rhizobia symbiosis. PLoS One. 2014;**9**:e93670

[39] Shantharam S, Mattoo AK. Enhancing biological nitrogen fixation: an appraisal of current and alternative technologies for N input into plants. Plant and Soil. 1997;**194**:205-216

[40] Boller BT, Nösberger J. Symbiotically fixed nitrogen from field-grown white and red clover mixed with ryegrasses at low levels of ^{15}N-fertilization. Plant and Soil. 1987; **104**:219-226

[41] Dubach M, Russelle MP. Forage legume roots and nodules and their role in nitrogen transfer. Agronomy Journal. 1994;**86**:259-266

[42] Lesuffleur F, Salon C, Jeudy C, Cliquet JB. Use of a $^{15}N_2$ labelling technique to estimate exudation by white clover and transfer to companion ryegrass of symbiotically fixed N. Plant and Soil. 2013;**369**:187-197

[43] Munroe JW, Isaac ME. N_2-fixing trees and the transfer of fixed-N for sustainable agroforestry: A review. Agronomy for Sustainable Development. 2014;**34**:417-427

[44] Fustec J, Lesuffleur F, Mahieu S, Cliquet JB. Nitrogen rhizodeposition of legumes. A review. Agronomy for Sustainable Development. 2010;**30**:57-66

[45] Adjesiwor AT, Islam MA. Rising nitrogen fertilizer prices and projected increase in maize ethanol production: The future of forage production and the potential of legumes in forage production systems. Grassland Science. 2016;**62**:203-212

[46] Trannin WS, Urquiaga S, Guerra G, Ibijbijen J, Cadisch G. Interspecies competition and N transfer in a tropical grass-legume mixture. Biology and Fertility of Soils. 2000;**32**:441-448

[47] Jalonen R, Nygren P, Sierra J. Transfer of nitrogen from a tropical legume tree to an associated fodder grass via root exudation and common mycelial networks. Plant, Cell and Environment. 2009;**32**:1366-1376

[48] Johansen A, Jensen E. Transfer of N and P from intact or decomposing roots of pea to barley interconnected by an arbuscular mycorrhizal fungus. Soil Biology and Biochemistry. 1996;**28**:73-81

[49] Li Y, Ran W, Zhang R, Sun S, Facilitated XG. Legume nodulation, phosphate uptake and nitrogen transfer by arbuscular inoculation in an upland rice and mung bean intercropping system. Plant and Soil. 2009;**315**:285-296

[50] He X-H, Critchley C, Bledsoe C. Nitrogen transfer within and between plants through common mycorrhizal networks (CMNs). Critical Reviews in Plant Sciences. 2003; **22**:531-567

[51] Nygren P, Leblanc HA. Dinitrogen fixation by legume shade trees and direct transfer of fixed N to associated cacao in a tropical agroforestry system. Tree Physiology. 2015;**35**:134-147

[52] Wahbi S, Maghraoui T, Hafidi M, Sanguin H, Oufdou K, Prin Y, Duponnois R, Galiana A. Enhanced transfer of biologically fixed N from faba bean to intercropped wheat through mycorrhizal symbiosis. Applied Soil Ecology. 2016;**107**:91-98

[53] Meng LB, Zhang AY, Wang F, Han XG, Wang DJ, Li SM. Arbuscular mycorrhizal fungi and rhizobium facilitate nitrogen uptake and transfer in soybean/maize intercropping system. Frontiers in Plant Science. 2015;**6**:1-10

[54] He X, Critchley C, Nara K, Southworth D, Bledsoe C. ^{15}N enrichment methods to quantify two-way nitrogen transfer between plants linked by mychorrhizal networks. In: Varma A, Kharkwal AC, editors. Symbiotic Fungi: Principles and Practices, Soil Biology. Vol. 18. Berlin: Springer; 2009. pp. 285-291

[55] Yong TW, Liu XM, Yang F, Song C, Wang XC, Liu WG, BY S, Zhou L, Yang WY. Characteristics of nitrogen uptake, use and transfer in a wheat-maize-soybean relay intercropping system. Plant Production Science. 2015;**18**:388-397

[56] Peoples M, Herridge D, Ladha J. Biological nitrogen fixation: An efficient source of nitrogen for sustainable agricultural production? Plant and Soil. 1995;**174**:3-28

[57] Mhadhbi H, Djébali N, Chihaoui S, Jebara M, Mhamdi R. Nodule senescence in *Medicago truncatula*–Sinorhizobium symbiosis under abiotic constraints: Biochemical and structural processes involved in maintaining nitrogen-fixing capacity. Journal of Plant Growth Regulation. 2011;**30**:480-489

[58] Ofosu-Budu KG, Ogata S, Fujita K. Temperature effects on root nodule activity and nitrogen release in some sub-tropical and temperate legumes. Soil Science & Plant Nutrition. 1992;**38**:717-726

[59] Guerra JCP, Coussens G, De Keyser A, De Rycke R, De Bodt S, Van De Velde W, Goormachtig S, Holsters M. Comparison of developmental and stress-induced nodule senescence in *Medicago truncatula*. Plant Physiology. 2010;**152**:1574-1584

[60] zu Schweinsberg-Mickan MS, Joergensen RG, Müller T. Fate of ^{13}C- and ^{15}N-labelled rhizodeposition of *Lolium perenne* as function of the distance to the root surface. Soil Biology and Biochemistry. 2010;**42**:910-918

[61] Jalonen R, Nygren P, Sierra J. Root exudates of a legume tree as a nitrogen source for a tropical fodder grass. Nutrient Cycling in Agroecosystems. 2009;**85**:203-213

[62] Issah G, Kimaro AA, Kort J, Knight JD. Nitrogen transfer to forage crops from a Caragana shelterbelt. Forests. 2015;**6**:1922-1932

[63] Mendonca ED, de Lima PC, Guimaraes GP, Moura WD, Andrade FV. Biological nitrogen fixation by legumes and N uptake by coffee plants. Revista Brasileira de Ciencia do Solo. 2017;**41**:1-10

[64] Giambalvo D, Ruisi P, Di Miceli G, Frenda AS, Amato G. Forage production, N uptake, N_2 fixation, and N recovery of berseem clover grown in pure stand and in mixture with annual ryegrass under different managements. Plant and Soil. 2011;**342**:379-391

[65] Haby VA, Stout SA, Hons FM, Leonard AT. Nitrogen fixation and transfer in a mixed stand of alfalfa and bermudagrass. Agronomy Journal. 2006;**98**:890-898

[66] Louarn G, Pereira-Lopes E, Fustec J, Mary B, Voisin AS, Carvalho PCD, Gastal F. The amounts and dynamics of nitrogen transfer to grasses differ in alfalfa and white clover-based grass-legume mixtures as a result of rooting strategies and rhizodeposit quality. Plant and Soil. 2015;**389**:289-305

[67] Saia S, Urso V, Amato G, Frenda AS, Giambalvo D, Ruisi P, Di Miceli G. Mediterranean forage legumes grown alone or in mixture with annual ryegrass: Biomass production, N_2 fixation, and indices of intercrop efficiency. Plant and Soil. 2016;**402**:395-407

[68] Zang HD, Yang XC, Feng XM, Qian X, YG H, Ren CZ, Zeng ZH. Rhizodeposition of nitrogen and carbon by Mungbean (*Vigna radiata* L.) and its contribution to intercropped oats (*Avena nuda* L.). PLoS One. 2015;**10**:e0121132

[69] Clermont-Dauphin C, Suvannang N, Pongwichian P, Cheylan V, Hammecker C, Harmand JM. Dinitrogen fixation by the legume cover crop *Pueraria phaseoloides* and transfer of fixed N to *Hevea brasiliensis*—Impact on tree growth and vulnerability to drought. Agriculture Ecosystems & Environment. 2016;**217**:79-88

[70] Frankow-Lindberg BE, Dahlin AS. N_2 fixation, N transfer, and yield in grassland communities including a deep-rooted legume or non-legume species. Plant and Soil. 2013;**370**:567-581

[71] Suter M, Connolly J, Finn JA, Loges R, Kirwan L, Sebastia MT, Luscher A. Nitrogen yield advantage from grass-legume mixtures is robust over a wide range of legume proportions and environmental conditions. Global Change Biology. 2015;**21**:2424-2438

[72] Chapagain T, Riseman A. Barley-pea intercropping: Effects on land productivity, carbon and nitrogen transformations. Field Crops Research. 2014;**166**:18-25

12

An Overview of the Effects of Heat Treatments on the Quality of Organic Wastes as a Nitrogen Fertilizer

Naoki Moritsuka and Kaori Matsuoka

Abstract

Sewage sludge is often heat-dried to eliminate water and pathogens. However, heat-drying can also change the form of nitrogen (N). To improve our understanding of this phenomenon, we examined the heat-induced changes in the rate of N mineralization from soils and organic wastes. Published results revealed that the response to the heating temperature differed between soils and organic wastes. As the heating temperature increased to 200°C, the rate of N mineralization increased in soils but decreased in organic wastes. In organic wastes such as sewage sludge, the content of mineralized N tended to decrease sharply when heating temperatures increased to 150–200°C. Furthermore, our results obtained from heat-drying of sewage sludge at 180°C indicated that the rate of carbon (C) mineralization decreased with increasing heating period after the sludge temperature reached 180°C. The C in sewage sludge heated at 180°C for 120 hours after complete drying contained more humin and aromatic C than that in sludge that was heat-dried at 180°C without the additional heating period. These results suggest that the heat-drying treatment can be divided into the drying and denaturing periods and that the temperature of the sludge, not that of the reactor, affects the quality of the end-product.

Keywords: carbon, heat-drying, nitrogen mineralization, organic wastes, sewage sludge, soil, stabilization

1. Introduction

The treatment of raw organic wastes increases their potential range of beneficial uses. The most conventional treatment is composting, which is typically conducted at ambient temperatures. It is generally recognized that composted organic matter is a good soil amendment that releases inorganic nitrogen (N) relatively slowly. Besides composting, heating has become a popular method to enhance the quality of organic wastes.

To eliminate water and pathogens, sewage sludge is heat-dried by various methods such as convective drying, conductive drying, and solar drying [1]. Several researchers have also reported that heat-drying of sewage sludge can change the form of N [2]. For example, Smith and Durham [3] evaluated the content of N in five sewage sludges in paired treatments, both anaerobically digested and one additionally heat-dried for pelleting or granulation. During heat-drying, more than 75% of inorganic ammonium was lost by volatilization. However, during the aerobic incubation at 25°C for 56 days, the production of nitrate in soil amended with heat-dried sludge was similar to, or even larger than, that from the corresponding sludge without heat-drying. These results suggest that heat-drying of sewage sludge greatly increased the content of mineralizable organic N.

The occurrence of such heat-induced N transformations is not surprising, because a similar phenomenon has been seen for a variety of soils. In 1901, Daikuhara [4] reported that heating of soil samples in a pan for 20 minutes increased the content of N that could be extracted by dilute acid solutions. His forgotten research is summarized briefly in our recent paper [5].

In contrast with soils, little information is available for organic wastes. In particular, limited attention has been paid to the relationship between the heating temperature and the resulting changes in the form of N. Case et al. [6] recently reported that heat-drying of sewage sludge at temperatures ranging from 130 to 250°C significantly decreased the rate of N mineralization. Their findings differ from previous results including ours [7], which showed increased N mineralization by heat-drying.

In this chapter, we briefly review what is known about this topic by examining data from previous papers in which the heating temperature was specified. The focus is on the effect of heating on the quality of organic wastes as a source of N for crops. The content is not limited to sewage sludge but covers other organic wastes and soils. For a more comprehensive review of sewage sludge, please refer to Rigby et al. [2]. Since Daikuhara's pioneering work, there have been many relevant publications from Japan. We introduce some of these papers to make them available to the international research community and our own unpublished results.

2. Previous publications from Japan on heat-induced changes in the quality of soils and organic wastes

2.1. Soils

Around 1940, the early findings of Daikuhara [4] were re-evaluated by Mitsui [8]. A series of his experiments were carried out before and during the Pacific War (World War II), when supplies of inorganic fertilizer ran short [5]. By heat-treating two soils at several temperatures (from 65 to 400°C, for 4 hours), Mitsui found that the content of mineralized N (initial inorganic N plus mineralizable N) reached a maximum with heating around 200°C. To generalize this result, he collected 44 types of soils from paddy and upland fields throughout Japan and treated them at 130 or 200°C for 4 hours. He then evaluated the contents of inorganic N and mineralizable N by means of aerobic incubation at 26°C for 31 days.

Both forms of N were increased by the heat treatments (**Figure 1**). The average content of inorganic N was 39.8 mg kg^{-1} (130°C) and 199.3 mg kg^{-1} (200°C) as compared to 25.2 mg kg^{-1} (original). The average content of mineralizable N was 89.4 mg kg^{-1} (130°C) and 111.1 mg kg^{-1} (200°C) as compared to 28.0 mg kg^{-1} (original). The amount of N mineralized by heating at 200°C and subsequent incubation was positively correlated with the content of total N in the original soil ($r = 0.55^{**}$), suggesting that the soil heating effect was greater for humus-rich soils.

The findings of Mitsui [8] were extended by Sakamoto et al. [9]. They tried to reveal the origin of the N mineralized by heating and measuring the amount of N mineralized from two soils fumigated with chloroform or heated at different temperatures. Chloroform fumigation can kill most of soil microbes, so it has been used to extract the elements such as carbon (C), nitrogen (N), and phosphorus (P) in soil microbial biomass.

The amount of N mineralized from the fumigated soil was similar to that from the soil heated at 50 and 100°C, but it was much less than that from the soil heated at 150 and 200°C (**Figure 2**).

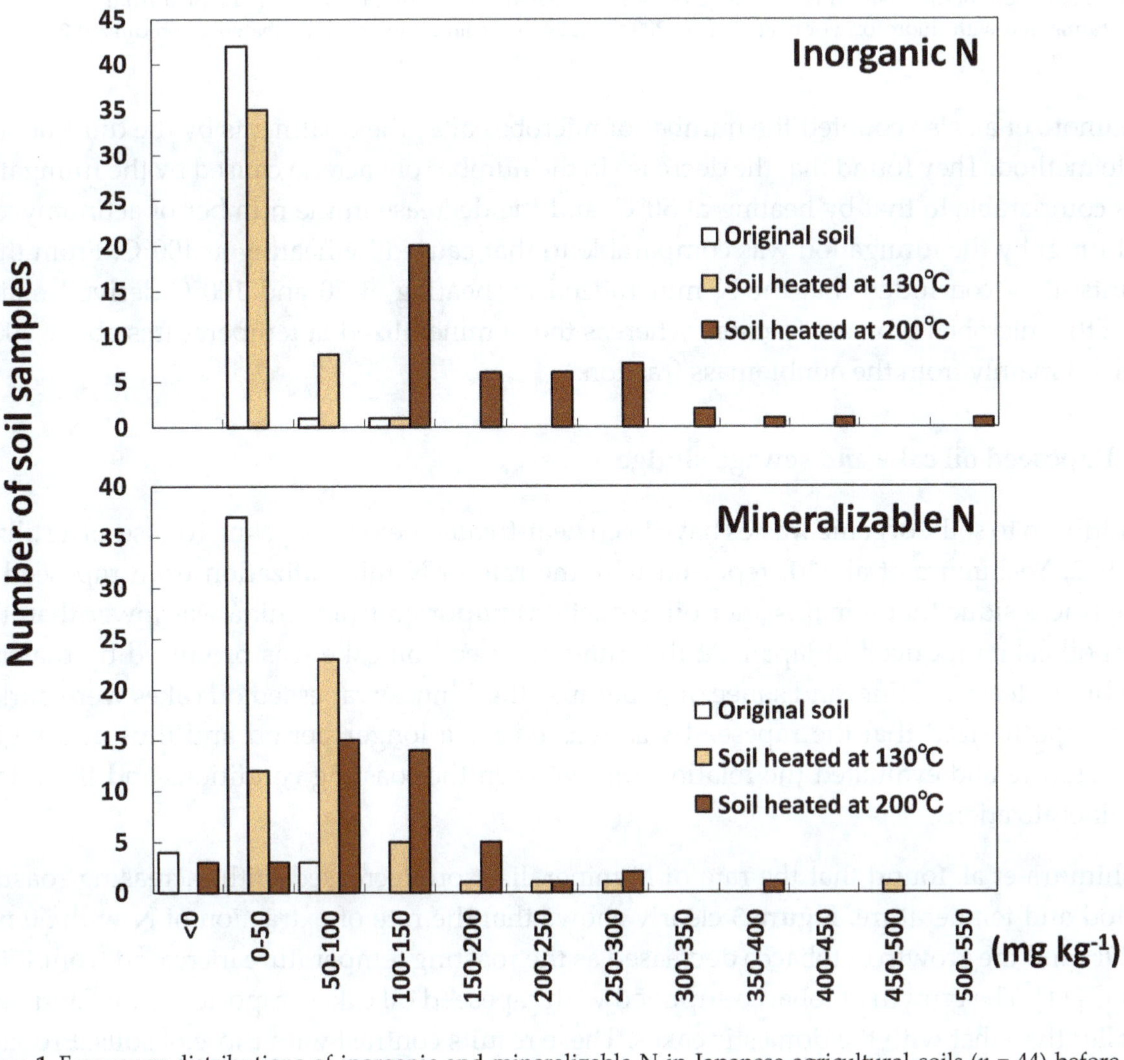

Figure 1. Frequency distributions of inorganic and mineralizable N in Japanese agricultural soils (n = 44) before and after heating at 130 or 200°C for 4 hours (adapted from Ref. [8]). Mineralizable N was evaluated by aerobic incubation at 26°C for 31 days.

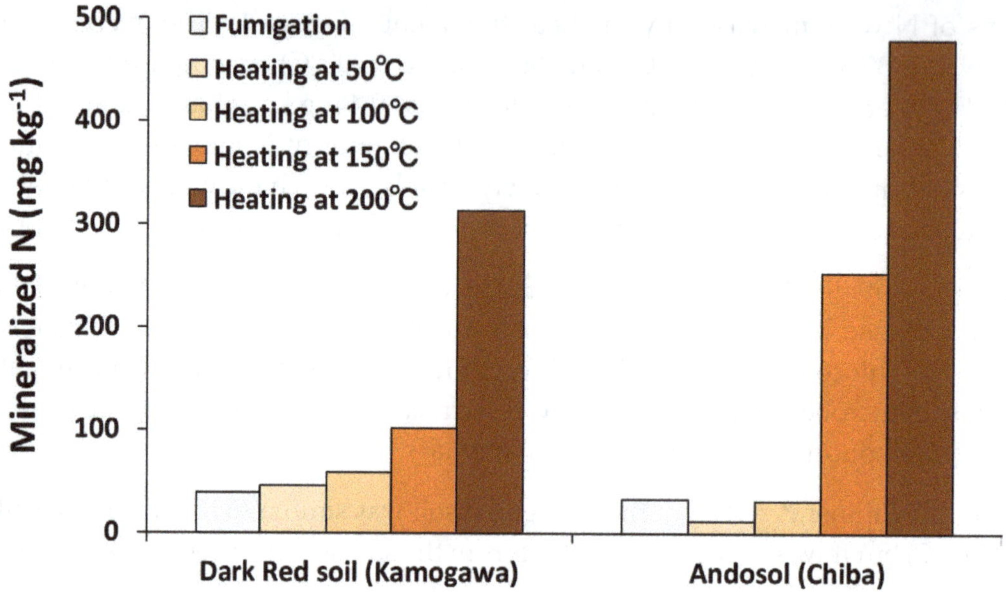

Figure 2. Amount of mineralized N in fumigated or heat-treated soils during incubation (adapted from Ref. [9]). Soils were fumigated with chloroform or heated at 50–200°C for 24 hours and then were incubated aerobically at 25°C for 20 days.

Sakamoto et al. also counted the number of microbes after the treatments by the dilution agar plate method. They found that the decrease in the number of bacteria caused by the fumigation was comparable to that by heating at 50°C, and the decrease in the number of actinomycetes and fungi by the fumigation was comparable to that caused by heating at 100°C. From these results, they concluded that the N mineralized by heating at 50 and 100°C derived mainly from the microbial biomass fraction, whereas the N mineralized at temperatures above 100°C derived mainly from the nonbiomass fraction.

2.2. Rapeseed oil cake and sewage sludge

In addition to soils, organic wastes have been heat-treated before recycling for use as fertilizer. In 1932, Yoshimura et al. [10] reported why the rate of N mineralization from rapeseed oil cake (the residue that remains after oil extraction) imported from China was lower than that from oil cake produced in Japan. At that time, rapeseed oil cake was produced by roasting, crushing, steam-heating, and squeezing. Because the Chinese rapeseed oil cakes were darker, they hypothesized that the rapeseed was roasted for a longer period and thus at a higher temperature and evaluated the relationship between the roasting conditions and the rate of N mineralization.

Yoshimura et al. found that the rate of N mineralization decreased with increasing roasting period and temperature. **Figure 3** clearly shows that the rate of extraction of N with 50 mM NaOH and the growth of tobacco decreased as the roasting temperature increased from 150 to 170°C [11]. The growth of tobacco supplied with rapeseed oil cakes imported from China was smaller than that with the domestic cakes. These results contrast with those of soils. Probably because of the uniqueness, we found no evidence that their paper had been cited until we rediscovered it recently [12].

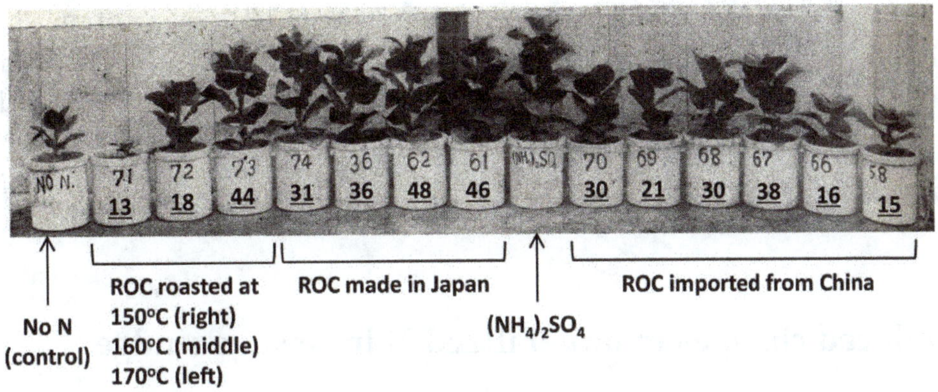

Figure 3. Effects of the application of rapeseed oil cake (ROC) on the growth of tobacco (reprinted from Ref. [11]). Rapeseed oil cake or $(NH_4)_2SO_4$ was applied at 1.5 gN pot^{-1}. Underlined values indicate the percentage of total N in the ROC that was extractable with 50 mM NaOH.

Yoshida [13] confirmed and extended those findings in 1970s. We are not sure if he was aware of Yoshimura's results, as he did not cite them. But whether by chance or not the materials that he selected were rapeseed oil cake and food sludge. Food sludge was dewatered by centrifugation, and both materials were dried at 90–95°C and crushed. They were then heat-treated at 150°C for 2 hours, 165°C for 1 hour, 175°C for 1 hour, or 200°C for 0.5 hour. These materials were subjected to the aerobic incubation at 30°C for 50 days, and the content of mineralized N produced during the incubation was evaluated.

When rapeseed oil cake was heat-dried at 175°C for 1 hour or at 200°C for 0.5 hour, the rate of N mineralization became much lower than that of the control (**Figure 4**). The results were similar for food sludge, indicating that the decrease of N mineralization caused by high temperatures was not limited to rapeseed oil cake. The influence of heating at 175°C was quite different from that of heating at 165°C for both materials. This difference suggests that there is a threshold temperature between 165 and 175°C after which the rate of N mineralization decreases with increasing temperature.

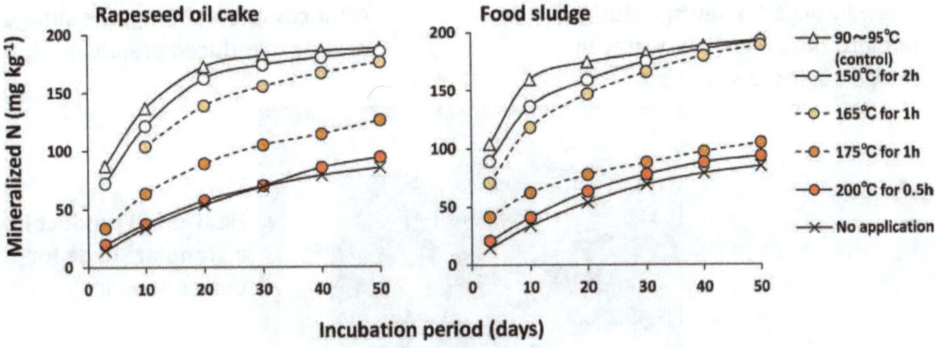

Figure 4. Patterns of N mineralization from a sandy loam soil to which rapeseed oil cake and food sludge treated at different temperatures were applied at a rate of 0.2 gN kg^{-1} soil (created by the authors from the data in Ref. [13]). Aerobic incubation was carried out at 30°C for 50 days. "No application" represents the original soil without application.

In the same year, Kurihara and Watanabe [14] reported that heat-drying of sewage sludge at 130°C for 1 hour could increase the rate of N mineralization by around 30%. In their experiment, mineralizable N was evaluated by the aerobic incubation at 29°C for 35 days. Their result is similar to the findings of Smith and Durham [3]. In addition to heat-drying, Kurihara and Watanabe reported that freeze-drying of sewage sludge decreased the rate of N mineralization by around 30%.

3. Heat-induced changes in mineralized N in sewage sludge

We have also examined the effect of heating of sewage sludge on the rate of N mineralization [7] and the growth of komatsuna (*Brassica campestris* L. var. *rapa*), a leafy vegetable [15]. We introduce our results in this section. For heating of sewage sludge, we used a pilot-scale conductive dryer ([16]; Krosaki Harima, K-10, Fukuoka, Japan). Sewage sludge in the reactor is heated indirectly by means of a surrounding oil heater in which the temperature can be regulated up to 200°C (**Figure 5**). During the treatment, the sludge is also mixed and crushed by a rotary stirrer to homogenize and granulate the products.

Dewatered sewage sludge collected from a wastewater treatment plant in Shimane Prefecture, Japan, was used in the experiment. We compared two heat treatments such as dry-heating of air-dried sludge (AD) and heat-drying of moist sludge without a preliminary air-drying step (HD). Using these treatments, we prepared four types of materials: AD dry-heated at 120°C

Figure 5. Reactor used in the pilot-scale conductive dryer system in our previous research [7, 16].

(AD-120), AD dry-heated at 180°C (AD-180), moist sludge heat-dried at 120°C (HD-120), and moist sludge heat-dried at 180°C (HD-180). Dry-heating requires an air-drying pretreatment and is not practical in terms of time and cost. We nonetheless used it to evaluate the effect of heating on the sludge properties more directly by comparing AD-120 and AD-180 with AD. The heating period was fixed at 16 hours.

These materials were mixed with three soils (an Andosol, a Fluvisol and an Arenosol) at a rate of 1% w/w, and the soils were aerobically incubated at 30°C for 84 days. Regardless of the soil type, the rate of N mineralization was increased significantly by heating of the air-dried sludge at 120°C and it was decreased significantly by heating of the air-dried sludge at 180°C (**Figure 6**). The conventional treatment based on heat-drying of moist sludge at 120 or 180°C exerted similar but less pronounced effects. These heat-induced changes were attributed to the transformation of sludge organic N, because volatilization of N during the heating treatments was negligible [7].

Using the same combination of sludges and soils, we carried out two successive pot experiments. Komatsuna, a popular test vegetable in Japan, was grown after a basal application of sludges at a rate of 20 Mg ha^{-1} (dry matter base). Sludge application was carried out only once before the start of the first experiment.

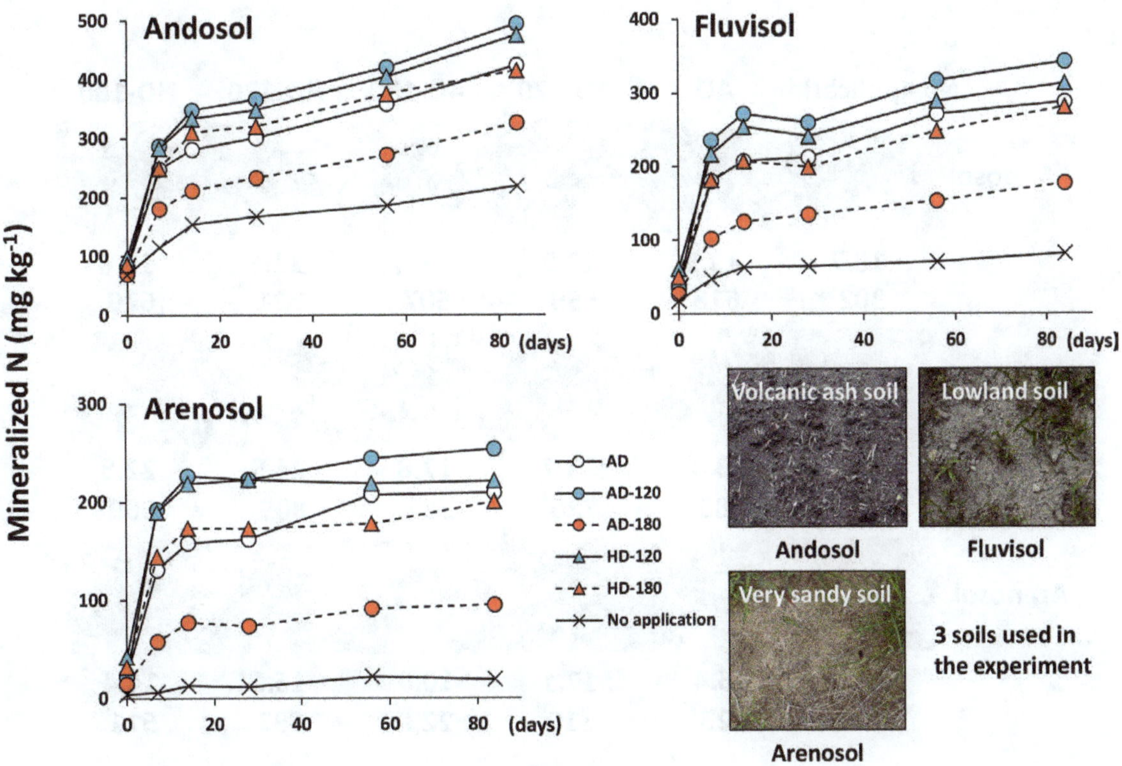

Figure 6. Patterns of N mineralization from three soil types amended with heat-treated sewage sludge (adapted from Ref. [7]). The heat treatment was followed by aerobic incubation at 30°C for 84 days. AD: air-drying, AD-120: dry-heating at 120°C, AD-180: dry-heating at 180°C, HD-120: heat-drying at 120°C, and HD-180: heat-drying at 180°C. Each treatment was duplicated.

In the first experiment, the amount of N uptake by the plants increased significantly by heating air-dried sludge at 120°C and decreased significantly by heating at 180°C, when the sludge was applied to the Fluvisol or Arenosol (**Figure 7**). Heat-drying of the sludge at 120 or 180°C also increased N uptake significantly. In the Andosol, on the other hand, N uptake was not so much influenced by the rate of N mineralization from sewage sludge as that observed in the other two soils (**Figure 6**). Because both Andosol and sewage sludge (which contained ferrous polysulfate as a coagulant) adsorbed P in soil solution (data not shown), we considered that the beneficial effect of N supply from sewage sludge on plant growth was offset by the limited supply of P [15].

The second pot experiment was carried out without an additional sludge application after complete removal of the plants from the first experiment. The plant growth in the second experiment became smaller than that in the first one, and the difference among the treatments decreased (**Figure 8**). The supply of N from the sludge did not last long. Thus, at least for the komatsuna plants used in our experiment, frequent application of sludge is required to sustain plant growth. However, it will also lead to the accumulation of sludge-derived N and P in soil, because less than 40 and 15% of the sludge N and P, respectively, were apparently recovered by two harvests of the plants [15]. From these results, we concluded that heated sludges can act as an effective organic N fertilizer, provided that they are applied to a suitable type of soil and that the short-term effects on soil productivity are balanced with the long-term effects on environmental quality.

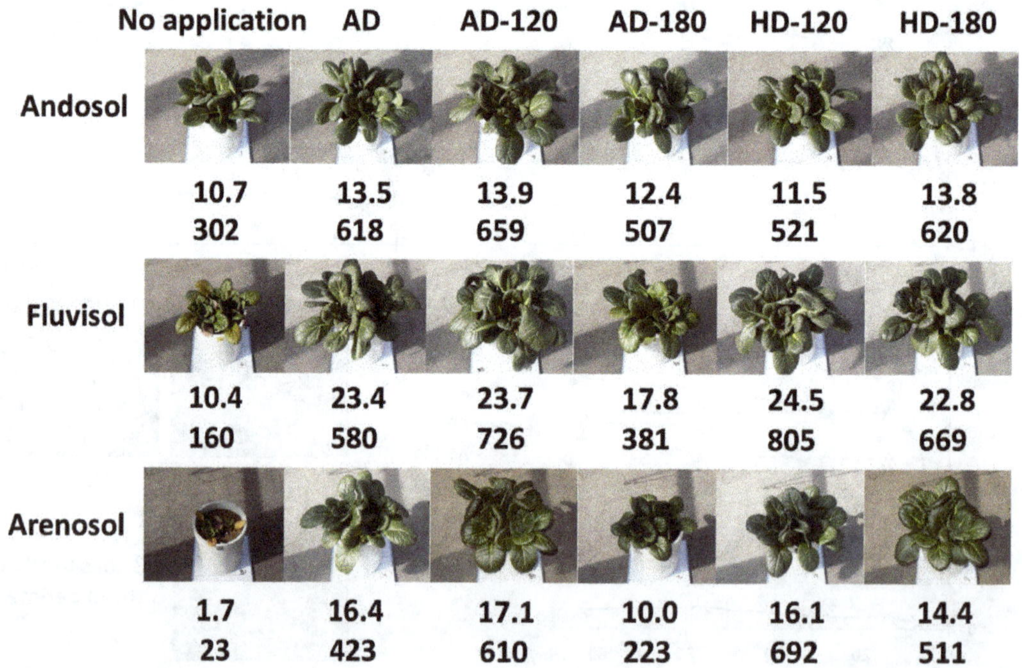

Figure 7. Effect of the application of heat-treated sewage sludge on the growth of a leafy vegetable (komatsuna) grown in three types of soils (adapted from Ref. [15]). This first experiment was carried out for 88 days in the winter. AD: air-drying, AD-120: dry-heating at 120°C, AD-180: dry-heating at 180°C, HD-120: heat-drying at 120°C, and HD-180: heat-drying at 180°C. Upper and lower values indicate the dry matter weight (g pot^{-1}, $n = 3$) and N uptake (mg pot^{-1}, $n = 3$), respectively. All pictures were taken from the same angle.

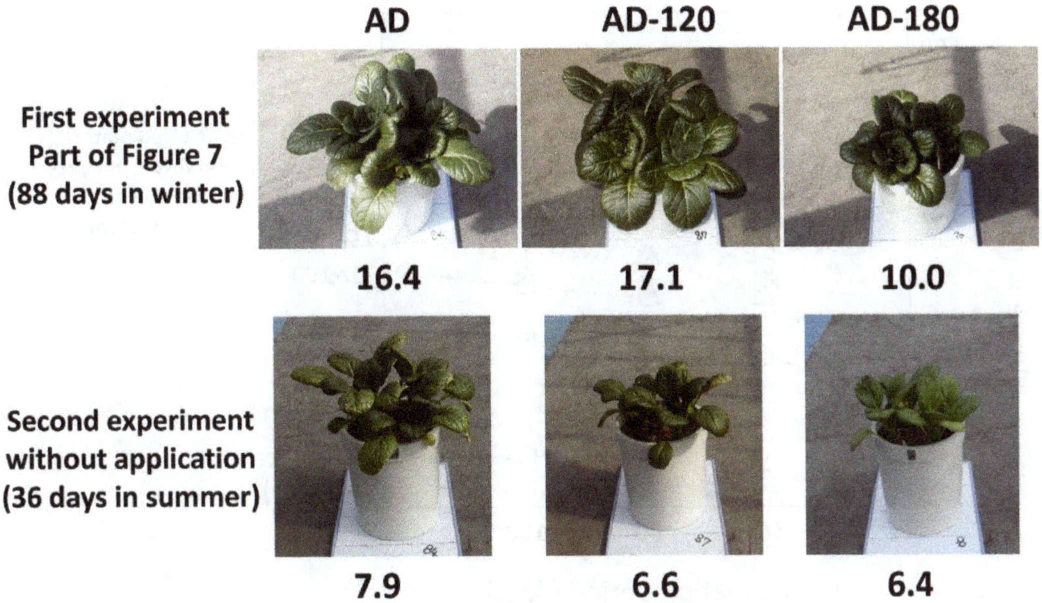

Figure 8. Sustainability of the effects of an initial application of sewage sludge on the growth of a leafy vegetable (komatsuna) grown in an Arenosol (adapted from Ref. [15]). The second experiment was carried out without sludge application. AD: air-drying, AD-120: dry-heating at 120°C and AD-180: dry-heating at 180°C. Values are the dry matter weight (g pot^{-1}, $n = 3$).

Case et al. [6] re-evaluated previous results, including ours. On the hypothesis that the effect of heat-drying on N mineralization from sewage sludge would differ among heating temperatures, they heat-dried anaerobically digested sewage sludge at different temperatures (70, 130, 190, or 250°C) until the water content reached less than 5%. Heat-drying treatment was carried out with a laboratory oven (laboratory-drying), and the product was abbreviated as LD. The sludge materials were applied to a sandy loam soil (Luvisol) and incubated aerobically at 15°C for 160 days. During the incubation, the production of exchangeable ammonium, nitrate, and carbon dioxide (CO_2) was monitored.

The production of nitrate during the whole incubation period was largest for the original sewage sludge (Original), followed by LD-70, LD-130 > LD-190 > LD-250 (**Figure 9**). Thus, nitrate production decreased with increasing temperature from 130 to 250°C. Because the content of exchangeable ammonium was almost zero at 80 days after the incubation (data not shown), the amount of nitrate produced after day 80 can be regarded as the amount of mineralized N (initial inorganic N plus mineralizable N).

The overall production of CO_2 was also largest in Original, LD-70, and LD-130, followed by LD-190 and LD-250 (data not shown). But the initial rate of CO_2 emission from Original was lower than that from LD-70 and LD-130, indicating that microbial decomposition of LD-70 and LD-130 occurred more rapidly. From these results, Case et al. concluded that heat-drying temperature significantly influenced the rate of N mineralization from sewage sludge but that heat-drying did not improve the rate of N mineralization at any temperatures they examined. They also emphasized that the heat-drying treatment in the laboratory produced different results from the treatment in a large-scale wastewater treatment plant (**Figure 9**) and the heat-drying temperature is one of the several factors that potentially affect the rate of N mineralization.

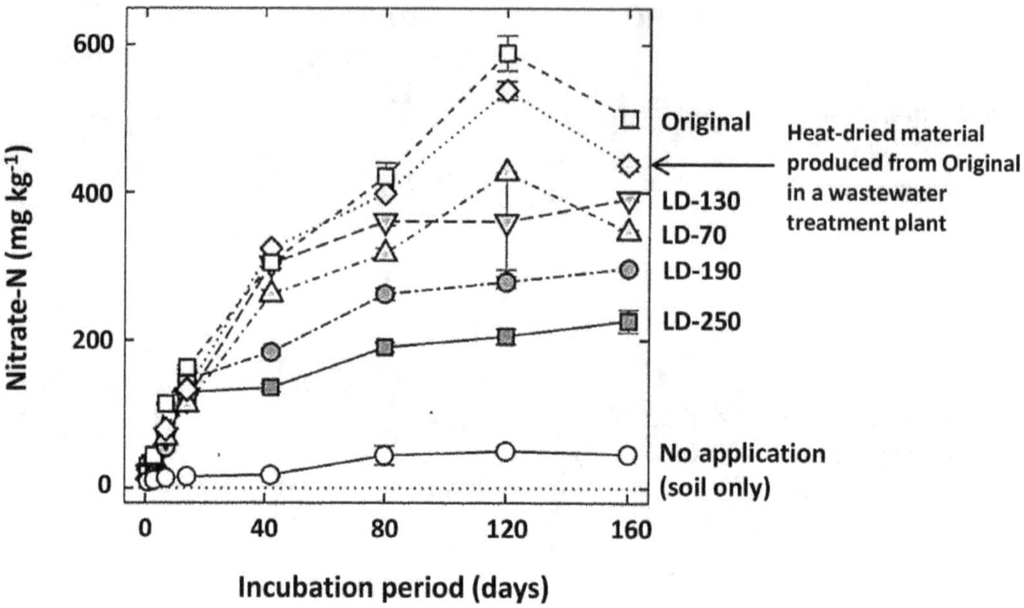

Figure 9. Patterns of nitrate production from a sandy loam soil (a Luvisol) amended with heat-treated sewage sludge at 2% (w/w). Aerobic (pF 2) incubation was carried out at 15°C for 160 days. Anaerobically digested sewage sludge (Original) was heated in a laboratory oven at 70, 130, 190, or 250°C until the water content was less than 5% (LD-70, LD-130, LD-190, and LD-250, respectively). Reproduced with permission from Case et al. [6], *Environ. Sci. Pollut. Res.*, Springer International Publishing AG.

In addition to Case et al. [6], the positive ([17]; 60°C for 13 hours) or negative ([18]; 250°C until reaching constant weight) effects of heat-drying on the amount of mineralized N in sewage sludge have been reported by comparing dewatered sludge with sewage sludge heat-dried at a single temperature. Maki et al. [19] also found that the content of chemically extractable organic N in cow dung manure was decreased by heating at temperatures from 80 to 180°C for 2 hours in a laboratory oven. As summarized by Rigby et al. [2] and Case et al. [6], several researchers have reported N mineralization from heat-dried sludge without specifying the heating conditions. According to the literature review by Rigby et al. [2], the percentage of mineralizable N to total organic N in heat-dried sewage sludge ranged from 26 to 71% (40% in average). This value was similar to aerobically digested sewage sludge whose value ranged from 32 to 58% (47% in average).

4. Summary of the heating effects on mineralized N in organic wastes

By compiling the abovementioned reports, we plotted the heat-induced changes of mineralized N (initial inorganic N plus mineralizable N) in organic wastes as a function of the heating temperature (**Figure 10**). It should be noted that only a rough comparison of these results among the studies is possible, because the materials used as a control, the heating conditions and the incubation conditions differed.

Nevertheless, the figure suggests certain trends. First, the response to the heating temperature differed between soils and organic wastes. As the heating temperature increased from around

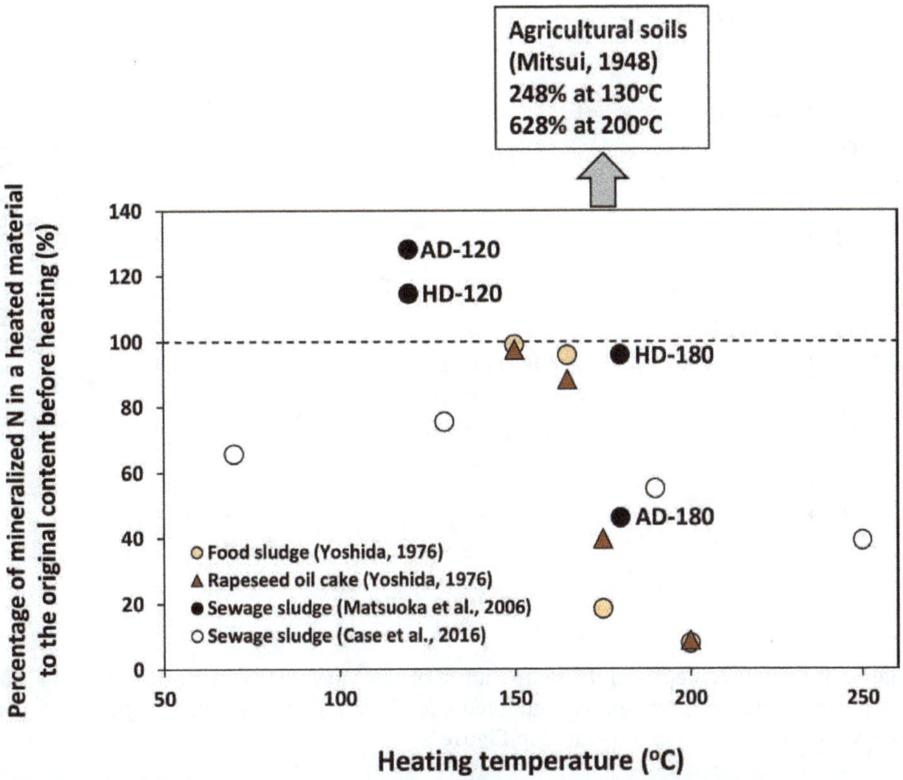

Figure 10. Relationship between temperature at which organic wastes were heated and heat-induced changes in mineralized N (initial inorganic N plus mineralizable N). The data from Mitsui [8] represent the average value obtained from 44 soils. Our data [7] also represent the average value obtained from three soils (**Figure 6**).

120 to 200°C, N mineralization increased in soil but decreased in organic wastes. The reason for the difference remains to be elucidated. Second, the content of mineralized N in organic wastes decreased sharply by heating at 150–200°C except for Case et al. [6] who reported a more gradual decrease. The reason for this difference is also uncertain. The release of CO_2 from the sludge of Case et al. during the incubation decreased slightly with heating at 130°C, but decreased sharply with heating at 190°C [6], which indicates that the threshold temperature for the stabilization of sludge C did exist between 130 and 190°C. On the other hand, we could not estimate the suitable temperature to increase N mineralization from organic wastes due to the insufficient number of available data. Lastly, the heating effect on N mineralization was influenced by both the heating temperature and by the initial water content of the sewage sludge. A typical example can be found in the difference between our samples at 180°C (AD-180 and HD-180).

We will have a closer look at the last point. **Figure 11** shows the temporal changes in the water content and the temperature of sewage sludge during the production of AD-180 and HD-180. Samples were heated for 16 hours with the reactor temperature set at 180°C. For both materials, the heat-drying process can be divided into two periods: the drying period (sludge < 100°C), in which most of the heat is consumed to dry the sludge, and the denaturing period (sludge > 100°C). Because of the difference in the initial water content, it took about 5 hours for AD-180 to reach 180°C, and about 14 hours for HD-180. As a result, HD-180 stayed at 180°C for only 2–3 hours. This clearly indicates that the temperature of sewage sludge during the heat-drying

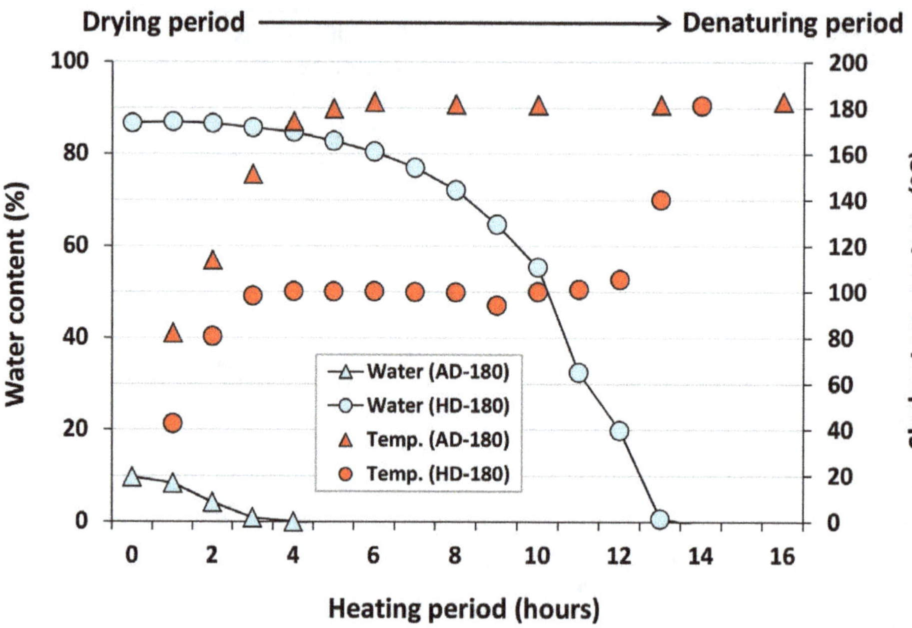

Figure 11. Changes in the water content and the temperature of sewage sludge during heating at 180°C with (AD-180) or without (HD-180) air-drying as a pretreatment (adapted from Ref. [7]). The heating treatment was carried out for 16 hours in the pilot-scale conductive dryer illustrated in **Figure 5**.

treatment rather than the temperature of the reactor is a most important factor that affects the availability of N in the heat-dried products. In the case of heat-drying of sewage sludge at temperatures higher than 150°C, it is plausible that the presence and the length of the denaturing period are key factors that determine whether the rate of N mineralization decreases or not.

Different from our samples treated beyond the drying period, the water content of the heat-dried sewage sludge produced at wastewater treatment plants varies from less than 10% [3, 6, 20] to more than 60% [21]. If sewage sludge was heat-dried homogeneously and the content of water in the end-product was relatively high, we can assume that the sludge temperature during the heat-drying treatment did not exceed 100°C.

5. Heat-induced stabilization of C in sewage sludge

5.1. Preliminary experiments

We carried out additional experiments to better understand why N mineralization was decreased significantly by dry-heating at 180°C (**Figure 6**). Since N mineralization is a part of the biological decomposition of organic C, we focused on the chemical form of the C. The heat-induced denaturation of C in sewage sludge (stabilization in the case of AD-180) was evaluated by using biological and chemical methods as follows; aerobic incubation, chemical extraction, and ^{13}C-NMR.

To extend the stabilization period, sewage sludge was heated at 180°C for a much longer period than in our previous experiment. Dewatered sewage sludge made from human waste was sampled from a wastewater treatment plant in Chiba Prefecture, Japan. The sludge was air-dried at room temperature or heat-dried at 180°C by the conductive dryer in **Figure 5**. The air-dried sludge without heating (AD) contained C and N at 36 and 5.6%, respectively. The sludge was heated for 14 hours until its temperature reached 180°C, or it was continued for an additional 24 hours (a total of 38 hours) or 120 hours (a total of 134 hours). The heat-dried materials treated at 180°C for the additional 0, 24, or 120 hours are referred to as HD-180 (0 hour), HD-180 (24 hours), or HD-180 (120 hours), respectively.

The difference between HD-180 (0 hour) and other two HD-180 materials is originated solely from the dry-heating process at 180°C. The total C content was about 37% in all HD-180 materials. However, the rate of C mineralization decreased significantly with increasing dry-heating period (**Figure 12**). The percentage of total C that was mineralized during the 35-days incubation was 12.5% in HD-180 (0 hour), which was slightly lower than AD (13.1%). It decreased to 5.8% in HD-180 (24 hours) and to 4.1% for HD-180 (120 hours). These results indicate that the stabilization of C occurred mainly during the initial 24 hours of the dry-heating period and proceeded slowly thereafter.

The color of the heat-dried sewage sludge was also quite different from that of the air-dried sludge. For example, HD-180 (30 hours) was darker than AD (**Figure 13**). This suggests that the stabilization of C and N in sewage sludge was concomitant with the Maillard reaction, which is a series of nonenzymatic browning reactions occurring when virtually all foods are heated [22]. This reaction starts from the condensation of the carbonyl group of a reducing sugar with a free amino group of an amino acid [22]. The color of powdered materials can be measured easily and precisely with a color sensor [23]. Thus, the color of the end-products can be a useful proxy for the temperature of the sludge throughout the heat-drying treatment.

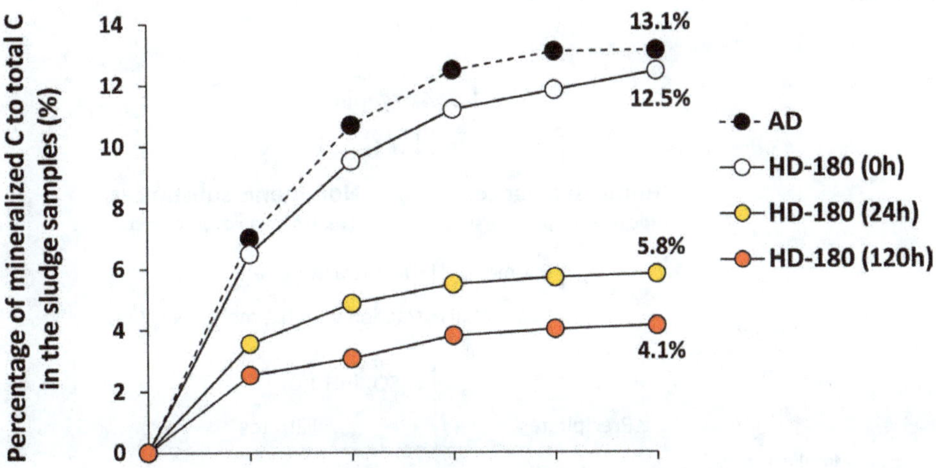

Figure 12. Average percentage of sludge C mineralized during aerobic incubation at 30°C for 35 days. The sludge materials were applied to an Acrisol from Shimane University at Honjo at 1% (v/v), and the soil was incubated at a moisture content of 60% of maximum water holding capacity. Each treatment was triplicated.

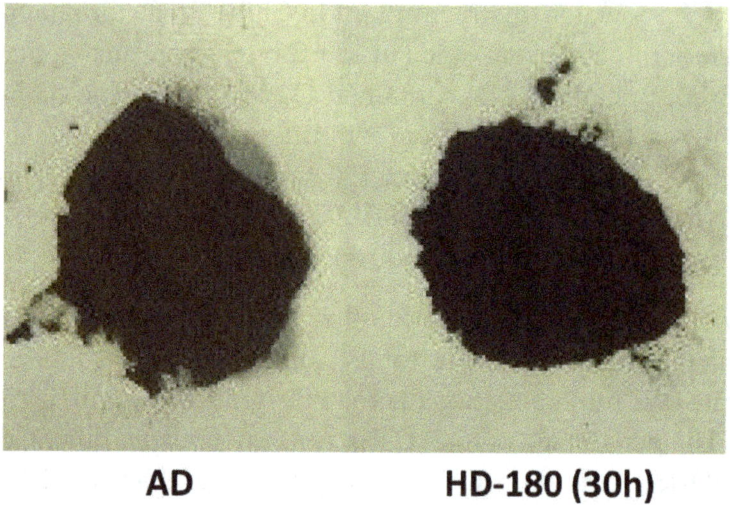

Figure 13. Color differences between two samples of finely ground sewage sludge. AD: air-drying, HD-180 (30h): heat-drying at 180°C with a stabilization period of 30 hours.

5.2. Fractionation of organic matter by chemical extraction

Based on these preliminary observations, we analyzed the composition of organic matter in three samples, i.e., AD, HD-180 (0 hour), and HD-180 (120 hours). The samples were crushed with an agate mortar to pass a 70-mesh sieve with an opening of 0.212 mm for the following analyses. Analyses were performed without replicates.

We used an extraction method known as Nagoya method [24], which is based on the different solubilities of organic matter fractions in alkaline and acid solutions. This method was originally designed to classify humic substances in soils, and we applied it to sewage sludge. **Figure 14** shows the procedures for separating humic substances into fulvic acid, humic acid, and humin fractions.

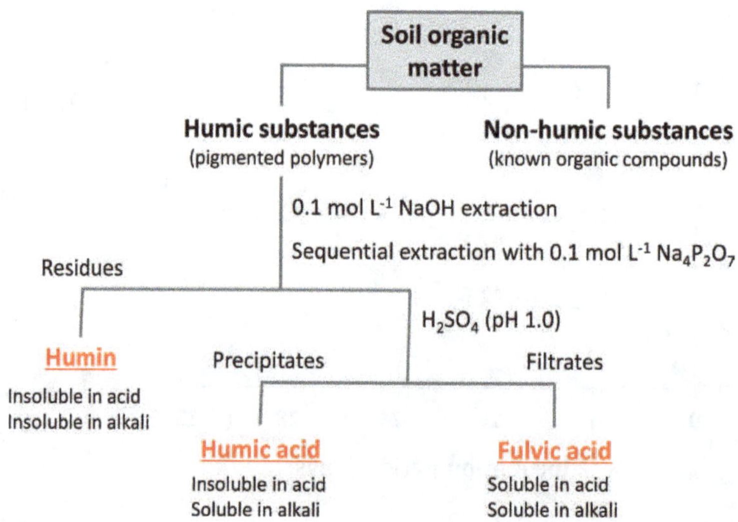

Figure 14. Procedures for the fractionation of soil organic matter into fulvic acid, humic acid, and humin according to the Nagoya method [24].

Figure 15 shows the relative contents of the fractions in our samples. The results of Collard et al. [20] were added to the figure, which were obtained from heat-drying of sewage sludge at 85 and 120°C. However, only a rough comparison between two datasets is possible, because Collard et al. used a slightly different extraction method (the IHSS method) and also removed lipids with a mixture of dichloromethane and methanol before extracting humic substances. In contrast, we did not remove nonhumic substances beforehand (the original method), so our extractable fractions (fulvic and humic acids) may have been overestimated as a result of contamination of nonhumic substances such as lipids and carbohydrates. Here we pay attention to the percentage of humin as an index of the stability of sludge C, since the humin fraction was considered to be least affected by such contamination.

The percentage of humin in the original sludge without heating suggested that our sample was less stabilized than that of Collard et al. In our samples, the heat-drying at 180°C greatly increased the percentage of humin; 12% in AD, 37% in HD-180 (0 hour), and 50% in HD-180 (120 hours). In the samples of Collard et al. [20], on the other hand, heat-drying at 120°C decreased the percentage of humin; 59% in Original, 59% in HD-85, and 30% in HD-120. These results indicate that the stability of sludge C was unaffected by heat-drying at 85°C, decreased at 120°C and increased at 180°C. The effects of heating on the stability of C differed between 120 and 180°C, which agreed with our results of N mineralization [7].

Comparing among our samples, however, the rate of C mineralization from AD and HD-180 (0 hour) was not so much different as the percentage of humin (**Figures 12** and **15**). This indicates that the increase of the percentage of humin was not proportional to the decrease of C mineralization during the incubation. In our previous study [7], the chemical forms of organic N in heat-treated materials were also evaluated by sequential extraction. Although organic N in AD-180 was most recalcitrant to chemical extractions, the results could not quantitatively explain the very low rate of N mineralization. These results suggest that mineralization of sludge

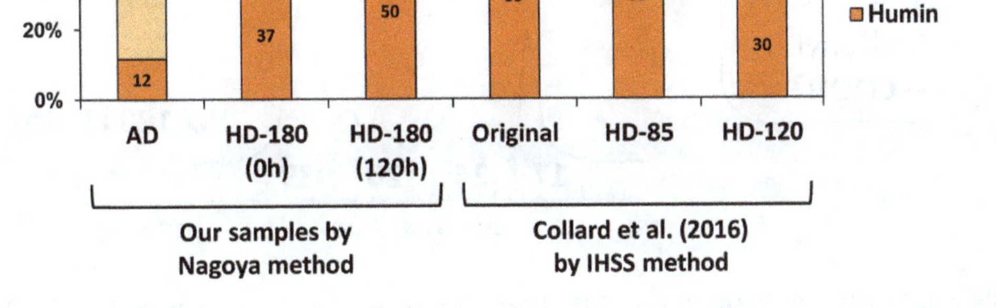

Figure 15. Relative percentages of fulvic acid, humic acid, and humin in sewage sludge samples. The percentage of lipid is also presented for the results of Collard et al. [20] after adaptation by the authors. In their paper, the activated sludge (Original) was heat-dried at 85°C (HD-85) and then pelleted at 120°C (HD-120).

C and N was determined not only by its chemical solubility but also by other physical factors such as microbial accessibility.

5.3. Analysis of the functional groups of C by ^{13}C-NMR

We also analyzed the C forms in AD, HD-180 (0 hour), and HD-180 (120 hours) by the solid-state ^{13}C nuclear magnetic resonance (^{13}C-NMR) spectroscopy. The ^{13}C-NMR method provides estimates of the relative percentages of the main functional groups of C. The analytical procedures and machine conditions were the same as those described by Hiradate et al. [25]. When our samples were analyzed without any pretreatment, the ^{13}C-NMR spectra were divided into the four regions. According to Hiradate et al. [25], these regions were assigned to the functional groups as follows (**Figure 16**); alkyl C (0–45 ppm), O-alkyl C (45–110 ppm), aromatic C (110–165 ppm), and carbonyl C (165–210 ppm).

In AD, the percentage of alkyl C plus O-alkyl C exceeded 60%. Fernández et al. [26] also reported the predominance of these functional groups (75%) in sewage sludge heat-dried at 70–80°C. During the drying period from AD to HD-180 (0 hour), the percentage of alkyl

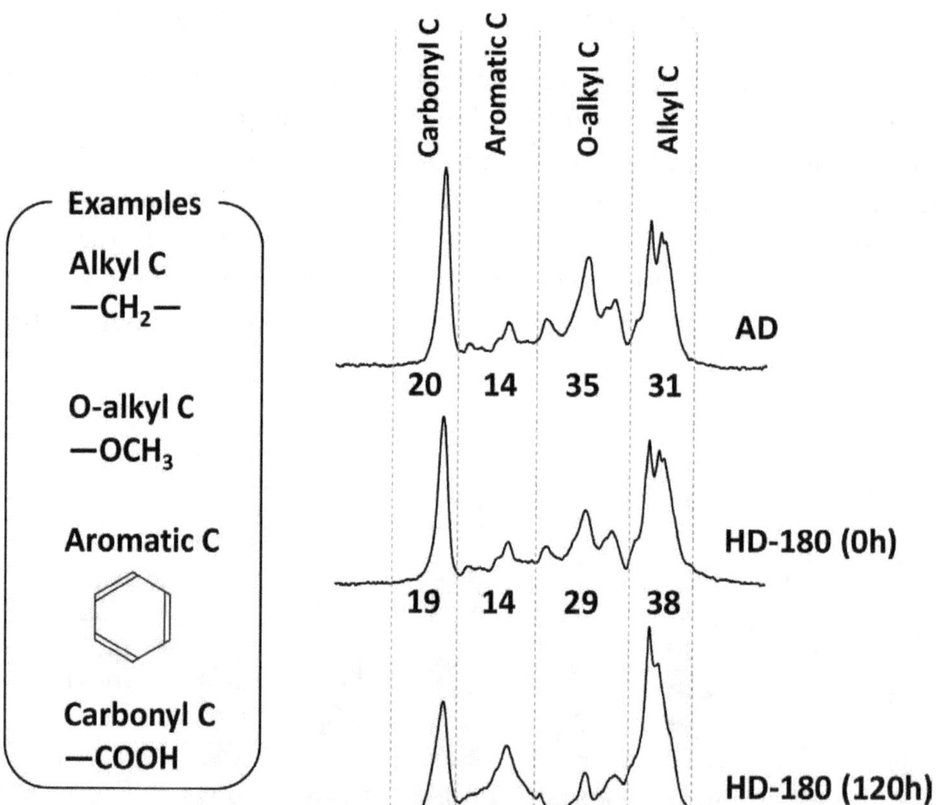

Figure 16. Solid-state ^{13}C-NMR (CP/MAS-TOSS, 75.45 MHz) spectra of sewage sludge samples. The ^{13}C-NMR spectrum was divided into the four chemical shift regions: 0–45 ppm (alkyl C), 45–110 ppm (O-alkyl C), 110–165 ppm (aromatic C), and 165–210 ppm (carbonyl C). The values in the spectra indicate the relative percentage of each type of C calculated by integrating the signal intensity.

C increased from 31 to 38%, whereas that of O-alkyl C decreased from 35 to 29%. During the stabilization period caused by dry-heating for 120 hours, the percentage of aromatic C increased from 14 to 24%, whereas that of O-alkyl C decreased from 29 to 20%. The proportion of carbonyl C was relatively constant throughout the treatments.

The transformation of C during the treatments at 180°C could be detected by applying the ^{13}C-NMR method to our crude samples. The transformation occurred not only during the stabilization period but also during the initial drying period, but the functional groups affected by the transformation differed between the periods. These results indicate that the reactions in the drying period differed from those in the stabilization period. A slight decrease of carbonyl C from 19 to 17% during the stabilization period implied the occurrence of the Maillard reaction. However, it is uncertain to what extent the changes observed in the ^{13}C-NMR spectra were responsible for the different rates of mineralization of C between HD-180 (0 hour) and HD-180 (120 hours). The higher proportion of aromatic C in HD-180 (120 hours) may have contributed to the lower rate of C mineralization, because the aromatic polymers in plant residues and microbial products are regarded as more recalcitrant to biological decomposition [27].

Similar to our results, the heating of Susuki (*Miscanthus sinensis* A.) leaves at 250°C for 1 hour in a laboratory oven increased the percentage of aromatic C from 17 to 66% and decreased that of O-alkyl C from 70 to 15% [28]. Heating at 250°C burned the leaves. On the other hand, the ^{13}C-NMR spectrum of the 200°C-heated samples remained unchanged from that of the untreated control. Heating of the leaves at 200°C for 1 hour did not cause burning but turned them brown possibly owing to the Maillard reaction.

As for sewage sludge, Fernández et al. [29] compared the properties of heat-dried sewage sludge with those of composted sludge. Sewage sludge that was heat-dried at the maximum sludge temperature of 75°C contained more fulvic acid than sewage sludge composted in windrows for 3 months. The solid-state ^{13}C-NMR spectroscopy indicated that the humic acid fraction of the heat-dried sludge contained more alkyl C and less aromatic C than that of the composted sludge. These results indicated that heat-drying of sewage sludge at 75°C (sludge temperature) did not increase the stability of C as much as did windrow composting. In their study, the heat-drying was carried out by indirect convection with air heated between 380 and 450°C. The temperature of the sludge during the treatment (<75°C) was therefore much lower than the hot air supplied for drying. This reemphasizes the importance of the temperature of sewage sludge during the drying period as a factor that affects the quality of the end-product.

6. Conclusions

By reviewing literature, we suggested that the heat-induced changes in N mineralization differ between soils and organic wastes. As the heating temperature increased to 200°C, the rate of N mineralization increased in soils but decreased in organic wastes. The rate of N mineralization from organic wastes tended to decrease sharply when heating temperatures increased to 150–200°C. Since the materials examined, heating conditions and analytical methods differed among the researchers, our findings may have been biased by these artifacts. More comparative studies are required to confirm these findings and reveal the processes involved. Furthermore, the results obtained from heat-dried sewage sludge indicated that the solid-state ^{13}C-NMR spectroscopy can be a powerful tool to characterize the heat-induced stabilization of sludge C in addition to the incubation and extraction methods that have been used by soil scientists.

Acknowledgements

We thank Dr. Noriko Iwashima (Shimane University) for the analysis of mineralization of C by the incubation method and Dr. Syuntaro Hiradate (Institute for Agro-Environmental Sciences, NARO) for the analysis of C forms by the solid-state ^{13}C-NMR spectroscopy. We also thank Dr. Toshiyuki Wakatsuki, Dr. Tsugiyuki Masunaga, Dr. Shingo Matsumoto, and Dr. Hisao Nakano; the faculty members of Shimane University who gave us valuable comments and technical supports during our research activities at Shimane University. This work was supported by the JSPS KAKENHI Grant Number JP15K00634.

Author details

Naoki Moritsuka[1*] and Kaori Matsuoka[2]

*Address all correspondence to: morituka@kais.kyoto-u.ac.jp

1 Graduate School of Agriculture, Kyoto University, Kyoto, Japan

2 Institute of Fruit Tree and Tea Science, NARO, Tsukuba, Japan

References

[1] Bennamoun, L., Arlabosse, P., Léonard, A. (2013) Review on fundamental aspect of application of drying process to wastewater sludge. *Renew. Sust. Energ. Rev.* **28**, 29-43. DOI: 10.1016/j.rser.2013.07.043

[2] Rigby, H., Clarke, B. O., Pritchard, D. L., Meehan, B., Beshah, F., Smith, S. R., Porter, N. A. (2016) A critical review of nitrogen mineralization in biosolids-amended soil, associated fertilizer value for crop production and potential for emissions to the environment. *Sci. Total Environ.* **541**, 1310-1338. DOI: 10.1016/j.scitotenv.2015.08.089

[3] Smith, S. R., Durham, E. (2002) Nitrogen release and fertilizer value of thermally-dried biosolids. *Water Environ. J.* **16**, 121-126. DOI: 10.1111/j.1747-6593.2002.tb00382.x

[4] Daikuhara, G. (1901) Studies on heated soil fertilizer. *The Bulletin of the Imp. Agri. Exp. Stat. Jpn.* **20**, 1-46. (In Japanese).

[5] Moritsuka, N., Matsuoka, K. (2016) An overview of the studies on biochar fertilizer carried out at the beginning of the twentieth century in Japan. *In* Larramendy, M and Soloneski, S., editors. *Organic Fertilizers – From Basic Concepts to Applied Outcomes.* pp. 203-221, InTech, Croatia. DOI: 10.5772/62526

[6] Case, S. D. C., Gómez-Muñoz, B., Magid, J., Jensen, L.S. (2016) Increasing thermal drying temperature of biosolids reduced nitrogen mineralization and soil N_2O emissions. *Environ. Sci. Pollut. Res.* **23**, 14383-14392. DOI: 10.1007/s11356-016-6607-3

[7] Matsuoka, K., Moritsuka, N., Masunaga, T., Matsui, K., Wakatsuki, T. (2006) Effect of heating treatments on nitrogen mineralization from sewage sludge. *Soil Sci. Plant Nutr.* **52**, 519-527. DOI: 10.1111/j.1747-0765.2006.00061.x

[8] Mitsui, S. (1948) Studies on the effect of soil heating. *Nat. Agri. Exp. Stat. Tokyo Bull.* **62**, 1-48. (In Japanese with an English summary).

[9] Sakamoto, K., Yoshida, T., Satoh, M. (1992) Comparison of carbon and nitrogen mineralization between fumigation and heating treatments. *Soil Sci. Plant Nutr.* **38**, 133-140. DOI: 10.1080/00380768.1992.10416960

[10] Yoshimura, K., Murata, H., Takase, Y. (1932) Studies on the fertilizer values of rapeseed oil cakes produced in Japan and China (part 1). *Jpn. J. Soil Sci. Plant Nutr.* **6**, 283-300. (In Japanese).

[11] Murata, K., Sonoda, T. (1934) Influence of thermal changes of nitrogenous compounds in oil-cake during the production on the fertilizer values. *Jpn. J. Soil Sci. Plant Nutr.* **8**, 37-38. (In Japanese).

[12] Moritsuka, N., Matsuoka, K. (2008) Past and present studies on the agronomic effects of soil heating. *Jpn. J. Soil Sci. Plant Nutr.* **79**, 505-510. (In Japanese).

[13] Yoshida, T. (1976) Utilization of activated wastewater sludge as fertilizers (2). Effect of treating conditions on fertilizer values. *J. Water Waste.* **18**, 419-426 (In Japanese).

[14] Kurihara, K., Watanabe, M. (1976) Factors affecting the N mineralization of activated sludges. *Ann. Rep. Div. Fert. Chem. Nat. Inst. Agri. Sci.* **189**, 2-15. (In Japanese).

[15] Moritsuka, N., Matsuoka, K., Matsumoto, S., Masunaga, T., Matsui, K., Wakatsuki, T. (2006) Effects of the application of heated sewage sludge on soil nutrient supply to plants. *Soil Sci. Plant Nutr.*, **52**, 528-539. DOI: 10.1111/j.1747-0765.2006.00062.x

[16] Matsui, K., Shibata, K. (2002) Recycling of sewage sludge in agriculture by using sludge thermal treatment. *J. Jpn Waste Manage. Assoc.* **55**, 292-296 (In Japanese).

[17] Silva-Leal, J., Torres-Lozada, P., Cardoza, Y. J. (2013) Thermal drying and alkaline treatment of biosolids: effects on nitrogen mineralization. *Clean Soil Air Water.* **41**, 298-303. DOI: 10.1002/clen.201100681

[18] Corrêa, R. S., White, R. E., Weatherley, A. J. (2012) Effects of sewage sludge stabilization on organic-N mineralization in two soils. *Soil Use Manage.* **28**, 12-18. DOI: 10.1111/j.1475-2743.2012.00387.x

[19] Maki, H., Kono, S., Nagai, K. (2009) Change in nutrient content and solubility by carbonization or heat treatment of cattle manure. *Jpn. J. Soil Sci. Plant Nutr.* **80**, 257-262. (In Japanese with an English summary).

[20] Collard, M., Teychené, B., Lemée, L. (in press) Comparison of three different wastewater sludge and their respective drying processes: Solar, thermal and reed beds – Impact on organic characteristics. *J. Env. Manage.* DOI: 10.1016/j.jenvman.2016.05.070

[21] Yoshida, H., Nielsen, M. P., Scheutz, C., Jensen, L. S., Christensen, T. H., Nielsen, S., Bruun, S. (2015) Effects of sewage sludge stabilization on fertilizer value and greenhouse gas emissions after soil application. *Acta Agri. Scand. B.* **65**, 506-516. DOI: 10.1080/09064710.2015.1027730

[22] Ames, J. M. (1990) Control of the Maillard reaction in food systems. *Trends Food Sci. Technol.* **1**, 150-154. DOI: 10.1016/0924-2244(90)90113-D

[23] Moritsuka, N., Matsuoka, K., Katsura, K., Sano, S., Yanai, J. (2014) Soil color analysis for statistically estimating total carbon, total nitrogen and active iron contents in Japanese agricultural soils. *Soil Sci. Plant Nutr.* **60**, 475-485. DOI: 10.1080/00380768.2014.906295

[24] Kumada, K., Sato, O., Ohsumi, Y., Ohta, S. (1967) Humus composition of mountain soils in Central Japan with special reference to the distribution of P type humic acid. *Soil Sci. Plant Nutr.* **13**, 151-158. DOI: 10.1080/00380768.1967.10431990

[25] Hiradate, S., Nakadai, T., Shindo, H., Yoneyama, T. (2004) Carbon source of humic substances in some Japanese volcanic ash soils determined by carbon stable isotopic ratio, $\delta^{13}C$. *Geoderma*, **119**, 133-141. DOI:10.1016/S0016-7061(03)00257-X

[26] Fernández, J. M., Plaza, C., Polo, A., Plante, A. (2012) Use of thermal analysis techniques (TG-DSC) for the characterization of diverse organic municipal wastes streams to predict biological stability prior to land application. *Waste Manage.* **32**, 158-164. DOI:10.1016/S0016-7061(03)00257-X

[27] Lützow, M. V., Kögel-Knabner, I., Ekschmitt, K., Matzner, E., Guggenberger, G., Marschner, B., Flessa, H. (2006) Stabilization of organic matter in temperate soils: mechanisms and their relevance under different soil conditions – a review. *Euro. J. Soil Sci.* **57**, 426-445. DOI: 10.1111/j.1365-2389.2006.00809.x

[28] Honma, H., Hiradate, S., Shindo, H. (1999) Analysis of the burning process of Susuki (*Miscanthus sinensis* A.) plants by ^{13}C-NMR, FT-IR and XRD. *Jpn. J. Soil Sci. Plant Nutr.* **70**, 277-282. (In Japanese with an English summary).

[29] Fernández, J. M., Hockaday, W. C., Plaza, C., Polo, A., Hatcher, P. (2008) Effects of long-term soil amendment with sewage sludges on soil humic acid thermal and molecular properties. *Chemosphere.* **73**, 1838-1844. Doi:10.1016/j.chemosphere.2008.08.001

Permissions

All chapters in this book were first published in NAU, by InTech Open; hereby published with permission under the Creative Commons Attribution License or equivalent. Every chapter published in this book has been scrutinized by our experts. Their significance has been extensively debated. The topics covered herein carry significant findings which will fuel the growth of the discipline. They may even be implemented as practical applications or may be referred to as a beginning point for another development.

The contributors of this book come from diverse backgrounds, making this book a truly international effort. This book will bring forth new frontiers with its revolutionizing research information and detailed analysis of the nascent developments around the world.

We would like to thank all the contributing authors for lending their expertise to make the book truly unique. They have played a crucial role in the development of this book. Without their invaluable contributions this book wouldn't have been possible. They have made vital efforts to compile up to date information on the varied aspects of this subject to make this book a valuable addition to the collection of many professionals and students.

This book was conceptualized with the vision of imparting up-to-date information and advanced data in this field. To ensure the same, a matchless editorial board was set up. Every individual on the board went through rigorous rounds of assessment to prove their worth. After which they invested a large part of their time researching and compiling the most relevant data for our readers.

The editorial board has been involved in producing this book since its inception. They have spent rigorous hours researching and exploring the diverse topics which have resulted in the successful publishing of this book. They have passed on their knowledge of decades through this book. To expedite this challenging task, the publisher supported the team at every step. A small team of assistant editors was also appointed to further simplify the editing procedure and attain best results for the readers.

Apart from the editorial board, the designing team has also invested a significant amount of their time in understanding the subject and creating the most relevant covers. They scrutinized every image to scout for the most suitable representation of the subject and create an appropriate cover for the book.

The publishing team has been an ardent support to the editorial, designing and production team. Their endless efforts to recruit the best for this project, has resulted in the accomplishment of this book. They are a veteran in the field of academics and their pool of knowledge is as vast as their experience in printing. Their expertise and guidance has proved useful at every step. Their uncompromising quality standards have made this book an exceptional effort. Their encouragement from time to time has been an inspiration for everyone.

The publisher and the editorial board hope that this book will prove to be a valuable piece of knowledge for researchers, students, practitioners and scholars across the globe.

List of Contributors

Ling Zhang and Xiaojun Liu
Jiangxi Key Laboratory of Silviculture, Co-Innovation Center of Jiangxi Typical Trees Cultivation and Utilization, College of Forestry, Jiangxi Agricultural University, Nanchang, China

Jianjun Chen
University of Florida, Institute of Food and Agricultural Sciences, Mid-Florida Research and Education Center, Apopka, FL, USA

Xiangying Wei
University of Florida, Institute of Food and Agricultural Sciences, Mid-Florida Research and Education Center, Apopka, FL, USA
Fujian University Key Laboratory of Plant-Microbe Interaction, College of Life Sciences, Fujian Agriculture and Forestry University, Fuzhou, China

Babar Mahmood
Department of Civil Engineering, Unitec Institute of Technology, Auckland, New Zealand

Felix Nchu, Yonela Matanzima and Charles P. Laubscher
Department of Horticultural Sciences, Faculty of Applied Sciences, Cape Peninsula University of Technology, Bellville, Cape Town, South Africa

Marcelo Carvalho Minhoto Teixeira Filho, Fernando Shintate Galindo, Salatiér Buzetti and Eduardo Henrique Marcandalli Boleta
Universidade Estadual Paulista (UNESP), Campus de Ilha Solteira, Ilha Solteira, São PauloState, Brazil

Akshit Puri and Kiran Preet Padda
Department of Soil Science, Faculty of Land and Food Systems, The University of British Columbia, Vancouver, Canada

Chris P. Chanway
Department of Soil Science, Faculty of Land and Food Systems, The University of British Columbia, Vancouver, Canada
Department of Forest and Conservation Sciences, Faculty of Forestry, The University of British Columbia, Vancouver, Canada

M. Luz Rodríguez-Blanco, M. Mercedes Taboada-Castro, Ricardo Arias and M. Teresa Taboada-Castro
Faculty of Sciences, Centre for Advanced Scientific Research (CICA), University of A Coruna, A Coruña, Spain

Shuangjie Huang, Chunfang Zhao and Cailin Wang
Institute of Food Crops of Jiangsu Academy of Agricultural Sciences, Nanjing, China

Yali Zhang
State Key Laboratory of Crop Genetics and Germplasm Enhancement, Key Laboratory of Plant Nutrition and Fertilization in Low-Middle Reaches of the Yangtze River, Ministry of Agriculture, Nanjing Agricultural University, China

Isabel Moura
Plant Stress & Biodiversity Lab, Linking Landscape, Environment, Agriculture and Food (LEAF), School of Agriculture (ISA), University of Lisbon (ULisboa), Portugal

Ana I. Ribeiro-Barros, Maria J. Silva and José C. Ramalho
Plant Stress & Biodiversity Lab, Linking Landscape, Environment, Agriculture and Food (LEAF), School of Agriculture (ISA), University of Lisbon (ULisboa), Portugal
GeoBioTec, Faculdade de Ciências e Tecnologia (FCT), Universidade NOVA de Lisboa (UNL), Portugal

Cristina Máguas-Hanson
Center for Ecology, Evolution and Environmental Changes (cE3c), Faculdade de Ciências, Universidade de Lisboa, Portugal

Natasha S. Ribeiro
Faculty of Agronomy and Forest Engineering, Eduardo Mondlane University, Maputo, Mozambique

Luigi Tedone, Salem Alhajj Ali and Giuseppe De Mastro
Department of Agricultural and Environmental Science, Bari, Italy

M. Anowarul Islam and Albert Tetteh Adjesiwor
Department of Plant Sciences, University of Wyoming, Laramie, WY, USA

Naoki Moritsuka
Graduate School of Agriculture, Kyoto University, Kyoto, Japan

Kaori Matsuoka
Institute of Fruit Tree and Tea Science, NARO, Tsukuba, Japan

Index

www.ingramcontent.com/pod-product-compliance
Lightning Source LLC
LaVergne TN
LVHW080912210926
848485LV00004B/634
* 9 7 8 1 6 3 9 8 9 3 8 4 3 *